Solutions to Climate Change Challenges in the Built Environment

Edited by

Colin A. Booth
Reader, Built Environment
School of Technology
University of Wolverhampton

Felix N. Hammond
Senior Lecturer, Built Environment
School of Technology
University of Wolverhampton

Jessica E. Lamond
Senior Research Fellow, Construction and Property
Faculty of Environment and Technology
University of the West of England

&

David G. Proverbs
Professor and Head of Department, Construction and Property
Faculty of Environment and Technology
University of the West of England

A John Wiley & Sons, Ltd., Publication

This edition first published 2012
© 2012 by Blackwell Publishing Ltd.

Blackwell Publishing was acquired by John Wiley & Sons in February 2007. Blackwell's publishing programme has been merged with Wiley's global Scientific, Technical, and Medical business to form Wiley-Blackwell.

Registered Office
John Wiley & Sons, Ltd, The Atrium, Southern Gate, Chichester, West Sussex, PO19 8SQ, United Kingdom

Editorial Offices
9600 Garsington Road, Oxford, OX4 2DQ, United Kingdom
2121 State Avenue, Ames, Iowa 50014-8300, USA

For details of our global editorial offices, for customer services and for information about how to apply for permission to reuse the copyright material in this book please see our website at www.wiley.com/wiley-blackwell.

The right of the author to be identified as the author of this work has been asserted in accordance with the UK Copyright, Designs and Patents Act 1988.

All rights reserved. No part of this publication may be reproduced, stored in a retrieval system, or transmitted, in any form or by any means, electronic, mechanical, photocopying, recording or otherwise, except as permitted by the UK Copyright, Designs and Patents Act 1988, without the prior permission of the publisher.

Wiley also publishes its books in a variety of electronic formats. Some content that appears in print may not be available in electronic books.

Designations used by companies to distinguish their products are often claimed as trademarks. All brand names and product names used in this book are trade names, service marks, trademarks or registered trademarks of their respective owners. The publisher is not associated with any product or vendor mentioned in this book. This publication is designed to provide accurate and authoritative information in regard to the subject matter covered. It is sold on the understanding that the publisher is not engaged in rendering professional services. If professional advice or other expert assistance is required, the services of a competent professional should be sought.

Library of Congress Cataloging-in-Publication Data

Solutions for climate change challenges in the built environment / edited by Colin Booth ... [et al.].
   p.   cm. – (Innovation in the built environment)
 Includes bibliographical references and index.
 ISBN 978-1-4051-9507-2 (hardback)
1. Sustainable construction.   2. Sustainable development.   3. Cities and towns–Environmental aspects.   4. Environmental protection.   5. Climatic changes–Prevention.   I. Booth, Colin (Colin A.)
 TH880.S655 2012
 720'.47–dc23
                                                                          2011035050

A catalogue record for this book is available from the British Library.

Wiley also publishes its books in a variety of electronic formats. Some content that appears in print may not be available in electronic books.

Set in 10/12pt Sabon by SPi Publisher Services, Pondicherry, India
Printed and bound in Malaysia by Vivar Printing Sdn Bhd

Series advisors

Carolyn Hayles, Queen's University, Belfast
Richard Kirkham, University of Manchester
Andrew Knight, Nottingham Trent University
Stephen Pryke, University College London
Steve Rowlinson, University of Hong Kong
Derek Thomson, Loughborough University
Sara Wilkinson, Deakin University

Innovation in the Built Environment (IBE) is a new book series for the construction industry published jointly by the Royal Institute of Chartered Surveyors and Wiley-Blackwell. It addresses issues of current research and practitioner relevance and takes an international perspective, drawing from research applications and case studies worldwide.

- Presents the latest thinking on the processes that influence the design, construction and management of the built environment

- Based on strong theoretical concepts and draws on both established techniques for analysing the processes that shape the built environment – and on those from other disciplines

- Embraces a comparative approach, allowing best practice to be put forward

- Demonstrates the contribution that effective management of built environment processes can make

Published and forthcoming books in the IBE series

Senaratne & Sexton, *Managing Change in Construction Projects: A Knowledge-Based Approach*
Lu & Sexton, *Innovation in Small Professional Practices in the Built Environment*
Pryke, *Construction Supply Chain Management: Concepts and Case Studies*
Kirkham, *Whole Life-Cycle Costing*
Booth, Hammond, Lamond & Proverbs, *Solutions for Climate Change Challenges in the Built Environment*
Roper & Borello, *Facility Management: International Perspectives, Best Practice and Case Studies*

We welcome proposals for new, high-quality, research-based books which are academically rigorous and informed by the latest thinking; please contact Madeleine Metcalfe.

Madeleine Metcalfe
Senior Commissioning Editor
Construction
Wiley-Blackwell
9600 Garsington Road
Oxford OX4 2DQ
mmetcalfe@wiley.com

This book is dedicated to those future generations who will experience the realities of climate change, in particular:

Esmée
Annabelle, Christabel, Ethan and Nathan
Francesca and William
Charlotte and Ella

Contents

Contributors	xiii

Chapter 1 Introductory Insights to Climate Change Challenges — 1
Felix N. Hammond, Colin A. Booth, Jessica E. Lamond and David G. Proverbs

1.1	Introduction	1
1.2	Climate Change Theory	2
1.3	The Controversy and Context	5
1.4	Organisation of the Book	8

Chapter 2 Climate Change: Nature and Emerging Trends — 11
Mark McCarthy

2.1	Introduction to the Climate System and Changes to the Radiative Forcing	11
2.2	Emerging Trends (Global)	13
2.3	Emerging Trends (UK)	17
2.4	Climate Trends and the Built Environment	19
2.5	Solutions	20

Chapter 3 Regional Implications — 23
Ana Lopez

3.1	Introduction	23
3.2	Climate Modelling	23
3.3	Projections of Future Climate Change	27
3.4	Solutions to the Challenges of Interpreting Climate Change Projections for the Characterisation of Climatic Risks	30

Chapter 4 Urbanization and Climate Change — 33
Felix N. Hammond, Kwasi Baffour Awuah Gyau and Stanislaus Y. Adiaba

4.1	Introduction	33
4.2	State of the World's Urbanization	33
4.3	Impact of Urbanization on Climate Change	35
4.4	How Does Urbanization Affect Climate Change?	36
4.5	Solutions for Change	39
4.6	Conclusion	41

Chapter 5 Global Political Initiatives and Overtones 45
Jean-Luc Salagnac

5.1	Introduction	45
5.2	Climate and the Built Environment	45
5.3	Background to Political Initiatives	47
5.4	Mitigation and Adaptation Policies	50
5.5	Solutions to Climate Change Challenges for the Built Environment	53

Chapter 6 Green Economics Dialogue and the Built Environment 57
Miriam Kennet

6.1	Introduction	57
6.2	Examples of Unsustainable Building Practices	59
6.3	The Choices We Face	62
6.4	Green Projects in Action	68
6.5	Conclusions	72

Chapter 7 Strategic Environmental Impact Assessment 75
Joseph Somevi

7.1	Introduction	75
7.2	Strategic Environmental Assessment	76
7.3	Contributions of SEA to Climate Change Solutions	77
7.4	Concluding Remarks	84

Chapter 8 Methods for Valuing Preferences for Environmental and Natural Resources: An Overview 87
Jessica E. Lamond and Ian Bateman

8.1	Introduction	87
8.2	Monetary Evaluation of Environmental Preferences: Theory	89
8.3	Methods for Monetary Evaluation of Environmental Preferences	91
8.4	Solutions to Valuation of Environmental and Natural Resources	95

Chapter 9 Ecological Value of Urban Environments 99
Ian C. Trueman and Christopher H. Young

9.1	Introduction	99
9.2	Ecological Value	100
9.3	Urban Habitats	101
9.4	Landscape Scales and Urban Areas	106
9.5	Ecological Implications of Climate Change	107
9.6	Implications of Climate Change for Urban Ecology	108
9.7	Solutions to Climate Change Challenges for the Built Environment	108

Chapter 10   The Pedological Value of Urban Landscapes   113
Jim Webb, Michael A. Fullen and Winfried E.H. Blum

10.1   Introduction   113
10.2   Urban Soils: The 'Grey Areas' on Soil Maps   113
10.3   Policy Responses for Urban Soils   119

**Chapter 11   Insights and Perceptions of Sustainable Design
and Construction**   127
David W. Beddoes and Colin A. Booth

11.1   Introduction   127
11.2   Sustainable Construction   128
11.3   Drivers for Sustainable Construction   130
11.4   Rethinking Construction   135
11.5   Thoughts for Change   137
11.6   Concluding Remarks   138

**Chapter 12   Progress in Eco and Resilient Construction
Materials Development**   141
Jamal M. Khatib

12.1   Introduction   141
12.2   Concrete   142
12.3   Brick and Masonry   144
12.4   Glass   145
12.5   Timber and Bamboo   145
12.6   Steel   146
12.7   Polymer-based Materials   147
12.8   Nanotechnology   148
12.9   Future Trends   149

Chapter 13   Energy Efficiency: Alternative Routes to Mitigation   153
David Coley

13.1   Introduction   153
13.2   Energy Efficiency   153
13.3   Carbon Sequestration and Climate Engineering   155
13.4   A Sustainable, Low-Carbon Future?   157
13.5   Solutions: Abatement Costs   161

Chapter 14   The Benefits of Green Infrastructure in Towns and Cities   163
Susanne M. Charlesworth and Colin A. Booth

14.1   Introduction   163
14.2   Integrating Vegetation into the Built Environment   163
14.3   Intercepting Rainfall and Reducing Flood Risk   167
14.4   Enhancing Urban Biodiversity   168
14.5   Limiting the Overheating of Buildings   169
14.6   Improving Human Health and Wellbeing   171
14.7   Sequestering Carbon to Offset CO_2 Emissions   172

14.8 'Green Infrastructure' Solutions for Climate
 Change Challenges 173
14.9 Conclusions 176

**Chapter 15 Particulate-Induced Soiling on Historic Limestone
Buildings: Insights and the Effects of Climate Change** 181
David E. Searle

15.1 Introduction 181
15.2 Urban Particulate Pollution 182
15.3 Soiling of Buildings 182
15.4 The Bath Study 184
15.5 Insights from the Bath Study 186
15.6 Effects of Climate Change on the Soiling of Buildings 188
15.7 Conclusions 188

Chapter 16 Sustainable Transportation 193
Panagiotis Georgakis and Christopher Nwagboso

16.1 Introduction 193
16.2 Climate Change and Sustainable Transportation 194
16.3 Perspectives of Sustainable Transportation 195
16.4 Development of Sustainable Transportation Systems 197
16.5 Solutions for Sustainable Transportation 203

**Chapter 17 Linkages of Waste Management Strategies
and Climate Change Issues** 207
Kim Tannahill and Colin A. Booth

17.1 Introduction 207
17.2 Integrated Solutions Approach 209
17.3 Key Policy Drivers: A European Perspective 216
17.4 Solutions for the Waste Management Sector 219

**Chapter 18 Climate Change and the Geotechnical
Stability of 'Engineered' Landfill Sites** 223
Robert W. Sarsby

18.1 Introduction 223
18.2 Ground Instability Effects 224
18.3 Stability of Soil Slopes 225
18.4 Soil Shear Strength 229
18.5 Landfill Sites 231
18.6 Insights and Solutions 235

**Chapter 19 Water Resources Issues and Solutions for the
Built Environment: Too Little versus Too Much** 237
Susanne M. Charlesworth and Colin A. Booth

19.1 Introduction 237
19.2 Too Little Water: Water Supply Shortages 238
19.3 Too Much Water: Urban Flooding 242

19.4	Property-Level Flood Resistance versus Resilience Measures	246
19.5	Present and Future Water Resources Solutions	247
19.6	Conclusions	248

Chapter 20   Organisational Culture and Climate Change Driven Construction — 251
Nii A. Ankrah and Patrick A. Manu

20.1	Introduction	251
20.2	Climate Change and Construction	251
20.3	Climate Change Driven Construction	252
20.4	The Role of Culture	253
20.5	The Culture of the UK Construction Industry	256
20.6	Achieving and Sustaining a Culture of Sustainability	257
20.7	Theory to Practice – Case Study	263
20.8	Emerging Solutions for a More Responsive Climate Change Culture in Construction	265

Chapter 21   Preparing for Extreme Weather Events: A Risk Assessment Approach — 269
Keith Jones

21.1	Introduction	269
21.2	What Is an Extreme Weather Event?	270
21.3	Relationship between Vulnerability, Resilience and Adaptive Capacity	271
21.4	A Risk Assessment Framework Model	276
21.5	Solutions: A New Risk Framework Model	277
21.6	Final Thoughts	279

Chapter 22   The Socio-environmental Vulnerability Assessment Approach to Mapping Vulnerability to Climate — 283
Fiifi Amoako Johnson, Craig W. Hutton and Mike J. Clarke

22.1	Introduction	283
22.2	The SEVA Approach	284
22.3	Results	291
22.4	Conclusions	299

Chapter 23   Mitigation via Renewables — 303
David Coley

23.1	Introduction	303
23.2	Current World Sustainable Energy Provision	304
23.3	Solar Power	305
23.4	Photovoltaics	308
23.5	Wind Power	308
23.6	Wave Power	310
23.7	Large-Scale Hydropower	310
23.8	Tidal Power	313
23.9	Biomass	314

23.10	Geothermal	316
23.11	Nuclear: Fast Breeders and Fusion	317
23.12	The Hydrogen Economy and Fuel Cells	318
23.13	Solutions	319

Chapter 24   Complexities and Approaches to Managing the Adaptation of Climate Change by Coastal Communities   **321**
Annie T. Worsley, Vanessa J.C. Holden, Jennifer A. Millington and Colin A. Booth

24.1	Introduction – What's Special about the Coast?	321
24.2	Coastal Landforms and Process	322
24.3	Challenges Facing Coastal Communities	323
24.4	Ways of Managing Coastal Challenges	324
24.5	Shoreline Management Plans	325
24.6	Case Study: The North Sefton Coast	326
24.7	Solutions for Coastal Communities	333

Chapter 25   Lessons for the Future   **337**
Jessica E. Lamond, David G. Proverbs, Colin A. Booth and Felix N. Hammond

25.1	Introduction	337
25.2	Technological Solutions	338
25.3	Working with the Natural Environment	339
25.4	Enabling Change	340
25.5	Final Remarks	342

Index   343

Contributors

Stanislaus Y. Adiaba
School of Technology
University of Wolverhampton
Wulfruna Street
Wolverhampton, WV1 1LY
UK

Nii A. Ankrah
School of Technology
University of Wolverhampton
Wulfruna Street
Wolverhampton, WV1 1LY
UK

Ian Bateman
School of Environmental Sciences
University of East Anglia
Norwich, NR4 7TJ
UK

David W. Beddoes
School of Technology
University of Wolverhampton
Wulfruna Street
Wolverhampton, WV1 1LY
UK

Winfried E.H. Blum
University of Natural Resources and Applied Life Sciences
Institute of Soil Research
Peter Jordan Str. 82
1190 Vienna
Austria

Colin A. Booth
School of Engineering and the Built Environment
University of Wolverhampton
Wulfruna Street
Wolverhampton, WV1 1LY
UK

Susanne M. Charlesworth
Faculty of Business, Environment and Society
George Eliot Building
Coventry University
Priory Street
Coventry, CV1 5FB
UK

Mike J. Clarke
GeoData Institute
School of Geography
University of Southampton
Highfield
Southampton, SO17 1BJ
UK

David Coley
Department of Architecture and Civil Engineering
University of Bath
Bath, BA2 7AY
UK

Michael A. Fullen
School of Applied Sciences
University of Wolverhampton
Wulfruna Street
Wolverhampton, WV1 1LY
UK

Panagiotis Georgakis
School of Technology
University of Wolverhampton
Wulfruna Street
Wolverhampton, WV1 1LY
UK

Kwasi Baffour Awuah Gyau
School of Technology
University of Wolverhampton
Wulfruna Street
Wolverhampton, WV1 1LY
UK

Felix N. Hammond
School of Technology
University of Wolverhampton
Wulfruna Street
Wolverhampton, WV1 1LY
UK

Vanessa J.C. Holden
Strata Environmental
PO Box 249
Hexham
NE46 9FL
UK

Craig W. Hutton
GeoData Institute
School of Geography
University of Southampton
Highfield
Southampton, SO17 1BJ
UK

Fiifi Amoako Johnson
GeoData Institute
School of Geography
University of Southampton
Highfield
Southampton, SO17 1BJ
UK

Keith Jones
School of Architecture
and Construction
University of Greenwich
Old Royal Naval College
Park Row
Greenwich
London, SE10 9LS
UK

Miriam Kennet
The Green Economics Institute
6 Strachey Close
Tidmarsh
Reading
Berkshire, RG8 8EP
UK

Jamal M. Khatib
School of Technology
University of Wolverhampton
Wulfruna Street
Wolverhampton, WV1 1LY
UK

Jessica E. Lamond
Faculty of Environment and Technology
University of the West of England
Frenchay Campus
Bristol, BS16 1QY
UK

Ana Lopez
School of Geography and the Environment
University of Oxford
South Parks Road
Oxford, OX1 3QY
UK

Patrick A. Manu
School of Technology
University of Wolverhampton
Wulfruna Street
Wolverhampton, WV1 1LY
UK

Mark McCarthy
The Meteorological Office
Fitzroy Road
Exeter, EX1 3PB
UK

Jennifer A. Millington
School of Technology
University of Wolverhampton
Wulfruna Street
Wolverhampton, WV1 1LY
UK

Christopher Nwagboso
School of Technology
University of Wolverhampton
Wulfruna Street
Wolverhampton, WV1 1LY
UK

David G. Proverbs
Faculty of Environment and Technology
University of the West of England
Frenchay Campus
Bristol, BS16 1QY
UK

Jean-Luc Salagnac
Economics and Human Sciences Department
Centre Scientifique et Technique
du Bâtiment (CSTB)
4 avenue Poincaré 75782
PARIS cedex 16
France

Robert W. Sarsby
Department of the Built Environment
Anglia Ruskin University
Bishop Hall Lane
Chelmsford, CM1 1SQ
UK

David E. Searle
School of Technology
University of Wolverhampton
Wulfruna Street
Wolverhampton, WV1 1LY
UK

Joseph Somevi
Sustainability Research, Information Planning and Environmental Services
Aberdeenshire Council
Woodhill House
Westburn Road
Aberdeen, AB16 5GB
UK

Kim Tannahill
School of Technology
University of Wolverhampton
Wulfruna Street
Wolverhampton, WV1 1LY
UK

Ian C. Trueman
School of Applied Sciences
University of Wolverhampton
Wulfruna Street
Wolverhampton, WV1 1LY
UK

Jim Webb
AEA Technology Plc.
Gemini Building
Harwell Business Centre
Didcot
Oxfordshire, OX11 0QR
UK

Annie T. Worsley
Strata Environmental
PO Box 249
Hexham
NE46 9FL
UK

Christopher H. Young
School of Applied Sciences
University of Wolverhampton
Wulfruna Street
Wolverhampton, WV1 1LY
UK

Introductory Insights to Climate Change Challenges

Felix N. Hammond, Colin A. Booth, Jessica E. Lamond
and David G. Proverbs

1.1 Introduction

The epic phenomenon of the 21st century, with which this book is concerned – climate change – was originally designated, or rather started, as a concern for global warming. A distinction is now maintained between the two terminologies. *Global warming* is restricted to the measurable rapid warming of the Earth's surface identified from a study of worldwide temperature records since 1880 attributable to human activities (Pielke *et al.*, 2004; Pielke, 2005; Nodvin, 2010; Riebeek, 2010). *Climate change* conversely now signifies 'changes in the state of the climate that can be identified by changes in the average and/or the variability of its properties… that persists [*sic*] for an extended period, typically decades or longer' (Nodvin, 2010). Climatic events associated with global warming include volatility and extremities of climatic events such as rainfall, sea level rise, drought, volcanic activities, hurricanes, loss of biodiversity, heightened storm intensity, frequent heat waves, altered precipitation patterns, reversal of ocean current and flooding, amongst others (Goulder, 2006; Tamirisa, 2007; American Institute of Physics [AIP], 2010). Because the climate of the earth is driven by the surface temperature of the earth (Lindsey, 2009), global warming is the prime cause of variation in global climate. To this end, the two phenomena cannot actually be decoupled in any serious sense; anything that influences global warming ultimately influences climate change.

There are areas of substantial uncertainty about climate change. What scientists agree on is that climate change is real and that if not curbed could result in catastrophic consequences (Stern *et al.*, 2006). As a result, the past two decades have seen unprecedented concern about the consequences of climate change and the cost of reducing its long-term impact. Climate

---

Solutions to Climate Change Challenges in the Built Environment, First Edition.
Edited by Colin A. Booth, Felix N. Hammond, Jessica E. Lamond and David G. Proverbs.
© 2012 Blackwell Publishing Ltd. Published 2012 by Blackwell Publishing Ltd.

change is now very high on the worldwide political agenda. This has led to major international initiatives such as the United Nations Framework Convention on Climate Change (UNFCCC), the Kyoto Protocol, the establishment of the United Nations Intergovernmental Panel on Climate Change and the 1992 Rio de Janeiro Conference. These are aimed at achieving a globally coordinated accord on adapting and mitigating climate change. 'Changing land cover and land use' have been implicated as a 'major underlying cause of' global warming (Intergovernmental Panel on Climate Change, Task Group on Data and Scenario Support for Impact and Climate Assessment [IPCC-TGICA], 2007). Changing land cover and land use are built environment and agricultural activities such as the construction of buildings, roads and highways and other infrastructure. The objective of this volume is to improve understanding of how built environment activities potentially induce global warming and climate change but, moreover, to highlight solutions to these challenges.

As the ensuing chapters show, climate change is a very intricate phenomenon, the understanding and handling of which involve mathematics, biology, physics, politics, economics, industrial science, climatology and so forth almost in equal measure (Lawson, 2006; AIP, 2010). That said, the complex thrust of the ongoing climate change debate and research can be distilled into five answerable focal questions: (1) Is the Earth temperature increasing with the possibility of reaching an intolerable limit at some point? (2) What is the threat that this poses to the very survival of life on Earth? (3) What are the fundamental cause(s)? (4) For our present purpose, are activities of the built environment elemental contributors to the warming of the Earth? And (5) what can be done, if anything, to forestall or cope with the problem of rising Earth temperature?

This volume contains 25 chapters that have attempted to address aspects of these questions from diverse perspectives, primarily by looking at the questions from the interdependent relations between climate change and built environment endeavours. For the sake of the uninitiated reader, it is useful to provide a brief overview of climate change theory before exploring its connections with the built environment.

1.2 Climate Change Theory

To understand climate change we must begin with an understanding of the term *climate*. In the text, climate has been used in two main senses, the narrow and broader sense. According to the IPCC,

> Climate in a narrow sense is usually defined as the "average weather," or more rigorously, as the statistical description in terms of the mean and variability of relevant quantities over a period of time ranging from months to thousands or millions of years. The classical period is 30 years, as defined by the World Meteorological Organization (WMO). These quantities are most often surface variables such as temperature, precipitation, and wind. Climate in a wider sense is the state, including a statistical description, of the climate system. (IPCC, 2007b)

Whereas:

> In a broader sense, [however] climate is the status of the climate system which comprises the atmosphere, the hydrosphere, the cryosphere, the surface lithosphere and the biosphere. These elements determine the state and dynamics of the Earth's climate. ([WMO], 2010)

In this book we are concerned with climate in the narrow sense.

1.2.1 Primary Cause of Climate Change

The problems of global warming and climate change have arisen because of the Earth's climate dependence on solar radiation (energy from the sun) as its primary source of power (Trenberth *et al.*, 2009; Congressional Budget Office [CBO], 2005). The solar radiation from the Sun is generated by its surface heat which is about 5500°C (Lindsey, 2009). The Sun transmits this heat or solar energy towards the Earth to power the Earth's climate. For at least two reasons, only a fraction of the solar radiation from the Sun ultimately reaches the Earth surface. The intensity of thermal energy reduces with distance. Thus, being some 150 million kilometres (93 million miles) away from the Sun, the intensity of the solar energy reduces drastically by the time it travels this distance to make contact with the surface of the Earth (AIP, 2010). Further reduction in the intensity of the solar energy from the Sun occurs as it penetrates the atmosphere, which encircles the Earth in order to reach the Earth's surface. By the time the solar energy reaches the top surface of the atmosphere, its intensity has diminished substantially. The atmosphere consists of nongreenhouse gases such as nitrogen and oxygen; water vapour; and greenhouse gases carbon dioxide, methane and others. On contact with the top surface of the atmosphere, about 30% of ultraviolet light is re-radiated by the atmosphere back to space. Of the remaining 70% that manages to penetrate the surface of the atmosphere, 19% is trapped (absorbed) by the greenhouse gases in the atmosphere. Approximately 51% of the ultraviolet light then passes through the atmosphere onto the Earth system – land surface or ocean (AIP, 2010). Whilst this is an on-going process, the Earth's temperature does not increase endlessly because energy is also dissipated away from the Earth.

The ultraviolet light from the Sun that manages to penetrate the atmosphere warms up objects on the Earth's surface. The warmed Earth emits heat energy in the form of infrared radiation back into space that cools down the Earth. The intensity of the infrared radiation emitted by the Earth is equal to that of the ultraviolet radiation it receives from the Sun. Without the intervention of the atmosphere, the process of radiations from the Sun to the Earth, and from the Earth to the Sun, would leave the temperature of the Earth unaltered or constant at −18°C (similar to that of the moon which is approximately the same distance from the Sun as the Earth). However, not all of the infrared radiation from the Earth reaches space. Indeed, NASA estimates that only about 6% of the infrared radiation from the Earth does so. This is because whilst the greenhouse gases in the atmosphere permit a considerable volume of the ultraviolet radiation from the Sun to penetrate the atmosphere to reach the Earth, they are not that transparent to the infrared radiation from the Earth. The greenhouse gases absorb the infrared radiation from the Earth and re-radiate a significant proportion back to the Earth's surface. This is then reflected back to the atmosphere and then back again onto the Earth surface and so forth. This process is called the greenhouse effect and it ultimately increases the surface temperature of the Earth. As a result, the average temperature of the earth surface hovers around 15°C, some 33°C warmer than a body without an atmosphere. Carbon dioxide (CO_2) is the main greenhouse gas implicated for the greenhouse effect which is causing global warming and hence climate change (Oregon Wild, 2007). Although CO_2 is not the most impenetrable of the gases, it is seen as most important because its levels have increased the most and hence it has influenced global warming the most. As Svante Arrhenius (the Swedish chemist who made the earliest effort to estimate the actual effect of greenhouse gases on climate in 1895) found, removing all CO_2 from the atmosphere would lower global temperature by 31°C (Warwick and Wilcoxen, 2002).

Though available data show that the greenhouse effect has caused the Earth temperature to fluctuate over time, its net effect in the twentieth century is a warming of the earth surface above that of the pre-industrial era. It is estimated that over the last century, the global Earth surface temperature has increased by about between 0.6°C and 0.8°C and is set to increase further in the next century. Though the forecasts may not all be identical owing to the variety of bodies involved in this research, there may safely be considered a overwhelming majority of experts predicting the warming of the earth resulting from greenhouse effects.

Since CO_2 is seen as the main culprit of climate change, it is useful to appreciate the source of atmospheric CO_2 and how its concentration can be regulated to generate a favourable Earth surface temperature. CO_2 is emitted through natural processes as well as through human actions. The planet Earth has a fixed volume of carbon (Oregon Wild, 2007). This can, however, be circulated and stored (that is, can be taken out of the atmosphere). Carbon is generally stored in all living things: rocks, sediments and the air (Holmes, 2008; Sedjo, 1993). Through a combination of natural activities (such as volcanic activities, death and decomposition of organic matter and living plants, leaves, animals and humans) and human actions (such as soil excavations and combustion of fossils through the use of fossil fuels – high carbon containing fuels – such as coal, natural gas, gasoline and oil for heating, transportation and electricity), CO_2 is released and shifted into the atmosphere (Oregon Wild, 2007).

The built environment is crucial in the climate change and global warming dialogue. It is estimated that nearly half (50%) of UK CO_2 emissions are buildings related, and 27% of UK CO_2 emissions come from housing (Department of Trade and Industry [DTI], 2006). As implied in the name, *built environment* refers to the aspects of the physical environment that have been built upon by humans. This mainly consists of the construction of buildings and infrastructure. This human-driven activity results in the release of considerable quantities of CO_2 into the atmosphere in many ways. Firstly construction involves excavation of topsoils, a process that releases stored carbons in the soil into the atmosphere. Additionally, construction is heavily dependent on the use of machines that are powered by either electricity or fossil fuels such as gasoline, oil or coal. Besides the extraction, manufacturing and transporting construction materials contribute some 10% of UK CO_2 emissions (DTI, 2006). Then again, the use and management of the constructed facilities involve substantial reliance of fossil fuels for heating, lighting and the operation of facilities such as computers, lifts and projectors. Built environment transportation systems such as trains, motor vehicles and air transport also make heavy use of fossil fuels. There is little doubt, therefore, that the built environment contributes to the quantity of CO_2 in the atmosphere and hence to global warming and climate change.

1.3 The Controversy and Context

Climate change is not without its own raging controversies. That climate change is influenced by greenhouse effect, and that human activities have increased the concentration of the CO_2 component in the greenhouse gases, are well accepted by experts in the field. There are, however, many controversial areas and uncertainties surrounding the science of climate change which are impeding progress in finding cost-effective solutions to the issue. The starting point of this debate is the anthropogenic climate change theory, or what is also sometimes referred to as *man-made catastrophic climate change theory*. The Intergovernmental Panel on Climate Change, comprising over 1000 scientists from over 100 countries, being the official mouthpiece of government across the globe on climate change and sponsored by the United Nations, remains the ardent proponent of this theory. Whilst CO_2 emissions may emanate from natural causes or human actions, this theory holds that the contributions from human activities alone are responsible for a significant increase in global warming. The IPCC concludes that:

> Global atmospheric concentrations of CO_2 [carbon dioxide], methane and nitrous oxide have increased markedly as a result of human activities since 1750 and now far exceed pre-industrial values determined from ice cores spanning many thousands of years.... The atmospheric concentrations of CO_2 and CH_4 in 2005 exceed by far the natural range over the last 650 000 years. Global increases in CO_2 concentrations are due primarily to fossil fuel use, with land-use change providing another significant but smaller contribution. It is *very likely that* the observed increase in CH_4 concentration is predominantly due to agriculture and fossil fuel use. The increase in N_2O concentration is primarily due to agriculture.... There is *very high confidence* that the global average net effect of human activities since 1750 has been one of warming, with a radiative forcing of +1.6. (2007b, 37)

In 2008, in affirmation of the above, the US Congressional Budget Office concluded that '[h]uman activities are producing increasingly large quantities of greenhouse gases, particularly carbon dioxide (CO_2), which accumulate in the atmosphere and create costly changes in regional climates throughout the World' (CBO, 2008). Based on the anthropogenic climate change theory, Sir Nicholas Stern *et al.* estimate that 'the overall cost and risks of climate change will be equivalent to losing at least 5% of global GDP each year, now and forever...these are risks of major disruption to economic and social activity, on a scale similar to those associated with the great wars and the economic depression of the first half of the 20th century' (2006). This brings the solution or at least moderation of the rate of global warming within the province of public policy; if it is caused by conscious human action, then it can be redressed through policy. After all, the aim of policy, whether economic or another, is to reform, shape or direct conscious human actions along lines that are compatible with the outcomes expected. Hence Stern *et al.* (2006), could assert that 'the benefit of strong early action outweigh the costs'. This probably explains why those in charge of policy – government policy advisers, government departments, United Nations, the World Bank and so on – have embraced the anthropogenic climate change theory and are working on policy solutions to climate change. At least by setting policy actions to regulate the volume and rate of human action contributions to climate change, policy makers and politicians are demonstrating to voters that they are performing their primary function – the reason why they exist – which is to promote the wellbeing of people. There is no presumption here that this is the motivation for the policy conclusion drawn. The main policy lines followed in controlling human actions for the benefit of climate change include: emission trading programs, emission taxes, performance standards and technology programs (Goulder, 2006).

There are great areas of uncertainty about the science of climate change. For instance, there is still uncertainty about the unique contribution of atmospheric water vapour, clouds and aerosols to climate change (Warwick and Wilcoxen, 2002). Besides these uncertainties, spells of global cooling have led to scepticism about whether the ongoing global warming can indeed reach the point of catastrophe.

Horner (2007) points out that the massive funding of climate change research was prompted by '"consensus" panic over "global cooling"'. As Bray (1999) asserts, 'before we take global warming as a scientific truth, we should note that the opposite theory was once scientific verity'. Along the same lines, Balling (1992) avers:

> Could the [cold] winters of the late 1970s be the signal that we were returning to yet another ice age? According to many outspoken climate scientists in the late 1970s, the answer was absolutely yes – and we needed action to cope with the coming changes....However, some scientists were sceptical, and they pointed to a future of global *warming*, not cooling, resulting from a continued build up of greenhouse gases. These scientists were in the minority at the time.

According to Crichton (2004):

"Just think how far we have come!" Henley said. "Back in the 1970s, all the climate scientists believed an ice age was coming. They thought the world was getting colder. But once the notion of global *warming* was raised, they immediately recognized the advantages. Global warming creates a crisis, a call to action. A crisis needs to be studied, it needs to be funded.

Michaels (2004) posits,

Thirty years ago there was much scientific discussion among those who believed that humans influenced the...reflectivity [which would] cool the Earth, more than...increasing carbon dioxide, causing warming. Back then, the "coolers" had the upper hand because, indeed, the planet was cooling. ...But nature quickly shifted gears. ...Needless to say, the abrupt shift in the climate caused almost as abrupt a shift in the balance of scientists who predictably followed the temperature.

The evidence of global cooling after the pre-industrial era brings into question the conclusiveness of anthropogenic climate change theory. For, if human activities contribute significantly to atmospheric CO_2 concentration, then since human activities have increased since the age of industrialisation, we should expect the Earth's temperature to head upwards only. But as the above quotes show, this has not been the case; there have been periods of severe cooling even to the point of causing concern about a possible global ice age. If the Earth has experienced so great a cooling in the past and it is now warming, it may well be that global cooling alternates with global warming and that the warming we experience today would eventually be replaced by a period of global cooling. This suggests at least that a better understanding of the underlying causes of the global temperature alternation is required in order to be able to make long-range predictions about global warming.

This lack of complete understanding has led to a conundrum. Subscribers to catastrophic anthropogenic climate change theory such as the IPCC have called for an immediate reduction of human activities that generate CO_2. The cost of doing so is phenomenal though believed to be trivial compared to the cost of the global warming-induced catastrophe that awaits us. Huge scarce resources are also now being directed towards enforcing such CO_2 reduction policies. Sceptics towards this theory hold the view that, since the contribution to temperature changes by human activities are actually very small and the current warming is likely to be replaced by cooling, there is no economic justification, or rather it will be a wasteful and unnecessary infraction on enjoyment to curtail certain human activities, spend all those resources and defer economic growth for a phenomena over which humans have no real control (Lawson, 2006). It is worth noting that this view is in the minority, and nearly all government and international development agencies hold tenaciously to the anthropogenic climate change theory.

Controversy aside and regardless of the degree of accuracy which climate models of future warming can claim, it is clear that the planet has warmed in the recent past. This is engendering extreme weather events that are damaging to the built environment. Also there is little doubt that man-made emissions have made some contribution to the warming and that the continued unchecked generation of greenhouse gases makes catastrophic warming scenarios more likely to occur. Climate change mitigation policy is here to stay for the foreseeable future. It is very important to note that alongside the threat of climate change there are also opportunities. These could represent new economic possibilities or the chance to improve the built or natural environment. New opportunities may offset the risks and reduce the need for society to defend existing buildings and livelihood. Built environment professionals and stakeholders should therefore seek to pursue climate mitigation and adaptation practices. To this end, this book presents solutions to some of the issues connected with climate change.

1.4 Organisation of the Book

Based on the challenges outlined above, this book comprises five main themes: (1) climate change experiences (Chapters 1, 2, 3, 5 and 6), (2) urban landscape development (Chapters 4, 9, 10, 11, 12 and 14), (3) urban management issues (Chapters 13, 15, 16, 17, 18, 19 and 20), (4) measurement of impact (Chapters 7, 8, 21 and 22) and (5) the future (Chapters 23, 24 and 25).

References

American Institute of Physics (2010) A hyperlinked history of climate change science. Spencer Weart and American Institute of Physics, College Park, MD. Available from http://www.aip.org/history/climate/summary.htm.

Balling, R.C., Jr. (1992) *The Heated Debate: Greenhouse Prediction versus Climate Reality*. Pacific Research Institute for Public Policy, San Francisco.

Bray, A.J. (1999) The Ice Age cometh. *Policy Review*, **58**, 82–84.

Congressional Budget Office (CBO) (2005) *Uncertainty in analyzing climate change: policy implications*. CBO Paper. Congressional Budget Office, Washington, DC.

Congressional Budget Office (CBO) (2008) *Policy options for reducing CO_2 emissions*. CBO Study. Washington, DC: Congressional Budget Office, Washington, DC.

Crichton, M. (2004) *State of Fear*. Avon Books, New York.

Department of Trade and Industry (DTI) (2006) *DTI Strategy for Sustainable Construction Consultation Events*. Department of Trade and Industry, London.

Goulder, L.H. & Pizer, W.A. (2006) *The economics of climate change*. Discussion paper RFF 06-06. Resources for the Future, Washington, DC.

Holmes, R. (2008) *The Age of Wonder*. Pantheon Books, New York.

Horner, C.C. (2007) *The Politically Incorrect Guide to Global Warming and Environmentalism*, Regnery Publishing, Washington, DC.

Intergovernmental Panel on Climate Change (IPCC) (eds.) (2007a) *Climate Change 2007: Synthesis Report: Contribution of Working Groups I, II and III to the*

Fourth Assessment Report of the Intergovernmental Panel on Climate Change. Geneva, IPCC.

Intergovernmental Panel on Climate Change (IPCC) (2007b) Frequently asked questions: what is climate? Available from http://www.wmo.int/pages/prog/wcp/ccl/faqs.html.

Intergovernmental Panel on Climate Change, Task Group on Data and Scenario Support for Impact and Climate Assessment (IPCC-TGICA) (2007) *General Guidelines on the Use of Scenario Data for Climate Impact and Adaptation Assessment.* Intergovernmental Panel on Climate Change, Task Group on Data and Scenario Support for Impact and Climate Assessment, Geneva.

Lawson, N. (2006) *The economics and politics of climate change: an appeal to reason.* Lecture to the Centre for Policy Studies. Centre for Policy Studies, London.

Lindsey, R. (2009) *Climate and Earth's Energy Budget.* NASA Earth Observatory, Greenbelt, MD.

Michaels, P.J. (2004) *Meltdown: The Predictable Distortion of Global Warming by Scientists, Politicians and the Media.* Cato Institute, Washington, DC.

Nodvin, S.C. (ed.) (2010) *Global Warming.* Environmental Information Coalition, National Council for Science and the Environment, Washington, DC.

Oregon Wild (2007) *The straight facts on forest, carbon and global warming.* Special Oregon Wild Report. Oregon Wild, Portland, OR.

Pielke, R.A. (2005) Heat storage within the Earth system. *Forum: American Meteorological Society,* 331–335.

Pielke, R.A., Davey, C. & Morgan, J.A. (2004) Assessing global warming with surface heat content. *EOS: American Geophysical Union,* **85**, 210–211.

Riebeek, H. (2010) Global warming. NASA Earth Observatory, Greenbelt, MD. Available from http://earthobservatory.nasa.gov/Features/GlobalWarming/.

Sedjo, R. (1993) The carbon cycle and global forest ecosystem. *Water, Air, and Soil Pollution,* **70**, 295–307.

Stern, N., Peters, S., Bakhshi, V., Bowen, A., Cameron, C., Catovsky, S., Crane, D., Cruickshank, S., Dietz, S., Edmonson, N., Garbett, S.L., Hamid L., Hoffman, G. & Ingram, T. (2006) *Stern Review: The Economics of Climate Change.* HM Treasury, London.

Tamirisa, N. (2007) Climate change and global economy. *Finance and Development,* **45** (1). Available from http://www.imf.org/external/pubs/ft/fandd/2008/03/tamirisa.htm.

Trenberth, K.E. & Fasullo, J.K. (2009) Earth's global energy budget. *Bulletin of the American Meteorological Society,* **90**, 311–323.

Warwick, J.M. & Wilcoxen, P.J. (2002) The role of economics in climate change policy. *Journal of Economic Perspectives,* **16**, 107–129.

World Meteorological Organisation (2010) *Understanding Climate.* World Meteorological Organisation, London.

Climate Change: Nature and Emerging Trends

Mark McCarthy

2.1 Introduction to the Climate System and Changes to the Radiative Forcing

An often used definition of climate is that it represents an 'average' of the weather. However, this is actually a manifestation only of the atmospheric component of the Earth's climate system. The climate system is also nonstationary, responding to both natural and anthropogenic (man-made) external drivers, and its own internal variability across timescales from days to millennia. For many applications within climate science, one considers the climate system as the representation of the processes maintaining the energy balance of the Earth, accounting for both the balance of incoming and outgoing radiation, and the redistribution of energy through the global ocean–land–atmosphere system.

As much as 99.97% of energy received by the Earth comes in the form of shortwave radiation from the Sun with small additional sources from geothermal and tidal (although these are generally considered negligible components of the overall energy balance). In order to maintain thermal equilibrium, the Earth must itself reflect and emit sufficient radiation to space in order to balance the incoming energy. The transparency of the atmosphere to the incoming solar (shortwave) and its opacity to terrestrial (longwave) radiation gives rise to the phenomenon commonly known as the greenhouse effect. The two most abundant atmospheric constituents of the Earth's atmosphere, oxygen and nitrogen, have a negligible contribution to the natural greenhouse effect. However, trace constituents, notably CO_2 and water vapour, act to absorb and re-emit upwelling longwave radiation from the surface and lower atmosphere, elevating temperatures at the surface and through the lowermost portion of the atmospheric column (the troposphere). The impact of a particular *forcing* on the climate is commonly described by

---

Solutions to Climate Change Challenges in the Built Environment, First Edition.
Edited by Colin A. Booth, Felix N. Hammond, Jessica E. Lamond and David G. Proverbs.
© 2012 Blackwell Publishing Ltd. Published 2012 by Blackwell Publishing Ltd.

the equivalent *radiative forcing* representing the instantaneous net change in irradiance (in Wm^{-2}) at the top of the troposphere. A positive radiative forcing will be associated with an increase in the heat content of the climate system, while a negative forcing is associated with a decrease.

A change in the energy balance of the climate system can therefore occur as a result of changes in the incident solar radiation (natural solar cycles, or changes in the Earth's orbit), a change in the fraction of reflected solar radiation (through changes in cloud cover, atmosphere composition or land cover change) or changes in the opacity of the atmosphere to longwave radiation (through changes in greenhouse gas concentrations). Current concentrations of atmospheric CO_2 and some other notable greenhouse gases such as methane (CH_4) and nitrous oxide (N_2O) are higher than at any point in the ice-core records of the past 650 000 years (e.g. Siegenthaler *et al.*, 2005). Emissions of CO_2 from fossil fuel burning and cement manufacture are likely to contribute to about two-thirds of this observed increase (at a rate of 1.9 ppm yr^{-1} between 1995 and 2005), and supported by analysis of the reduction in the ratio between the two dominant carbon isotopes ($^{13}C/^{12}C$) being consistent with expectation of CO_2 release from fossil fuel burning (Keeling *et al.*, 2010). Increases in atmospheric CO_2 between 1750 and 2005 results in the largest radiative forcing of the climate system equivalent to approximately 1.7 Wm^{-2}, but this figure rises to 2.6 Wm^{-2} if contributions from all other long-lived greenhouse gases are accounted for. There is high confidence that this net forcing is unprecedented in over 10 000 years, and it is considered extremely likely that human activity has contributed to a net warming influence on the climate (IPCC, 2007). Additional forcings from land cover change and atmospheric aerosol effects reduce the net forcing estimate to 1.6 Wm^{-2} which is still an order of magnitude larger than the equivalent estimate of forcing from changes in solar irradiance over the same period estimated at 0.12 Wm^{-2}.

The above discussion provides a convenient summary for assessing and comparing the net impact of both man-made and natural forcings on the climate system, but it is far from a complete description of a changing climate and cannot be easily related to an equivalent rise in surface temperature. The atmosphere, ocean, cryosphere (land and sea-ice) and terrestrial biosphere are all components of the Earth's climate system and will react to external forcings. For example, a majority of any additional heat provided to the climate system will go into the large ocean heat store, while climate feedbacks such as changes in cloud cover, cloud properties and changes in ice and snow cover will in turn influence the net energy balance and redistribution of energy through the climate system. Meanwhile, components of the terrestrial and ocean biosphere affect the natural exchanges of carbon between surface stores and the atmosphere. Approximately half of the man-made emissions of CO_2 have gone into the land and ocean carbon stores, thus the ecosystem has provided a service to limit the full potential impact of greenhouse gas emissions to date. The importance of terrestrial ecosystem services in regulating CO_2 concentrations is now widely recognised, as is their sensitivity to climate change, and many climate models now explicitly include representation of the carbon cycle. The impacts of climate change are expected to have a negative impact on the ability of the Earth's biosphere to remove anthropogenic greenhouse gas emissions in the future (Friedlingstein et al., 2006). An ever increasing understanding of the importance of the various components of the Earth's climate system is reflected in the ever increasing complexity of global climate and earth system models and continued active development since the 1970s. However, early experiments to test the climate response to elevated CO_2 conducted by Svante Arrhenius (1896) and George Callendar (1938) provided estimates of surface temperature responses to a doubling of CO_2 that are still within the uncertainty range of modern numerical climate modelling methods, highlighting that the development of our understanding has a strong foundation in fundamental and robust physical explanation of the role of greenhouse gases in the definition of climate and climate change.

2.2 Emerging Trends (Global)

Warming of the climate over the twentieth century is unequivocal (IPCC, 2007). Global mean temperatures have increased by close to 0.8°C between the mid-nineteenth century and the first decade of the twenty-first century. The three decades of the 1980s, 1990s and 2000s have each been warmer than the preceding decade, and each has set a statistically significant record compared to any other decade in the 150-year observational record (Arndt et al., 2010). Global mean temperature is a useful and long-term gauge of climate change, and it is far from the only strand of evidence for a changing climate. Monitoring of key variables relating to the state of the oceans, the atmosphere and the cryosphere all exhibit significant trends consistent with warming of the global climate system, some (but not all) of which are discussed below. In addition multiple analyses of each essential climate variable exist that are based on independent data platforms and data analysis methods. These diverse but consistent strands of data provide high confidence in the continued warming of the global climate system. In a report entitled 'State of the Climate in 2009' by Arndt et al. (2010), published at http://www1.ncdc.noaa.gov/pub/data/cmb/bams-sotc/climate-assessment-2009-lo-rez.pdf (and adapted from Kennedy et al., 2010), the figure available at page 26 shows time series from a range of indicators that would be expected to correlate strongly with the surface record. Note that the stratospheric cooling is an expected consequence of greenhouse gas increases. This figure demonstrates the warming of land, oceans and atmosphere and the retreat of snow and ice in the northern hemisphere. In each case the magnitude of the emerging signal is greater than the uncertainty resulting from observational or analysis methods, and comparison to multiple paleo-climate reconstructions from ice cores and tree rings suggests that the global warming trend is unusual in the context of the last millennia (Jansen et al., 2007).

The observational evidence for climate change is very strong, as is the evidence of significant perturbations to the energy budget discussed previously. But these strands of information in isolation do not provide sufficient evidence of a direct man-made contribution to the observed warming trend through greenhouse gas emissions. Detection and attribution methods (Stott et al., 2010) provide a means to assess the significance of change (detection) and determine the most likely underlying causes (attribution). In the case of detection we must determine whether or not any observed signal is outside the expectation of internal variability of the system. Multicentury simulations of climate provide a suitably stable and lengthy dataset with which to determine the statistical significance of changes in the observational records.

Considerable speculation and controversy have been sparked by an apparent levelling off of global mean temperatures over the period 1998 to 2010, with little or no apparent warming during this period. Decade-long periods with little warming, or even cooling, are to be expected under long-term radiatively forced climate change. Climate models have sufficient natural variability to also exhibit similar periods of no trend or even slight cooling (Knight et al., 2009). Therefore, detection of climate change must consider at the very least multidecadal, and preferably centennial-scale, observational records.

Attribution is the identification of 'fingerprints' of change in the climate system in response to specific forcing mechanisms, and provides an assessment of there being a detectible manifestation of specific fingerprints within the observational record (for a summary, refer to Stott et al., 2010). Anthropogenically forced temperature change has been detected over the last 50 years across all continents, and is summarised in Figure 2.1 (reproduced from IPCC, 2007); consistent changes in temperature-related indices such as frost days, growing season and heatwaves can also be explained only by including anthropogenic forcing (e.g. Meehl et al., 2007; Christidis et al., 2005). There is growing interest in the potential for the attribution of specific extreme events to climate change. It would be inappropriate to make statements linking a specific event to climate change directly, but it is possible to attribute changes in the likelihood of events occurring. For example, human influences on climate have more than doubled the risk of European summer heatwaves of a magnitude similar to that of the unprecedented heatwave of the summer of 2003 (Stott et al., 2004).

Temperature trends through the depth of the troposphere (the lowest portion of the atmosphere up to approximately 15–20 km) show a consistent warming (Arndt et al., 2010, upper-right panel of Figure 2.1 adapted from Kennedy et al., 2010); however, greater uncertainty in the various satellite and weather balloon data records resulting from significant technological advancements, and sampling strategies, over the twentieth century have limited, but not prevented, the detection of specific forcing responses in these data. This figure demonstrates that temperatures in the stratosphere are cooling significantly, which is again an expected consequence of increased greenhouse gas concentrations warming the troposphere below. Decreases in ozone concentration further contribute to lowering of stratospheric temperatures. In contrast, increase in solar irradiance would be expected to contribute to elevated stratospheric temperatures due to the additional absorption of incoming radiation by ozone and other trace gas constituents at this level. Natural climate forcing resulting from volcanic activity is also clearly apparent in the historical stratospheric temperature records.

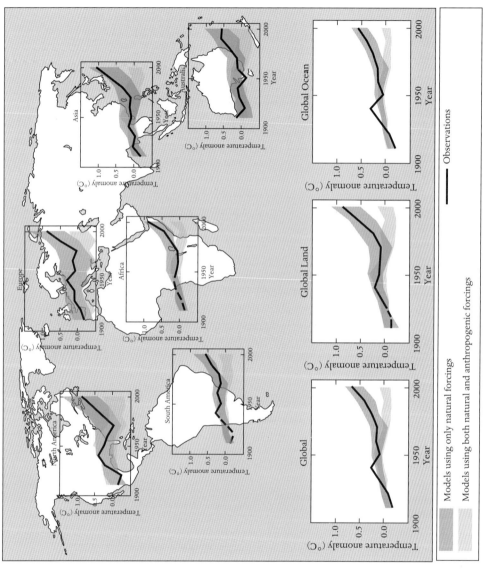

Figure 2.1 Comparison of observed continental and global scale changes in surface temperature with results simulated by climate models using natural and anthropogenic forcings. From *Climate Change 2007: The Physical Science Basis*. Working Group I Contribution to the Fourth Assessment Report of the Intergovernmental Panel on Climate Change, Figure SPM.4. Cambridge University Press. Reproduced by permission of the IPCC.

Although there is a general cooling trend in the lower stratosphere, warming events lasting several years following the major eruptions of the tropical volcanoes Mount Agung (1963–64), El Chichon (1982) and Mount Pinatubo (1991) are apparent. These result from the mass injection of sulphate aerosols into the upper atmosphere where they have a residence time of several years; they also reduce the solar radiation reaching the surface and act to cool the surface and troposphere below. This again illustrates how different forcing mechanisms exhibit distinct fingerprints of change on different aspects of the climate system, and contribute to the continued development of our understanding of the underlying physical processes involved.

An important natural climate feedback of the atmosphere arises from water vapour feedback. As the atmospheric temperature increases, the water-holding capacity of the column also increases, and evaporation from the ocean increases. Water vapour is itself a strong natural greenhouse gas, and therefore the increased water content of the atmospheric column increases the opacity to terrestrial radiation. The net forcing from this feedback is partly offset by changes in the vertical profile of temperature (also known as the *lapse rate feedback*), but it remains the single most important natural feedback to greenhouse gas-induced climate change. Anthropogenic signals have been detected in both surface- and satellite-based datasets of atmospheric humidity for the latter part of the twentieth century (Willett *et al.*, 2007; Santer *et al.*, 2007).

A large proportion of excess heat resulting from anthropogenic forcing will be partitioned into the ocean heat store. Therefore, in order to obtain a more complete description of the state of the climate system, we must account for this. The subsurface ocean has been sparsely observed in many parts of the world, resulting in considerable sampling uncertainties in the historical observational record. However, recent analysis of specific properties of the oceans heat content have been shown to be very well produced by climate models that include anthropogenic forcing (Palmer *et al.*, 2009).

Global mean sea levels have been observed to rise at a rate of approximately 1.8 mm/year, consistent with estimated contributions from a combination of thermal expansion and the melting of land water stores in glaciers and icecaps. Melting of the polar ice sheets are a small secondary contribution of twentieth-century sea level rise, but contribute to a significant uncertainty in projections of future rises. The potential impact of small apparent sea level rises will be most devastating when they are combined with storm surge events, posing additional risks for low-lying coastal and estuary communities and ecosystems. Significant reductions in northern hemisphere snow cover and sea ice are also consistent with the large-scale warming trends.

The wealth of observational data and attribution studies highlights the ever growing evidence basis for human-induced climate change. The development of attribution methods utilising climate models has furthered our understanding of the climate system, and directly contributed to the identification of where the uncertainties remain. The overwhelming picture from the observational evidence and emerging trends across a range of global and regional indices is one of significant warming of the climate system that is consistent in its pattern and magnitude with expectation of the response to the observed increase in atmospheric concentrations of important greenhouse gases.

Our confidence in global climate models (GCMs) as a tool for quantifying future projections of climate change is supported by their ability to reproduce observed changes in historical climate including the natural variability. They can provide realistic representations of known atmospheric and oceanic processes, based on a foundation in physical and mathematical principles and the fundamental laws that govern the conservation of mass, energy and momentum. However, future projections of climate change carry a large uncertainty range resulting from a combination of the natural variability in the climate system, socioeconomic uncertainty affecting future greenhouse gas emission scenarios and modelling uncertainty relating to the necessary numerical approximations (parameterisations) and incomplete or missing physical processes.

Regardless of the inherent uncertainties robust findings are provided by comprehensive multimodel analyses such as those contributing to the IPCC assessments (IPCC, 2007). Even if concentrations of greenhouse gases are rapidly stabilised, further warming and related climate change would be expected to occur as a result of the thermal inertia of the climate system. This is often referred to as *committed warming*, creating a requirement for some adaptive responses to future climate change irrespective of current and future emission rates. Recent research has shown that the peak warming associated with man-made forcing is related to the total cumulative emissions of greenhouse gases, rather than the emission rates (Allen et al., 2009; Meinshausen et al., 2009). If emission of greenhouse gases continues at or above current rates, then it is very likely that twenty-first-century climate change would exceed that observed in the twentieth century. Warming is greatest at northern latitudes over land, and heat-related extremes will become more frequent and longer lasting. Snow cover and sea ice extent will reduce, and glaciers will continue to recede. Precipitation patterns will be altered with generally increasing precipitation in the tropical rain band and in the higher latitudes, and decreases in precipitation in the subtropics. Increased water holding capacity is also expected to result in increased intensity of rainfall events. Furthermore, future warming would tend to reduce the capacity of the land and ocean biospheres to absorb anthropogenic CO_2 resulting in an ever increasing fraction of emitted CO_2 remaining in the atmosphere.

2.3　Emerging Trends (UK)

The Central England Temperature record (CET) is the longest continuous climate record of anywhere in the world, extending back to 1659 (Manley, 1974; Parker, 1992). It is currently comprised of Rothamstead (Hertfordshire), Pershore (Worcestershire) and Stonyhurst (Lancashire) and is representative of the temperature variability across the United Kingdom (Croxton et al., 2006). The CET has risen by approximately 1°C since 1950, with a significant warming since 1980. There is considerable variability on inter-annual and interdecadal timescales in the CET relating to natural variations in both atmospheric and oceanic circulations in the North Atlantic and Arctic. However, it is still considered likely that there has been a significant influence from human activity on the recent warming (Karoly and Stott, 2006). Nine of the ten warmest years in the CET record have occurred since 1989, and 2003 saw temperatures exceed 38°C for the first time recorded in the United Kingdom. Heating degree days (requirement for buildings to be heated) has seen close to an 18% reduction in southern parts of England over the period 1961–2006 (Jenkins et al., 2008), while cooling degree days (requirement for cooling of buildings) have increased significantly.

The longest running precipitation record for the United Kingdom is the England and Wales Precipitation dataset (EWP) extending back to 1766 (Alexander and Jones, 2001). There has been no significant trend in the total annual mean EWP, but there are seasonal trends showing increases in winter precipitation and high-intensity rain events, and decreases in summer, consistent with expectation of climate change. Even if the total amount of rain over the United Kingdom remains similar under a changing climate, the potential for changing seasonality could have impacts on water resource management. Furthermore, changes to the intensity of rainfall events or the risk of summer droughts (Burke et al., 2010) could pose specific challenges to the built environment in the United Kingdom.

Uncertainty is a fundamental aspect of climate change projection and can rapidly escalate as one moves from global to regional to local climate impacts. There have been significant advances in the methods and tools for both making and interpreting projections of future climate change that are not limited solely to the development of numerical global climate models. While the GCMs provide the basis of our physical understanding of the climate system, and underpin detection, attribution and projections of future change, there are important developing methods for making probabilistic projections, and for decision making in the face of large uncertainty, a problem that is not unique to climate research.

A shift from commonly used frequentist statistics of hypothesis testing to a Bayesian paradigm in which one tests the probability of an outcome given some prior distribution of the expected outcomes is presented in the UK climate change projections of Murphy et al. (2009) (UKCP09). The probabilities provided by the UKCP09 quantify the degree of confidence in projections of a range of essential climate variables accounting for natural variability, modelling uncertainty and future emissions. A key development of this method is that the climate projections are no longer extracted directly from the climate model simulations, but instead enter a more sophisticated Bayesian framework with which to define the relative likelihood of certain outcomes.

The UKCP09 methodology estimates that it is very unlikely (10% probability) that UK summer temperature will rise by much less than 1°C by the 2050s with a central estimate of 2°C to 3°C depending on location. The projections show robust signature of increasing winter precipitation and decreases in summer. However, the uncertainty range for summer rainfall changes does cross zero (i.e. we cannot reject the possibility of small increases in summer rain based on the available evidence).

For the United Kingdom, despite considerable natural variability of our climate, there are detectable signatures of man-made climate change. Continued rises in greenhouse gas concentrations and subsequent global climate change would be expected to continue these trends through the twenty-first century. Significant developments in the application of climate change results from GCMs have yielded probabilistic assessments of climate change for the United Kingdom that can provide a foundation for subsequent assessment of impacts and adaptation.

2.4 Climate Trends and the Built Environment

An often over-looked component of climate change is the role of the built environment. Urban areas account for only a tiny fraction of the available land surface of the world, but a majority of the world's population is now considered to be urban, with a projected growth of urban population to 6 billion by 2050 (United Nations, 2007) making urban areas focal points of potential vulnerability to climate change. World cities account for approximately two-thirds of global primary energy demand, and are therefore responsible for a major share of man-made greenhouse gas emissions.

It has long been known that cities generate their own microclimates, first reported by Luke Howard in his seminal nineteenth-century study of London (Howard, 1833). Long before scientists were able to associate greenhouse gas emissions with global climate change, we have been aware of the very localised man-made climate change associated with cities. The thermal, aerodynamic and air quality of the urban surface and atmosphere affects temperature, wind, clouds and precipitation both in the city itself and in the immediate surroundings and downwind of the city. The most apparent expression of the urban microclimate is the urban heat island in which cities are significantly warmer than their rural surroundings.

Climate research has gone to considerable lengths to ensure that the influence of urbanisation does not disproportionately affect the estimates of global mean temperature used to detect the signature of greenhouse gas-induced change. As such it is estimated that urbanisation influences are less than $0.006°C$ per decade since 1900 on the global land record and $0.002°C$ on the global land and ocean record (IPCC, 2007). However, the climate change to which many people are exposed will be a combination of climate change induced by the global radiative forcing from greenhouse gases, and the local climate modifications arising from regional and local land use change – in particular urbanisation. With dramatic urbanisation projected through the twenty-first century, the impacts of climate change on people and the built environment will be exacerbated by the continued impact of urbanisation on the local urban microclimate. For example, warming trends at and above the rate of global climate change have been detected in many Japanese cities over the twentieth century (Fujibe, 2009), with some locations exhibiting a trend of $10°C$ per century in extreme nocturnal minimum temperatures. Furthermore the response of the urban environment to global change can be significantly greater than for non-urban surfaces, particularly for night-time extremes (McCarthy *et al.*, 2010).

The influence of greenhouse gas emissions associated with urban areas is included in future assessments of global climate change. But a secondary consequence of urban energy use on climate is gaining increasing attention. Energy used for electricity, heating, cooling, hot water, transport and even human metabolism can be reasonably assumed to eventually be released to the environment as heat. We can treat energy resulting from the combustion of fossil fuels as an additional heat source to the climate system. At the global scale the forcing associated with this is of order $0.01\,\text{Wm}^{-2}$ and is generally considered to be a negligible contribution to the global energy balance. However, the potential for continental-scale impacts for future energy projections is possible (Flanner, 2009). Concentrated over the relatively small urban land surface, these heating rates can become significant exceeding 100s or even 1000s of Wm^{-2} (Ichinose *et al.*, 1999).

Global climate change, the urban environment and urban energy demands are all interdependent. The magnitude of the urban heat island is controlled by both the regional climate and the urban morphology, function and extent, which in turn impact on urban energy use, in turn influencing both local and global climate change. Urban areas represent focal points of greenhouse gas emissions, and are also focal points of vulnerability to climate change. However, this means that they can also become powerful focal points for adaptation and mitigation strategies. Interpreting climate change impacts on the built environment should therefore consider not only the influence of greenhouse gas induced change, but also the cumulative impacts from climate change and urbanisation. Rapid advancements in urban climatology in recent decades are providing an increasing number of numerical tools capable of simulating the complex urban microclimate, which in turn are increasingly suitable for coupling to dynamical atmosphere models, or building models, in order to better understand and represent the important processes and interactions that will operate across the building to street to neighbourhood to city to regional and global climate scales.

2.5 Solutions

The emerging trends in the meteorological, oceanic and biogeochemical indicators of change continue to provide overwhelming evidence of twentieth-century climate change that is very likely to be a direct consequence of the man-made emissions of important greenhouse gases including CO_2, methane and nitrous oxide. Further change is an inevitable consequence of the emissions we have made through greenhouse gas emission to date, and mitigation of climate change reductions is required to minimise the potential full extent of such change. Meanwhile adaptation strategies are required to help to reduce the potential impacts of committed climate change over the coming decades. Robust evidence of past, present and future climate change has been collated; there are inherent uncertainties associated with future projections of change, not least because of the dependence on socioeconomic development. However, developments in climate research relate not only to the continued improvement in the physical models of the climate system, but also in the ongoing development of more useful and intuitive analysis methods to provide, for example, probabilistic assessments of future projections.

The built environment has the potential to play a critical role in solutions to address both the problem of adapting society and infrastructure to a changing climate, and the issue of mitigating our emissions of greenhouse gases with the potential for mutually beneficial strategies that consider the interactions of the built environment with local-, regional- and global-scale climate change.

References

Alexander, L.V. & Jones, P.D. (2001) Updated precipitation series for the U.K. and discussion of recent extremes. *Atmospheric Science Letters*, DOI: 10.1006/asle.2001.0025.

Allen, M.R. *et al.* (2009) Warming caused by cumulative carbon emissions towards the trillionth tonne, *Nature*, DOI:10.1038/nature08019.

Arndt, D.S., Baringer, M.O. & Johnson, M.R. (eds.) (2010) State of the Climate in 2009. *Bull. Amer. Meteor. Soc.*, **91** (6), S1–S224.

Arrhenius, S. (1896) On the influence of carbonic acid in the air upon the temperature on the ground, *Philos. Mag.*, **41**, 237–276.

Burke, E.J., Perry, R.H.J. & Brown, S. (2010) An extreme value analysis of UK drought and projections of change in the future, *J. Hydrology*, DOI:10.1016/j.jhydrol.2010.04.035.

Callendar, G.S. (1938) The artificial production of carbon dioxide and its influence on temperature, *Q. J. R. Meteorol. Soc.*, **87**, 223–237.

Christidis, N. *et al.* (2005) Detection of changes in temperature extremes during the 20th Century. *Geophys. Res. Lett.*, DOI:10.1029/2005JD006280.

Croxton, P.J., Huber, K., Collinson, N. & Sparks, T.H. (2006) How well do the Central England temperature and the England and Wales precipitation series represent the climate of the UK?, *Int. J. Climatol.*, **26**, 2287–2292.

Flanner, M.G. (2009) Integrating anthropogenic heat flux with global climate models, *Geophys. Res. Lett.*, 36, L02801, doi:10.1029/2008GL036465.

Friedlingstein, P. *et al.* (2006) Climate–carbon cycle feedback analysis: results from the C4MIP model intercomparison. *J. Climate*, 19, 3337–3353, doi: 10.1175/JCLI3800.1.

Fujibe, F. (2009), Detection of urban warming in recent temperature trends in Japan, *Int. J. Climatol.*, DOI: 10.1002/joc.1822/.

Howard, L. (1833) *The climate of London, deduced from meteorological observations made in the metropolis and various places around it*. Harvey and Darton, London.

Ichinose, T., Shimodozono, K. & Hanaki, K. (1999), Impact of anthropogenic heat on urban climate in Tokyo. *Atmospheric Environment*, **33**, 3897–3909.

IPCC (2007) *Climate Change 2007: The Physical Science Basis. Contribution of working group I to the Fourth Assessment Report of the Intergovernmental Panel on Climate Change*, ed. Solomon, S., Qin, D., Manning, M., Chen, Z., Marquis, M., Averyt, K.B., Tignor, M. & Miller, H.L. Cambridge University Press, Cambridge.

Jansen, E. et al. (2007) Paleoclimate. In: *Climate Change 2007: The physical science basis. Contribution of working group I to the Fourth Assessment Report of the Intergovernmental Panel on Climate Change*, ed. Solomon, S., Qin, D., Manning, M., Chen, Z., Marquis, M., Averyt, K.B., Tignor, M. & Miller, H.L. Cambridge University Press, Cambridge.

Jenkins, G.J., Perry, M.C. & Prior, M.J. (2008) The climate of the United Kingdom and recent trends. Met Office Hadley Centre, Exeter, UK. http://ukclimateprojections.defra.gov.uk.

Karoly, D. & Stott, P. (2006) Anthropogenic warming of Central England temperature. *Atmospheric Science Letters*, 7, 81–85.

Keeling, R.F., Piper, S.C., Bollenbacher, A.F. & Walker, S.J. (2010) Monthly atmospheric $^{13}C/^{12}C$ isotopic ratios for 11 SIO stations. In *Trends: A Compendium of Data on Global Change*. Carbon Dioxide Information Analysis Center, Oak Ridge National Laboratory, U.S. Department of Energy, Oak Ridge, TN. http://cdiac.ornl.gov/trends/co2/iso-sio/iso-sio.html.

Kennedy, J.J., Thorne, P.W., Peterson, T.C., Ruedy, R.A., Stott, P.A., Parker, D.E., Good, S.A., Titchner, H.A. & Willett, K.M. (2010) How do we know the world has warmed? *Bull. Amer. Meteor. Soc.*, 91 (6), S26–S27.

Knight, J. et al. (2009) Do global temperature trends over the last decade falsify climate predictions? *Bull. Amer. Meteor. Soc.*, 90 (8), S22–S23.

Manley, G. (1974) Central England Temperature: monthly means 1659 to 1973. *Q. J. R. Meteorol. Soc.*, 100, 389–405.

McCarthy, M.P., Best, M.J. & Betts, R.A. (2010) Climate change in cities due to global warming and urban effects. *Geophys. Res. Lett.*, DOI:10.1029/2010GL042845.

Meehl, G., Arblaster, J. & Tebaldi, C. (2007) Contributions of natural and anthropogenic forcing to changes in temperature extremes over the United States. *Geophys. Res. Lett.*, DOI:10.1029/2007gl030948.

Meinshausen, M. et al. (2009) Greenhouse-gas emission targets for limiting global warming to 2°C. *Nature*, DOI:10.1038/nature08017.

Murphy, J.M. et al. (2009) UK Climate Projections Science Report: Climate change projections. Met Office, Hadley Centre, Exeter, UK. http://ukclimateprojections.defra.gov.uk.

Palmer, M.D., Good, S.A., Haines, K., Rayner, N.A. & Stott, P.A. (2009), A new perspective on warming of the global oceans. *Geophys. Res. Lett.*, 36, L20709, DOI:10.1029/2009GL039491.

Parker, D.E., Legg, T.P. & Folland, C.K. (1992) A new daily Central England Temperature Series, 1772–1991. *Int. J. Climatol.*, 12, 317–342.

Santer, B.D. et al. (2007) Identification of human-induced changes in atmospheric moisture content. *Proc. Natl. Acad. Sci.*, DOI: 10.1073/pnas.0702872104.

Siegenthaler, U. et al. (2005) Stable carbon cycle–climate relationship during the late Pleistocene. *Science*, 310 (5752), 1313–1317. DOI: 10.1126/science.1120130.

Stott, P.A., Gillett, N.P., Hegerl, G.C., Karoly, D.J., Stone, D.A., Zhang, X. & Zwiers, F. (2010) Detection and attribution of climate change: a regional perspective. *Wiley Interdisciplinary Reviews: Climate Change*, 1 (2), 192–211.

Stott, P.A, Stone, D.A. & Allen, M.R. (2004) Human contribution to the European heatwave of 2003. *Nature*, 432, 610–614.

United Nations (2007) *World Urbanisation Prospects: The 2007 Revision*. Department of Economic and Social Affairs, United Nations, New York. Available from www.un.org/esa/population/publications/wup2007/2007wup.htm.

Willett, K.M., Gillett, N.P., Jones, P.D. & Thorne, P.W. (2007) Attribution of observed surface humidity changes to human influence. *Nature*, 449, DOI:10.1038/nature06207.

3

Regional Implications

Ana Lopez

3.1 Introduction

The levels of greenhouse gases in the atmosphere are higher than they have been for hundreds of thousands of years and there is every indication that they will continue to rise. The effect of these increasing levels, and the consequent changes in the climate system, on food production, water availability, human health, the built infrastructure and energy demand, are expected to be wide ranging. In order to quantify the scale of these impacts, projections of changes at the relevant regional and local scales are needed. In this chapter we discuss what the currently available tools to generate these projections are and what they imply for regional climate change.

We start with a short description of climate models, including a discussion about the different factors that influence the interpretation of model results such as spatial and temporal scales of interest, climate variables under consideration and uncertainties in climate change projections. We then move on to summarize the robust projections of regional climate change as described in the IPCC *Fourth Assessment Report* (Solomon et al., 2007a, b). We close the chapter by discussing key issues that in our view should be taken into account when using climate modelling to evaluate the impacts of climate change.

3.2 Climate Modelling

Projections of changes in future climate are based on climate models simulations. Even though the physical and chemical processes in the climate system follow known scientific laws, the complexity of the system implies that many simplifications and approximations have to be made when modelling it. The choice of approximations creates a variety of physical

---

Solutions to Climate Change Challenges in the Built Environment, First Edition.
Edited by Colin A. Booth, Felix N. Hammond, Jessica E. Lamond and David G. Proverbs.
© 2012 Blackwell Publishing Ltd. Published 2012 by Blackwell Publishing Ltd.

climate models that can be broadly divided into two groups: simple climate models and general circulation models (GCMs).

The simplest possible models reduce the climate system to a zero-dimensional problem that describes the time evolution of the global mean temperature. These models have the advantage that they are easy to evaluate and interpret. On the other hand, they are subject to many simplifying assumptions, and only represent some parts of the climate system, typically global mean temperature, without any internally generated variability (i.e. weather noise). They are useful, for instance, to understand the effects of uncertainties in future emissions and changes in millennial scales.

At the other end of the spectrum, the general circulation models or global climate models are fully dynamical models that numerically calculate the exchanges of air, moisture, heat and radiation on a three-dimensional grid through time. The atmospheric component of the model interacts with a land surface model that represents storage and transport of soil moisture, river runoff and evapo-transpiration from the ground and vegetable layer. It also interacts with an ocean and a sea ice model interchanging heat, momentum, water and, in the newer models, chemical compounds (McGuffie and Henderson-Sellers, 2005; Randall et al., 2007).

The GCMs assume that everything within each grid cell or box is uniform, and they cannot explicitly calculate anything smaller. Due to limitations in current computer power, typically atmospheric grid boxes have a horizontal resolution of 1° to 3° in latitude and longitude. In the vertical direction there are around 30 levels whose dimensions range from a few hundred metres close to the surface, at the boundary layer, to several kilometres at the top layers. Processes that act at smaller spatial scales cannot be resolved explicitly. As a consequence of this, models cannot simulate local variations in climate. More importantly, since many of these unresolved processes also affect climate on larger scales, they have to be represented by simple empirical approximations called *parameterisations*. Analogously, the models run at discrete time steps, and cannot resolve anything happening at time intervals smaller than 10 to 20 minutes for current models.

Since climate processes occurring in a given area depend on phenomena occurring at smaller and larger spatial scales, the variability of weather over a GCM grid box depends on the parameterizations that represent the smaller scale processes and that are only known empirically.

Therefore, even though GCM output can be obtained at the grid box resolution (and similarly for regional climate model output; see below), model projections should not be taken literally at the spatial resolution provided. The general (not proven) assumption is that, when averaging over several grid boxes, the GCM output effectively models the smaller scale processes, and as a result GCM output should only be interpreted at scales greater than about 1000 km. A similar argument holds for time resolution[1].

---

[1] Moreover, it can be shown that the time step and the spatial resolution used to solve the GCM's equations numerically must satisfy certain relationships for the solution to exist (Courant et al., 1967).

However in this case, model output is provided at daily or monthly steps, and not at the 10–20-minute model step since the latest is not considered to provide any useful information over the former one (Stone, 2008; Stone and Knutti, 2010).

The interpretation of climate model output depends not only on the spatial and temporal scales of interest, but also on the climate variable under consideration (Stone and Knutti, 2010). Temperature for instance is a direct variable as simulated by the climate model, while precipitation is represented entirely by parameterizations. While temperature is a continuous variable whose changes vary smoothly and fit standard statistical models, precipitation is intermittent, can vary over very short distances and periods of time and does not entirely fit any statistical model. Even though models can differ by more than 1°C in their estimation of global mean surface air temperature, their estimations of warming over the last hundred years differ by only about a tenth of that. If the interest is on trends rather than absolute values, the disagreement in absolute values might not matter. However, for many questions, such as those related with the strength of the temperature-snow or ice feedbacks, it does matter which side of the freezing point the absolute value is. In the case of precipitation, long-term and large-scale averages compare better to observations because the episodic characteristics of precipitation averages out. There are exceptions to this, for instance in regions where topography is important or precipitation events are very rare; in these cases the location or timing of events cannot be considered as random. Humidity is also a direct variable that varies continuously over large scales, similarly to temperature. Surface radiation is a continuous variable, but it depends on the representation of clouds in the model and, in this sense, depends indirectly on the parameterizations. Surface winds depend very strongly on the local topography and vegetation, which are poorly represented in models.

GCMs forced by scenarios of greenhouse gases predict significant changes in the Earth's climate in this century (Solomon et al., 2007a). The range in the projections is in some cases very large, in particular at the regional and local scales relevant for the analysis of impacts and adaptation options under climate change. Several sources of uncertainty are present in GCM's simulations, including radiative forcing, initial conditions, model formulation (including resolution) and model inadequacy uncertainties (Stainforth et al., 2007).

In order to explore these types of uncertainties, different computational experiments are carried out. For instance, running climate models forced by different emissions scenarios (Nakicenovic et al., 2000) allows to quantify uncertainty in forcing of the climate system induced by human activities.

The analysis of multi-GCM ensembles, usually termed *ensembles of opportunity*, such as those available from the Program for Climate Model Diagnosis and Intercomparison (PCMDI) archives, is one approach that provides information on GCM structure uncertainties (Solomon et al., 2007a). An alternative philosophy makes use of perturbed physics ensembles (PPEs), which are specifically aimed at evaluating uncertainty in GCM formulation (Murphy et al., 2004, 2007; Stainforth et al., 2005; Frame *et al.*, 2009). These ensembles usually comprise a large number of runs of a state-of-the-art climate model, where each individual run uses a version of the model with the parameters representing various physical processes set to different values within their acceptable range. For each combination of parameter values, an initial condition ensemble is used so that the relative contribution of formulation and initial condition uncertainty can be evaluated.

In the following section we describe results obtained using multi-GCM ensembles. However, it is important to mention that large perturbed physics ensembles are currently becoming a common tool that, in combination with multi-GCM ensembles, are being used to explore and quantify uncertainties in projections of climate change (Stainforth *et al.*, 2005; Jenkins *et al.*, 2009), and to inform possible adaptation options to climate change (Lopez *et al.*, 2009).

As previously mentioned, the GCM spatial resolution is of the order of 1° to 3° in latitude and longitude, much larger than the scales relevant to evaluate impacts of climate change. Different approaches have been developed to downscale the GCM model data to local scales: dynamical and statistical downscaling (Christensen *et al.*, 2007). We close this section by briefly discussing them.

Dynamical downscaling is carried out using regional climate models (RCMs). These are regional versions of a GCM that have the characteristics of a particular region of interest and that are embedded within a GCM. The states of the GCM simulation are used as the boundary conditions for the RCM. Current RCMs run at resolutions of about 30 km. The main assumption underlying the use of RCMs is that subgrid processes in the GCM outside the region of interest do not affect the climate of this region. Conversely, since the coupling is only one way (not feedback from the RCM is passed back to the GCM), it is assumed that the climate of the region of interest does not affect the climate outside this region. These are both very strong assumptions that should be tested when using this approach. Clearly all the caveats regarding interpretation of model data at the grid box scale, characteristics of the climate variable of interest and different types of model uncertainties also hold for RCMs.

Statistical downscaling is based on the possible existence of robust statistical relationships between some large-scale meteorological quantity and the local variable of interest. In order to apply this approach, historical data are needed to find these statistical relationships. These are then applied to the large-scale climate model data in order to obtain future projections at higher spatial resolution (for a review, see Wilby and Fowler, 2010, and references therein). This approach requires long observational data to adequately train and validate the statistical models. The main assumption in this case is that the statistical relationships will not change in a future climate. This is not true in all cases, and this assumption becomes particularly limiting if future states move outside of the range of the states that have been observed in historical records and were used to develop the statistical model.

Up to this point, we have briefly described GCMs and the different factors that influence the interpretation of their projections such as spatial and temporal scales of interest, climate variables under consideration and uncertainties in climate change projections. In the next section we summarize the main features of the regional climate change projections reported in the IPCC *Fourth Assessment Report*.

3.3 Projections of Future Climate Change

When analysing climate model output, it is important to keep in mind that GCMs simulations are forced with different scenarios of future greenhouse gases and sulphate aerosol emissions. Therefore, estimates of future climate change are projections that apart from being conditioned by the model's approximations depend explicitly on the hypothetical scenarios used to force it.

Current projections of global mean temperature changes suggest that, even if emissions were drastically reduced now, at least in the short term the world would become warmer (Solomon *et al.*, 2007a, b). Projections of global mean temperature change for the highest emissions scenario (SRES A2) and the lowest emissions scenario (SRES B1) overlap until the 2020s. Even under the commitment scenario[2], the projections partially overlap with the simulations from the other two highest and lowest emission scenarios (Solomon *et al.*, 2007a, b; Stone, 2008). Physically this can be explained by the fact that the climate system has a large inertia, and consequently it responds slowly to the greenhouse gases already in the atmosphere.

For the 2020s, changes in global mean precipitation are masked by its natural variability in the short term, partly because precipitation is a noisier variable, and partly because its response to higher greenhouse gas concentrations is not as direct as for temperature. When looking at longer term projections, the picture is different. Beyond the short to medium term, the emissions path does matter as well as the different models' sensitivities. For instance, changes in global mean temperature for the A2 scenario and for the B1 scenario do not overlap at the end of the twenty-first century (Solomon *et al.*, 2007a, b). Changes in global mean precipitation become distinguishable from its natural variability, and some robust patterns emerge across the ensemble of GCMs, such as an increase in the tropical precipitation maxima, a decrease in the subtropics and increases at high latitudes. However, since precipitation is a variable that strongly depends on the way the models parameterise unresolved processes such as cloud formation, the

---

[2] The SRESA2 storyline represents a world of independently operating nations, continuously increasing population, regionally oriented economic development and slow and fragmented per capita economic growth and technological change. The SRESB1 storyline represents a world that foments local solutions to economic, social and environmental sustainability. Global population growth is slower than for A2, the levels of economic development are intermediate, and technological change is less rapid and more diverse than in other storylines. The commitment scenario is the one where the emissions are held constant at the year 2000 values.

confidence in precipitation response to greenhouse gas increases is much lower that the confidence in simulated temperature response (Stone, 2008).

Clearly, regional changes can be larger or smaller than global averages, and in general, the smaller the scale the less consistent the picture is across the ensemble of GCM projections, particularly for some climate variables. The IPCC *Fourth Assessment Report* (Solomon et al., 2007a), in its 'Regional Climate Projections' chapter (Christensen et al., 2007), presents projections at continental scales, and goes down to subcontinental scales (Giorgi regions) in order to provide quantitative information that is still robust even though it corresponds to spatial scales that are not as coarse grained as continents. One key feature of these regions is that they are not smaller than the horizontal scales on which current GCMs are useful for climate simulations (typically judged to be roughly 1000 km).

Christensen et al. (2007)[3] state that confidence in temperature projections has increased due to the fact that a larger number and variety of simulations are available, models have improved, and there is a better understanding of model deficiencies. Generally the criterion used to evaluate the robustness of a result is that, if a greater agreement in projections in changes is supported by strong physical arguments, then there is more confidence about the likelihood of a given regional change.

The greatest amount of warming is expected, and has been observed, over the land masses. In particular, it is expected that large warming will occur at higher latitudes.

Even though these are regions with the largest uncertainties in projections, and by the 2020s some GCMs project very small (or even slightly negative) temperature changes, by the 2080s all GCMs project warming of one or more degrees with respect to the 1997–2006 decade for the SRES A1B (Stone, 2008).

Precipitation changes are less consistent than temperature changes, partly because precipitation is much more variable than temperature, and partly because it does not respond as directly to increases in greenhouse gases concentrations. The changes in annual means projected for the 2020s indicate that the largest potential changes and simultaneously the largest uncertainties occur in areas where precipitation is low, such as deserts and polar regions. By the 2080s, projections are noisy but some patterns emerge, such as the fact that precipitation in polar regions is projected to increase. This is related to the fact that models project a retreat of sea and lake ice, allowing surface waters to evaporate directly (Stone, 2008)[4].

At the regional level, seasonality of changes has to be considered, since clearly changes in annual averages do not uniquely determine the way in which the frequency or intensity of extreme weather events might change in the future. For instance, the warming in the Arctic is projected to be largest

---

[3] For an update on the assessment of climate risks since the IPCC *Fourth Assessment Report*, see Fussel (2009).
[4] These projections may be a result of the limitations of the sea ice component of climate models; therefore, their robustness might be questioned (Eisenman et al., 2007).

in winter and smaller in the summer. Similarly, annual Arctic precipitation is projected to increase with the largest increase in the winter and the smallest in the summer. In Europe, where the annual mean temperature is likely to increase, the largest warming is projected to be likely in winter in northern Europe and in summer in the Mediterranean area. Annual precipitation is likely to increase in northern Europe and decrease in most of the Mediterranean area (Christensen *et al.*, 2007).

Levels of confidence in projections of changes in frequency and intensity of extreme events, in particular regional statements concerning heat waves, heavy precipitation and droughts, can be estimated using different sources of information including observational data and model simulations. It has been observed that extreme cold and warm daily temperatures in most land regions have increased faster than the change in average temperature since 1950. Modelling studies also project that extremely high temperatures will continue to rise due to increasing climate variability (see Fussel, 2009, and references therein). This result is consistent with the expectation that heat waves will become more common, as a consequence of the lack of moisture to cool the ground through evaporation during more frequent dry periods. For extreme rainfall events, it is expected that these will be unrelated to changes in average rainfall. Average rainfall amount depends on the vertical temperature gradient of the atmosphere, which in turn depends on how quickly the top of the atmosphere can radiate energy into space, which is expected to change only slightly with changes in carbon dioxide concentrations. On the other hand, extreme precipitation depends on how much water the air can hold, which increases exponentially with temperature. Thus it is reasonable to expect that in a warmer climate, short extreme rainfall events could become more intense and frequent, even in areas that become drier on average. Some studies have found that in regions that are relatively wet already, extreme precipitation will increase, while areas that are already dry are projected to become drier due to longer dry spells. The spatial extent of areas with severe soil moisture deficits and the frequency of short-terms drought is expected to double by the late twenty-first century, and long-term droughts are projected to become three times more common, in particular in the Mediterranean, West African, Central Asian and Central American regions (see Fussel, 2009, and references therein).

Projections of extreme events in the tropics are uncertain, due in part to the difficulty in projecting the distribution of tropical cyclones using current climate models with too coarse spatial resolution, but also due to the large uncertainties in observational cyclones datasets for the twentieth century. For instance, some studies suggest that the frequency of strong tropical cyclones has increased globally in recent decades in association with higher sea surface temperatures, consistently with the hypothesis that as the oceans warm there is more energy available to convert to tropical cyclone wind. However, the reliability of this sort of analysis based on estimating trends from observational datasets has been questioned based on the argument that improved satellite coverage, new analysis methods and operational changes in the tropical cyclone warning centres have contributed to discontinuities in the datasets and more frequent identification of extreme tropical cyclones after 1990 (Fussel, 2009).

3.4 Solutions to the Challenges of Interpreting Climate Change Projections for the Characterisation of Climatic Risks

In the previous sections, we discussed how future climate projections are developed. We also summarised the projections of changes in future climate that are considered to be robust as reported by the IPCC *Fourth Assessment Report*. We close this chapter by briefly discussing some important issues that should be taken into consideration when climate model information is used to quantify climate change impacts and inform adaptation to climate change[5].

Traditionally, the design of systems that are potentially vulnerable to climatic risks has relied on historical observations. For instance, the system is designed to perform adequately under the occurrence of a particular climatic event with a given recurrence period (extreme precipitation, flood and long drought), and the information about the possible magnitude of the event and the recurrence period is extracted from historical data. This approach assumes that the climate is stationary, that is, that it varies within an unchanged envelope of variability. However, mounting evidence against the stationary assumption (Solomon *et al.*, 2007a) suggests that other approaches should be used.

One possibility would be to use observational records long enough to guarantee that many different states of the climate system would have been explored during that trajectory. However, even if these records were available, their temporal and spatial resolutions would probably not be useful at the regional and local scales of interest for impacts studies.

A remaining option is to use climate model simulations to explore projections of future changes. In this case, the issues discussed in the previous sections regarding the characteristics of climate model output and how to interpret it must be taken into consideration. Some of the questions that need to be addressed are:

> *Does the climate model have any skill in simulating the phenomenon of interest? For instance, when analysing impacts of extreme precipitation in a given region, a basic requirement should be for the climate model to be able to simulate the correct position of the storm tracks in that region. If it doesn't, its projections will not be robust.*
> *Are processes not included in the climate model important to quantify the impact one is interested in? For example, if changes in land use will affect the local climate of the region of interest, the results should be*

---

[5] *Adaptation* here is defined as all the steps taken to reduce vulnerability to climate change, including climate variability and extremes.

> *interpreted considering that changes in land use are not included in current scenarios.*
>
> *What are the uncertainties that affect the climate projections? The relative and absolute importance of different sources of uncertainties depends on the spatial scale, the lead time of the projection, and the variable of interest. At shorter time scales, natural variability and other nonclimatic risks would have a higher impact than climate change for many systems. For example, during the next few years, changes in urbanization and developments in inadequate areas could increase significantly the risk of flooding independently of climate change.*
>
> *Is there a possibility that the system of interest could be affected by catastrophic changes due to nonlinear feedbacks and processes that are not known or have not yet been incorporated in the climate models?*[6]

In conclusion, climate modelling information is potentially a powerful tool to explore the impacts of climate change on different sectors, and to help inform adaptation decision making. However, an adequate interpretation of the projections fully understanding their limitations is crucial for the correct use of this information. Otherwise, there is the risk that over-interpretation of climate model information could lead to maladaptation.

References

Christensen, J.H., Hewitson, B. et al. (2007). Regional climate projections. In: *Climate Change 2007: The Physical Science Basis. Contribution of Working Group I to the Fourth Assessment Report of the Intergovernmental Panel on Climate Change*, ed. Solomon, S., Qin, D., Manning, M. et al. Cambridge University Press, Cambridge.

Courant, R., Friedrichs, K. & Lewy, H. (1967) On the partial difference equations of mathematical physics. *IBM Journal*, 215–234.

Eisenman, I., Untersteiner, N. & Wettlaufer, J.S. (2007) On the reliability of simulated Arctic sea ice in global climate models. *Geophysical Research Letters*, 34, L10501.

Frame, D.J., Aina, T., Christensen, C.M., Faull, N.E., Knight, S.H.E., Piani, C., Rosier, S.M., Yamazaki, K., Yamazaki, Y. & Allen, M.R. (2009) The climateprediction.net BBC climate change experiment: design of the coupled model ensemble. *Phil. Trans. R. Soc. A*, 367, 855–870.

Fussel, H.M. (2009). An updated assessment of the risks from climate change based on research published since the IPCC Fourth Assessment Report. *Climatic Change*, 97, 469–482.

Jenkins, G.J., Murphy, J.M. et al. (2009). *UK Climate Projections: Briefing Report*. Met Office Hadley Centre, Exeter, UK.

Lopez, A., Fung, F. et al. (2009). From climate model ensembles to climate change impacts: a case study of water resource management in the South West of England. *Water Resources Research*, 45, W08419.

---

[6] See Fussel (2009) and references therein for an up-to-date brief review about abrupt climate change and tipping elements.

McGuffie, K. & Henderson-Sellers, A. (2005). *A Climate Modelling Primer*. John Wiley & Sons, New York.

Nakicenovic, N., Alcamo, L. *et al.* (2000). *IPCC Special Report on Emissions Scenarios*. Cambridge University Press, Cambridge.

Randall, D.A., Wood, R.A. *et al.* (2007). Climate models and their evaluation. In: *Climate Change 2007: The Physical Science Basis. Contribution of Working Group I to the Fourth Assessment Report of the Intergovernmental Panel on Climate Change*, ed. Solomon, S., Qin, D., Manning, M. *et al.* Cambridge University Press, Cambridge.

Solomon, S., Qin, D. *et al.* (2007a). *Climate Change 2007: The Physical Science Basis. Contribution of Working Group 1 to the Fourth Assessment Report of the Intergovernmental Panel on Climate Change*. Cambridge University Press, Cambridge.

Solomon, S., Qin, D. *et al.* (2007b). Technical summary. In: *Climate Change 2007: The Physical Science Basis. Contribution of Working Group I to the Fourth Assessment Report of the Intergovernmental Panel on Climate Change*, ed. Solomon, S., Qin, D., Manning, M. *et al.* Cambridge University Press, Cambridge.

Stainforth, D.A., Aina, T. *et al.* (2005). Uncertainty in predictions of the climate response to rising levels of greenhouse gases. *Nature*, **433**, 403–406.

Stainforth, D.A., Allen, M.R. *et al.* (2007). Confidence, uncertainty and decision-support relevance in climate predictions. *Phil. Trans. R. Soc. A*, **365**, 2145–2161.

Stone, D.A. (2008). Predicted climate changes for the years to come and implications for disease impact studies. *Rev.Sci.Tech.Off.int.Epiz.*, **27** (2).

Stone, D.A. & Knutti, R. (2010). Weather and climate. In: *Modelling the Impact of Climate Change on Water Resources*, ed. Fung, F., Lopez, A. and New, M. Blackwell & Wiley, New York.

Wilby, R.L. & Fowler, H.J. (2010). Regional climate downscaling. In: *Modelling the Impact of Climate Change on Water Resources*, ed. Fung, F., Lopez, A. and New, M. Blackwell & Wiley, New York.

Urbanization and Climate Change

Felix N. Hammond, Kwasi Baffour Awuah Gyau and Stanislaus Y. Adiaba

4.1 Introduction

Urbanization and climate change are perhaps the two main environmental concerns in the twenty-first century (UN-Habitat, 2009; Seto and Shepherd, 2009). Whilst urbanization is perceived as unstoppable and economically productive, it is equally seen to have adverse impacts on environmental systems. This chapter discusses the relationship between urbanization and urban growth, and climate change. It commences with the state of the world's urbanization, followed by its impacts on climate change and how these impacts are generated and outlines some solutions.

4.2 State of the World's Urbanization

The transition of the world from rural to urban society through agglomeration of large numbers of people has been a continuous and unstoppable process which has characterized the world since the commencement of the western European Industrial Revolution (Mumford, 1961; Hall, 1975; Konadu-Agyemang, 1998; UN-Habitat, 2008). In 1950 only 733 million of the world population lived in urban areas. However, between 1950 and 2000 the world's urban population almost quadrupled to 2,857 million (Cohen, 2006). Indeed, over the last two decades the rates of growth and urbanization, as well as the number and size of the world's largest cities, have been unprecedented (Songsore, 2004; Cohen, 2006; UN-Habitat, 2008). The United Nations Population Fund (UNFPA, 2007) asserts that 50% of the world's population of over 6 billion live in urban areas and by 2025 two thirds of the world's population will be living in cities. Out of

---

Solutions to Climate Change Challenges in the Built Environment, First Edition.
Edited by Colin A. Booth, Felix N. Hammond, Jessica E. Lamond and David G. Proverbs.
© 2012 Blackwell Publishing Ltd. Published 2012 by Blackwell Publishing Ltd.

this estimate, 90% will be in developing countries. Other sources estimate that by 2050, 70% of the population will live in cities, with the greatest concentration in Asia and Africa (UN-Habitat, 2009; Watson, 2009; Seto and Shepherd, 2009).

Whilst at the commencement of the twentieth century, 16 cities, the majority of which were in advanced industrial countries, had populations of 1 million or more, at 2006, almost 400 cities had populations of 1 or more million, 75% of which were in developing countries. Furthermore, statistics indicate that Africa, the least urbanized region, has the highest rate of urbanization growing at an estimated rate of 4.58% per annum (UN-Habitat, 2007). However, it is also estimated one-third of all urban population will be living in Asia (Seto and Shepherd, 2009) which will have 18 of the 27 mega cities by 2050 (Roy, 2009), a situation pointing to massive urbanization and urban growth within the period under reference.

As espoused in urban economics discourse, the concentration of people and economic activities, a characteristic of urban areas, results in economies of agglomeration which lead to the attraction of larger and larger agglomeration of people and economic activities for socioeconomic development (Hirsch, 1973). The urban area's dominant role in global economy is as centres of production and consumption, and being at the forefront of socioeconomic development campaign is, thus, well known in the development literature (UN-Habitat, 2008; Hammond, 2006; Hammond and Adarkwah, 2010). It is argued that cities occupy 2% of the Earth surface but consume 75% (UNFPA, 2007; Roy, 2009) of world output and, therefore, in some cases have the highest capacity to act towards socioeconomic development (British Council, 2004). Indeed in South Africa, the cities of Cape Town, Durban and Johannesburg account for some 50% of gross domestic product (GDP). Lagos alone accounts for 60% of non-oil GDP in Nigeria. Similarly, Sao Paulo contributes approximately 25% of Brazil's GDP. It is further estimated that by 2020 China's three leading extended urban regions, comprising the Pearl River Delta, the Lower Yangtze River Delta and the Bohai Bay Region, will contribute about 80% of China's GDP (UN-Habitat, 2008).

Clearly the economic importance of urbanization and growth of urban areas cannot be overemphasized. In fact, upon proper management, urbanization offers important opportunities for socioeconomic development. For example, it is argued that high population may be appropriate for minimizing the effect of humans on local ecosystems and reduce the per capita cost in the provision of housing and environmental services (Songsore, 2004; Cohen, 2006). Again, China embarked on countrywide reforms in 1978. These reforms resulted in rapid urban growth and urbanization between that period and 2000, which saw an increase in the country's small towns from 2176 to 20 312, nearly double that of the world's average during that period. The number of cities also increased from 190 to 663, whilst the proportion of the population rose from 18% to 39%. As a consequence of this urban growth and urbanization during the aforementioned period, China recorded an average growth rate of 9.5% compared to the rates of 2.5% of the developed world and 5% of the developing world (Zhou et al., 2004).

4.3 Impact of Urbanization on Climate Change

In spite of the socioeconomic gains associated with properly managed urbanization and urban growth, the phenomenon has outpaced the capacity of most governments, particularly those of developing economies, where it is occurring rapidly and under fragile economic conditions, thereby posing environmental problems (Songsore, 2004; Cohen, 2006; Roy, 2009; Bart, 2010). McGranahan *et al.* (2005) submit that despite their relatively small land coverage, the physical transformation of landscapes associated with the process of urbanization has important consequences for global environmental change. The UNFPA (2007) also asserts that urban areas generate 75% of all waste with the effect that the function of urban areas has serious implications for the global environment. This undoubtedly poses a threat to sustainable management of cities.

One environmental issue that has attracted serious attention in recent times, and over which urbanization and urban growth of cities have an effect, is climate change. Indeed, the UN-Habitat (2009) points out that the most important environment concern facing the world in this twenty-first century is climate change. This recent concern stems from climate change impacts on basic elements of life for people around the world, which include: access to water, food production, health and the environment (UN-Habitat, 2009; Bosello *et al.*, 2006; Songsore, 2004).

The vulnerability to climate change impacts within the cities of developing economies is estimated as follows: approximately 0.8 million annual deaths from ambient air pollution, 1.9 million from physical inactivity and 1.5 million from indoor air pollution (Campbell-Lendrum and Corvalán, 2007). UN-Habitat (2009) argues further that hundreds of millions of people are likely to suffer hunger, water shortages and coastal flooding as global warming increases, and the most vulnerable groups are the poorest countries and their people. This has generated interest in studies on activities that impact on global climate change and, in particular, the relationship between urbanization and urban growth and climate change. Seto and Shepherd (2009), for example, report that even though the focus of research in prevailing climate change has been effects of anthropogenic greenhouse gases, the Fourth Assessment of the Intergovernmental Panel on Climate Change (IPCC) has noted the emerging interest in understanding the role of urbanization through land use on the climate systems (Satterthwaite, 2008).

Studies carried out in this environmental genre, as well as other related reports, point to a strong connection amounting to a direct effect of urbanization through the built environment on climate change. Even though disputed, many sources such as Munich Reinsurance, the United Nations Human Settlements and the Clinton Climate Initiative estimate that cities generate between 75% and 80% of global greenhouse gas emissions. Satterthwaite (2008), in disagreement with this estimate, did an envelope back assessment using data from Intergovernmental Panel on Climate Change (IPCC) and arrived at between 30% and 40%, a result that is lower by half than earlier reported figures but still worthy of consideration. Also, in their study of impacts of urbanization on climate change in the continental United States, Kalnay and Cai (2003) observed a decrease in diurnal temperature range and attributed half of the decrease to urban and other land use changes. Indeed, the results of their study through comparing trends in observed and reanalysis of surface temperature over the period from 1950 to 1999 showed an estimate of 0.27°C mean surface warming per century due to land use changes. This was at least twice as high as previous estimates based on urbanization alone. This 0.27°C mean surface warming per century is also a much larger estimate than the area-weighted average warming effect of urban heat islands (UHI) (see Chapter 14) over land during the twentieth century. This was estimated at less than 0.06°C per century globally and between 0.06°C and 0.15°C per century in the United States based on differences in temperature trends between rural and urban stations (Zhou et al., 2004).

Zhou et al. (2004) report, based on the study of impacts of land use change on climate change in 13 provinces and municipalities in southeast China, of warming surface temperature of 0.05°C per decade for the period from 1978 to 2001. This result is much higher than results from other studies; the earlier mentioned studies carried out in the United States and that of Liu et al. (2003) who carried out a similar study in southeast China between 1951 and 2001. Liu et al. (2003) reported 0.011°C of warming mean surface temperature per decade based on analysis of the rural and urban differences in annual mean temperature. The difference between the study undertaken by Zhou et al. (2004) and the other studies is attributed to the rapid urbanization experienced by China during the period 1978 to 2001, the period that underlies the Zhou et al. (2004) study. From the aforementioned, it is clear that urbanization and urban growth contribute to greenhouse gas emissions to the atmosphere, thereby contributing to climate change, but how does urbanization affect climate change?

4.4 How Does Urbanization Affect Climate Change?

Global warming resulting from emission of greenhouse gases into the atmosphere is a major determinant of climate change, but major changes in land use such as urbanization, agriculture and deforestation are also very significant (Kalnay and Cai, 2003). The influence of cities on global climate change is mainly through high-level energy consumption, concentration of economic activities and burning of fossil fuels (Roy, 2009; Ruth and Coelho, 2007; UN-Habitat, 2009; Watson, 2009). Seto and Shepherd (2009) point out that the built environment is a source of heat, has a poor storage system for water, represents an impediment to atmospheric motion and is a source of aerosols. From the foregoing, it is clear that urban areas consume vast energy through burning of fossil fuel, which in turn transports CO_2 emissions and other greenhouse gases into the atmosphere, thereby causing climate change. But the extent to which cities consume energy depends on their spatial configurations. This determines distances from home to work and other destinations and modes of transportation, which also influence the level of energy consumption through the burning of fossil fuels (Seto and Shepherd, 2009).

Unfortunately, the current spate of urbanization and urban growth in the world is associated with urban sprawl and compounded by poor land use and other land management policies, particularly in developing countries where urban growth and urbanization are occurring at a rapid pace (Cohen, 2006). This state of affairs, apart from causing traffic congestion, has led to long travelling time and ultimately consumption of more energy. Cohen (2006) further points out that congestion in many large cities can be very severe and air pollution is a serious environmental concern with concentrations of carbon monoxide, lead and suspended particulate matter greatly exceeding the World Health Organizations (WHO) guidelines. Roy (2009), describing the current urbanization in fast-growing cities in the world as *metropolisation*, also indicates that it has led to urban sprawl and that whilst work places are in the centre of these cities, people will have to commute long distances, which has serious environmental repercussions. Indeed, it is estimated that in the next 50 years per capita distance travelled will grow by 0.2–0.8% in the OECD countries and 6% and 5% in China and India, respectively (Fulton and Eads, 2004; Roy, 2009). In the United States, for example, environmental concerns including emissions from traffic congestion and travelling over long distances, following urban sprawl, have in recent times propelled state government to take an interest in land use issues (Holcombe and Staley, 2001; Pena, 2002; Schmidt & Buehler, 2007; Ihlanfeldt, 2009). Additionally, Bart (2010) investigates the relationship between transport emissions and urban lands within EU member states between the periods 1990 and 2000. The study found a much stronger relationship between transport CO_2 emissions and the artificial land area than between CO_2 emissions and GDP or population data. Further, Bart (2010) reports that transport emissions accounted for around 20% of all greenhouse gas emissions in the EU-25 in 2005; however, road transport alone accounted for 93%, and since 1990 transport emissions have increased by one-fifth, with Bart noting that it is a significant factor responsible for climate change. Also, in Bangladesh, the transport sector is responsible for 70% of that country's CO_2 emissions; its capital, Dhaka, is one of the fastest growing cities in the world, had a population of 12.4 million as of 2006 and contributes between 20% and 30% of such emissions (Roy, 2009).

Apart from transport emissions, urbanization and urban growth have led to changes in land uses with serious ramifications for climate change. As cities and urban centres grow, agricultural land use and the vegetation, which protect against UHI are reduced as part of the island is taken over by other land uses, such as residential, industrial and other commercial activities at the countryside. Upon the reduction of these green spaces, which hitherto served to absorb CO_2 emissions and other anthropogenic greenhouse gas emissions, the spatial pattern of the UHI correspondingly becomes less scattered and more intense (Seto and Shepherd, 2009; He *et al.*, 2007). Earlier, Carlson and Arthur (2000) writing on this subject relative to American cities, indicated that urban areas used to be isolated, but are now becoming sprawling with regions becoming interconnected in dendritic fashion, and lands devoted to agriculture and also serving as protection against the UHI effect are being replaced by broad and profound transformation of the countryside. This situation is attributed to lack of recognition by human beings of the effect of their activities on the environment and the desire of fortunate city dwellers for nature and imperium (Carlson and Arthur, 2000).

It has, however, been argued that the level of wealth of cities determines the degree or extent to which a particular city's urban growth and urbanization impinge on climate change (Roy, 2009). Indeed McGranahan and Songsore (1994) and McGranahan et al. (1996, 2001), using the urban environmental transitional conceptual model, demonstrate that the wealthiest cities in the developed world draw more heavily on the global resource base and generate a disproportionate share of global pollution and, thus, account for a greater proportion of global warming, acid rain and depletion of the ozone layer. The model generally posits that as a city's wealth level increases, home and neighbourhood environmental issues (such as sanitation facilities, poor housing conditions and so on) recede in importance, and city-wide environmental problems (such as ambient air pollution) become dominant in middle-income cities, whilst the wealthiest cities (such as those in the Global North) generate a large proportion of the global pollution. Drawing on the insights of the urban environmental transitional model, Roy (2009) cites the example of 'China-dominated Asia', thought of as the future of the global economy and estimated to have 18 of the 27 world mega-cities by 2020, to generate a large proportion of global pollution. In fact, the level of pollution in Asia, which used to be a small proportion of global emissions in the 1970s, has increased rapidly since the mid-1990s to surpass that of North America and Europe (Akimoto, 2003).

It is also imperative to note that this metropolisation of urbanization and urban growth, particularly in fast-growing poor cities in developing countries, is not matched with the provision of basic infrastructure resulting in poor physical environment and, thereby, contributing to emissions of greenhouse gases (Songsore, 2004; Cohen, 2006; Roy, 2009; UN-Habitat, 2009). The truth of the matter is that the phenomena of urbanization and urban growth are occurring under weak economic conditions (Kissides, 2005) characterised by shifting demand for labour and transitional unemployment and by high cost of food and nonfood items, such as transport, health, electricity and water. This has resulted in a reduction of real urban incomes, with people in low-income bracket being the worst affected. Indeed, the increasing levels of poverty associated with urbanizing cities in developing economies are well known. It is, for instance, estimated that 43% of the urban population in Africa are poor (UN-Habitat, 2008). With weak economies, both national and local governments are unable to provide basic amenities for the rising urban population, and the majority of the urban population are also unable to even pay for the limited available basic amenities. Waste management in African cities is, for example, described as a monster that has aborted most efforts made by city authorities, state and federal governments and professionals alike (Onibokun, 1999). Added to this point are high, unrealistic development standards and policies, and weak planning and other land management institutions (UN-Habitat, 2009; Roy, 2009). It is, for instance, estimated that about 90% of new developments in the urban environment are taking the form of slums (UN-Habitat, 2008). This has, therefore, resulted in poor housing conditions and deterioration of the physical environment with serious ramifications for the climatic conditions given the greenhouse gas emissions that result from it.

4.5 Solutions for Change

It is implied from the foregoing that urbanization and urban growth and climate change are connected (Seto and Shepherd, 2009). Whilst urbanization and urban growth through their several manifestations impact the physical environment and, ultimately, the global climatic conditions, changes in global climatic conditions also have adverse effects on humankind. These effects manifest themselves on basic elements of life for people around the world in the areas of access to water, food production, health and the environment, with the poor usually suffering much of the impacts (UN-Habitat, 2009; Bosello et al., 2006; Songsore, 2004). Indeed, UN-Habitat (2009) reports that sufferers of environmental challenges caused by urbanization, such as slope instability resulting in landslides and flooding, are the world's 1 billion urban slum dwellers who are not protected by planning regulations. The discussion also establishes that urbanization and urban growth are not necessarily bad and, if managed properly, could contribute immensely to the socioeconomic development of society. In fact, it has been argued that urban centres hold the key to the socioeconomic development and reduction of poverty and that given the trends of urbanization, particularly in developing economies, it is unstoppable (UN-Habitat, 2008). This, therefore, requires that the adverse effects of urbanization and urban growth on the environment are reduced to the barest minimum, as far as possible, or ways should be found to limit the impact of change on humankind. In other words, there should be a balance between urbanization and urban growth and the environment.

Consequent upon the foregoing, and taking into account how urbanization and urban growth affect climate change and vice versa, several measures have been suggested towards addressing the relationship between urbanization and climate change regarding societal welfare. These measures either have been adaptive or seek to mitigate the effects on climate change. Adger et al. (2003), cited in Roy (2009), submits that with regard to climate change, fast-growing cities require a number of adaptive measures to adjust the urban system to moderate the impacts of climate change, take advantage of new opportunities or bear the consequences. Also, applying the urban environment transitional model, researchers (McGranahan and Songsore, 1994; McGranahan et al., 1996, 2001) explain that wealthy cities in the Global North, despite their huge generation of greenhouse gas emissions, resulting from tremendous use of global resources, have been able to develop adaptive systems to counteract the adverse affects of climate change. Of course, it is through large-scale environment resource use concomitant to urbanization and urban growth that made these cities wealthy and, out of that, they have developed technologies to deal with the adverse effects of climate change. It is noted that the United Nations under the National Adaptive Programmes for Action (NAPA) has initiated a number of adaptive programmes particularly in developing economies (see http://unfccc.int).

The other aspect of addressing the relationship between urbanization and urban growth, and climate, centres on pollution abatement. These include the use of clean technology, energy-efficient equipments and modes of transport and appropriate land use planning policies to ensure the right spatial configuration of the urban system to reduce, for example, travel distance and consequently the reduction of energy use. Bart (2010), for example, recommends planning regulation to deal with CO_2 emission of the transport sector particularly the road transport sector in EU cities, which is a significant contributor of greenhouse gas emissions and stems from urban sprawl.

Literature on the subject indicates that adoption of any of these measures depends on a number of factors such as financial, institutional and ethical ones. Roy (2009) for example mentions four factors, namely, urban morphology, infrastructure aspects, individual lifestyle and consumer behaviour and policy intervention. Roy (2009) further identifies urban land use planning as one of the solutions through undertaking urban designs, which promote energy efficiency and maximization of location and density, to reduce reliance on private transport. But then, such planning should be part of a holistic concept of sustainable urbanization that can be achieved through regional, national and even international planning and policy frameworks. Indeed, the UN-Habitat (2009) also advocates for sustainable urbanization with the following goals among others:

- Greenhouse gas emissions are reduced and serious climate change mitigation and adaptation actions are implemented;
- Urban sprawl minimized and more compact towns and cities served by public transport are developed;
- Non-renewable resources are sensibly used and conserved;
- Renewal resources are not depleted;
- Energy used and the waste produced per unit of output or consumption is reduced; and
- Waste produced is recycled or disposed of in ways that do not damage the wider environment, and the ecological footprint of towns and cities is reduced.

Whilst these solutions are worthy of note, what is also critical to the efficacy of any of the solutions is to identify and tackle the causal factors of urbanization and urban growth and its manifestations of modification of land uses. For example, why are some cities in 'China-dominated Asia' growing at a faster rate with modification of vegetation into industrial land uses and emitting greenhouse gases? Appropriate answers to such questions will also go a long way to help address the issue.

Economic theory suggests that human beings act on the basis of incentives and expectations (Smith, 1776; Mises, 1949; Rothbard, 2004). Human beings need food, clothing, shelter and indeed socioeconomic development – general wellbeing. These factors impel individuals to embark on action, be it to undertake economic ventures or seek employment. As established earlier, cities and urban areas are centres of production and consumption (Cohen, 2006). Hirsch (1973) points out that cities are places where all the needs of humankind are provided. Consequently, people move to urban centres in anticipation that they will serve as a platform for the realization of their dreams of a better life. Indeed, the trends of ongoing urbanization indicate that migration is the greatest contributor to urban population increases (Cohen, 2006). In developing countries, particularly those of Sub-Saharan Africa, with weak economies (Songsore, 2004; Kissides, 2005), the kinds of conditions necessary for the realization of the dreams of the majority of individuals do not exist. Rather, these individuals are faced with austere economic conditions with limited resources, they are supposed to meet high unrealistic planning standards and they are expected to make do with high prices of land and utilities which are poorly managed. In the end, the majority of these urban dwellers settle at the countryside, where land prices are cheap and ahead of planning, or in slums and shantytowns with serious environmental consequences (UN-Habitat, 2009). Conversely, in the developed countries of the Global North, it is the individual desire to further improve welfare through drawing closer to nature at the countryside that creates urban sprawl with its attendant environmental consequences (Carlson and Arthur, 2000). Similarly, in emerging economies such as that of China-dominated Asia, it is the quest for rapid socioeconomic development that underpins current urban growth (Zhou *et al.*, 2004).

From the foregoing it is abundantly clear that different circumstances prevail in different parts of the world, which account for the current relationship between urbanization and climate change. It is also noted from literature that whilst many studies have been carried out on urbanization and climate change in areas like United States, Europe and China, the same cannot be said of developing regions like Africa and Latin America (Seto and Shepherd, 2009). This obviously requires comprehensive studies in these areas to inform policy. However, based on the discussions so far, this chapter submits that any proposal to find the right balance regarding the relationship between urbanization and climate change should take into account the need to address the incentives question.

4.6 Conclusion

The chapter set out to discuss the impact of urbanization on climate change. The discussions unearth that urbanization and climate change are very much connected and, whilst improperly managed urbanization causes climate change through modification of land use and combustion of energy through inappropriate spatial configuration of urban system, climate change also has adverse effects on the physical environment in which humans live. Consequently, since humans need suitable environmental conditions, and yet also depend on the environment for survival, there is the need for an appropriate balance in the relationship between urbanization, urban growth and climate change. Whilst this chapter acknowledges earlier recommendations, such as the prescriptions in sustainable urbanization, it is recommended that the incentives that lie behind people's quest to live in urban centres should be taken into account in any measure to foster the right mix between urbanization and climate change.

References

Akimoto, H. (2003) Global air quality and pollution. *Science*, **302**(5651), 1716–1719.

Bart, I.L. (2010) Urban sprawl and climate change: A statistical exploration of cause and effect, with policy options for the EU. *Land Use Policy*, **27**(2), 283–292.

Bosello, F., Roson, R. & Tol, R.S.J. (2006) Economy-wide estimates of the implications of climate change: human health. *Ecological Economics*, **58**(3), 579–591.

British Council (2004) *A briefing on climate change and cities*. Briefing sheet 30. Tyndall Centre for Climate Change Research, London.

Carlson, T.N. & Arthur, S.T. (2000) The impact of land use – land cover changes due to urbanization on surface microclimate and hydrology: A satellite perspective. *Global and Planetary Change*, **25**(1), 49–65.

Cohen, B. (2006) Urbanization in developing countries: Current trends, future projections, and key challenges for sustainability. *Technology in Society*, **28**, 63–80.

Campbell-Lendrum, D. & Corvalán, C. (2007) Climate change and developing country cities: implications for environmental health and equity. *Journal of Urban Health: Bulletin of the New York Academy of Medicine*, **84**(1), 109–i117.

Fulton, L. & Eads, G. (2004) *IEA/SMP Model Documentation and Reference Case Projection*. World Business Council for Sustainable Development's Sustainable Mobility Project, International Energy Agency (IEA), Paris.

Hall, P. (1975) *Urban and Regional Planning*. John Wiley, New York.

Hammond, F.N. (2006) *The Economic Impacts of Sub-Saharan Africa Urban Real Estate Policies*. PhD Thesis, University of Wolverhampton, UK.

He, J.F., Liu, J.Y., Zhuang, D.F., Zhang, W. & Liu, M.L. (2007) Assessing the effect of land use/land cover change on the change of urban heat intensity. *Theory Application Climatology*, **90**(3–4), 217–226.

Hirsch, W.Z. (1973) *Urban Economic Analysis*. McGraw-Hill, New York.

Holcombe, R.G. & Staley, S.R. (eds.) (2001) *Smarter Growth, Market-Based Strategies for Land use Planning in the 21st Century*. Greenwood Press, London.

Ihlanfeldt, K.R. (2009) Does comprehensive land use planning improve cities? *Land Economics*, **85**(1), 74–86.

Kalnay, K. & Cai, M. (2003) Impact of urbanization and land use change on climate. *Nature*, **423**, 528–531.

Kessides, C. (2005) *The Urban Transition in Sub-Saharan Africa: Implications for Economic Growth and Poverty Reduction*. Urban Development Unit, World Bank, New York.

Konadu-Agyemang, K. (1998) *The Political Economy of Housing and Urban Development in Africa: Ghana's Experience from Colonial Times to 1998*. Praeger, Westport, CT.

Liu, S., Li, X. & Zhang, M. (2003) *in Scenario Analysis on Urbanization and Rural–Urban Migration in China, Interim Report IR-03-036*. International Inst. for Applied Systems Analysis, Vienna, Austria. Available from http://www.iiasa.ac.at/Publications/Documents/IR-03-036.pdf on May 17, 2010.

McGranahan, G. & Songsore, J. (1994) Wealth, health and the urban household: weighing environmental burdens in Accra, Jakarta and Sao Paulo. *Environment*, **36**(6), 40–45.

McGranahan, G., Songsore, J. & Kjellen, M. (1996) Sustainability, poverty and urban environmental transactions. In: *Sustainability: The Environment and Urbanization*, ed. Pugh, C., London, Earthscan, 103–133.

McGranahan, G., Jacobi, P., Songsore, J., Surdjadi, C. & Kjellen, M. (2001) *The Citizens at Risk: From Urban Sanitation to Sustainable Cities*. London, Earthscan.

McGranahan, G., Marcotullio, P.J., Bai, X., Balk, D., Braga, T., Douglas, I., Elmqvist, T., Rees, W., Satterthwaite, D., Songsore, J., Zlotnik, H., Eades, J. & Ezcurra, E. (2005) Urban systems. In: *Millennium Ecosystems Assessment: Current State and Trends: Findings of the Condition and Trends Working Group. Ecosystems and Human Well-being*, vol. 1. Island Press, Washington, DC.

Mises, L.V. (1949) *Human Action: A Treatise on Economics*, scholar's ed. Ludwig von Mises Institute, Auburn, AL.

Mumford, L. (1961) *The City in History, Its Origins and Its Transformation*. London, Seeker and Warburg.

Onibokun, A.G. (1999) *Managing the Monster: Urban Waste and Governance in Africa*. International Development Research Centre, Ottawa.

Pena, S. (2002) Land use planning on the U.S.–Mexico border: a comparison of the legal framework. *Journal of Borderlands Studies*, **17**(1), 1–20.

Rothbard, M.N. (2004) *Man, Economy and State*, scholar's ed. Ludwig von Mises Institute, Auburn, AL.

Roy, M. (2009) Planning for sustainable urbanization in fast growing cities: mitigation and adaption issues addressed in Dhaka, Bangladesh. *Habitat International*, **33**(3), 276–286.

Ruth, M. & Coelho, D. (2007) Understanding and managing the complexity of urban systems under climate change. *Climate Policy*, **7**(4), 317–336.

Satterthwaite, D. (2008) Cities' contribution to global warming: notes on the allocation of greenhouse gas emissions. *Environment and Urbanization*, **20**(2), 539–550.

Schmidt, S. & Buehler, R. (2007) The planning process in the US and Germany: a comparative analysis. *International Planning Studies*, **12**(1), 55–75.

Seto, K.C. & Shepherd, J.M. (2009) Global land use trends and climate impacts. *Current Opinion in Environment Sustainability*, **1**(1), 89–95.

Songsore, J. (2004) *Urbanization and Health in Africa: Exploring the Interconnections between Poverty, Inequality and the Burden of Disease*. Ghana Universities Press, Accra.

United Nations Population Fund (UNFPA) (2007) *State of the World's Population 2007: Unleashing the Potential of Urban Growth*. United Nations Population Fund, New York.

UN-Habitat (2007) *Enhancing Urban Safety and Security*. UN-Habitat, Nairobi.

UN-Habitat (2008) *Securing Land Rights for All*. UN-Habitat, Nairobi.

UN-Habitat (2009) *Global Report on Human Settlements 2009, Planning, Sustainable Cities: Policy Directions*), abridged ed. Earthscan, London.

Watson, V. (2009) 'The planned city sweeps the poor away …': urban planning and 21st century urbanization. *Progress in Planning*, **72**(3), 151–193.

Zhou, L., Dickson, R.E., Tian, Y., Fang, J., Li, Q. & Kaufmann, R. (2004) Evidence for a significant urbanization effect on climate in China. *PNAS*, **101**(26), 9540–9544.

Global Political Initiatives and Overtones
Jean-Luc Salagnac

5.1 Introduction

The built environment and climate have always had a very close relationship. Until recently, constructors mainly had the lessons of the past to improve the adaptation of construction to local climatic conditions. A rapidly emerging new climatic context creates new circumstances in which to define adaptation rules. Political initiatives are now being implemented to reduce greenhouse gas (GHG) emissions. Climate projections confirm that changes may nevertheless happen during the coming decades, so future adaptation policies need to be defined. The coherence between mitigation and adaptation policies is crucial. This chapter first addresses the relations between climate and the built environment. The background to political initiatives is then addressed before a more detailed discussion of some of these initiatives.

5.2 Climate and the Built Environment

The built environment consists of buildings, towns and networks that societies have constructed for their own needs (housing, education, transport, commerce, industry etc.). These developments are the result of human decisions and actions. The great variety of design, technology and building types results from the need to satisfy universal functions, among which is the protection of persons and goods against climatic events. Descriptions by historians of ancient buildings and the first cities (about 6000 years ago) show that most of the technical and symbolic features of these early constructions are very similar to those of contemporary buildings. This is for instance obvious for building parts ensuring protection against wind, rain, heat, cold and the like. Walls, roofs and openings are universal solutions.

Solutions to Climate Change Challenges in the Built Environment, First Edition.
Edited by Colin A. Booth, Felix N. Hammond, Jessica E. Lamond and David G. Proverbs.
© 2012 Blackwell Publishing Ltd. Published 2012 by Blackwell Publishing Ltd.

Nevertheless, the detailed solutions imagined and developed by human groups are very diverse according to local climate, culture and resources. The relatively stable climate, though variable, over the last 6000 years has favoured the development of societies and the built environment. These anthropogenic developments have been huge and fast during the last hundred years. The associated production of GHG has been increasing as well as the concentration of these gases in the atmosphere.

Construction is indeed a major GHG contributor. In OECD (Organisation for Economic Co-operation Development) countries, buildings are responsible for 25–40% of total energy use and approximately 30% of GHG emissions. The share of the residential sector accounts for the major part of the energy consumed in buildings; in developing countries, this share can be over 90% (United Nations Environment Programme [UNEP], 2007). Construction is also one of the major economic activities as it provides 5–10% of employment at the national level and normally generates 5–15% of a country's GDP (UNEP, 2007). It also contributes to the majority of its gross fixed capital. As such, construction is a major domain in which to develop policies aimed both at mitigating GHG emission and at adapting the built environment to a new climatic context. A major issue in such a perspective is the development of buildings consuming less energy and emitting fewer GHG as there is no doubt that GHG concentration in the atmosphere will contribute to climate change during the coming decades and centuries as concluded by Intergovernmental Panel on Climate Change (IPCC) experts (IPCC, 2007). This strong requirement has to be fulfilled whilst easily accessible (and cheap) fossil energy reserves decrease rapidly (at the human scale of time: within decades or maybe within one or two centuries) and world population experiences an exponential growth. In short, power generation, mineral resource extraction and others all exhibit the exponential growth that parallels the forces that create them. As such perspectives are clearly not sustainable, 'sustainable development' has been perceived for some years as a major challenge for the twenty-first century.

In spite of national political decisions limiting the emissions of GHG, IPCC experts infer from their work that a new climatic context may nevertheless develop during the coming decades: 'It is very likely (likelihood of outcome between 90% and 99%) that anthropogenic greenhouse gas increases caused most of the observed increase in global average temperatures since the mid-20th century. Without the cooling effect of atmospheric aerosols, it is likely (likelihood of outcome between 66% and 90%) that greenhouse gases alone would have caused a greater global mean temperature rise than that observed during the last 50 years' (IPCC, 2007). There is still a great uncertainty concerning the local climate evolutions, but, as far as the protection of persons and goods against climate is concerned, it is wise to examine now how the built environment (both existing constructions and those built in the future) will be adapted to this new context. The December 2009 Copenhagen United Nations Framework Convention on Climate Change (UNFCC) conference confirmed that these issues are not just technical but are much more complex as geopolitical, cultural, economic and demographic issues are entangled. The absence of a committing decision after this conference, then, is not completely surprising.

Just before and still after this conference, the activity of climate sceptics clearly became stronger. Their argument, according to which climate has always been changing, is definitely true but consequences of climate change will also definitely be different from what human societies experienced in the past.

Some examples will confirm this assertion:

1. About 18 000 years ago, the sea level was around 120 metres lower than today but the global population was about 10 million (INED, 2003) and the stakes exposed to climatic hazards were infinitely less than now. A 1 metre sea level rise would be much more harmful in the current context than it could have been a millennia ago.
2. Intense and prolonged urban heat islands during heat wave periods may have immediate and tragic consequences as during the summer of 2003 in Europe (Salagnac, 2007). This threat may increase as the share of world population living in urban areas has increased from 13% in 1900, to slightly less than 30% in 1950 and close to 50% in 2005, and is expected to reach 60% in 2030 (United Nations [UN], 2007).
3. Clay soil drying shrinking is known as a major cause of costly building defects (e.g. foundation subsidence and wall cracks). The record of such events in France jumped from about 20 000 a year over the 1989–2002 period to 100 000 in 2003 after the heat wave (Taillefer and Salagnac, 2010). The very likely frequency increase of such events announced by IPCC may significantly increase this particular risk.

To face such an unprecedented climatic context shift under an exponential demographic growth, political initiatives have already been taken and others will be taken.

5.3 Background to Political Initiatives

Vernacular buildings were for many centuries built using techniques pragmatically developed through trial-and-error processes. Clever, efficient and locally adapted solutions were developed that are still in use in many parts of the world (e.g. wooden structures and masonry). The strong development of the sciences since the eighteenth century allowed a better understanding of construction physics. This knowledge, together with the unprecedented industrial development that started in the nineteenth century, produced innovative ways to ensure building functions. The following innovations are now used daily by professionals or occupants in industrialised countries. It is nevertheless important to remember that most of these innovations did not even exist a century ago or were reserved for a wealthy minority of the population:

- Materials such as steel, reinforced concrete, prestressed concrete, plaster boards, float glass and thermal insulation;
- Equipment such as central heating, air conditioning and double flux air ventilation; and
- Services such as drinkable tap water, domestic hot water, electricity and gas.

48   Solutions to Climate Change Challenges in the Built Environment

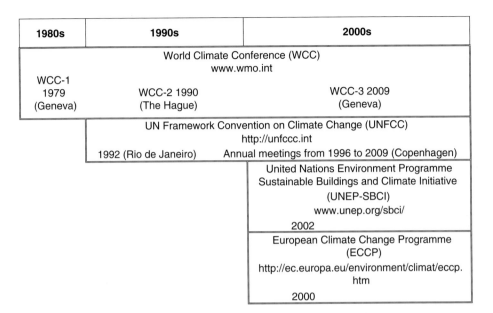

Figure 5.1   Main climate-related political initiatives taken at international and European levels.

The conjunction of these innovations together with the availability of cheap energy resources dramatically changed the building industry during the 20th century. Demands for all-season indoor comfort requiring both heating and cooling equipment have for instance been constantly rising.

The resulting complexity of the construction activity called for the codification of construction rules. The first buildings technical codes occurred very recently on a human society timescale. These codes concerned construction structural issues (in France, the first code for reinforced concrete building structures was enacted in 1906, and first code of wind and snow effects on building structures in 1946). The codification of buildings thermal performances is even more recent and was triggered by the first oil shock in 1973 (in 1974, France passed its first building energy regulation). The first and second oil shocks were determinant events that influenced policies from the early 1970s until now. Energy concern slowly evolved into environmental concern during the 1990s and into sustainable and climatic concern from the early 2000s onwards.

The main climate-related political initiatives at international and European levels follow this time pattern, as illustrated in Figure 5.1. Though the World Climate Conference does not directly address built environment issues, it is interesting to mention that WCC-3 decided to establish a Global Framework for Climate that will be crucial for building climate-resilient societies: 'Through strengthened observations, research and information systems, as well as new interaction mechanisms for climate information users and providers, the Framework will ensure that all sectors of society have user-friendly climate products that enable them to plan ahead in the face of changing climate conditions' (World Meteorological Organization [WMO], 2009). This quotation addresses both mitigation and adaptation issues. The WMO helps establishing the information background needed to elaborate political initiatives regarding climate and the built environment. The UNFCC focuses on GHG emission mitigation in all sectors. It provides reference data and methodology to assess mitigation actions but does not specifically target the built environment. By supporting sustainable development, it nevertheless encompasses built environment issues.

The United Nations Environment Programme Sustainable Buildings and Climate Initiative (UNEP-SBCI) works to promote sustainable building practices worldwide. This initiative was launched in 2002 with the final objective to promote a worldwide adoption of sustainable buildings and construction practices with a first focus on energy efficiency and CO_2 emissions.

The European Climate Change Programme (ECCP) was launched in 2000 by the European Commission. The goal of the ECCP is to identify and develop all the necessary elements of an EU strategy to implement the Kyoto Protocol. The ECCP white paper published in 2009 also addresses adaptation issues (ECCP, 2009).

Other political initiatives aiming at developing a low-carbon and resource-efficient economy could be added to this list, particularly at the European level, such as:

- The 2002 directive on the energy performance of buildings (Directive 2002/91/EC; European Parliament and the Council of the European Union, 2003)
- The 2006 directive on energy end-use efficiency and energy services (Directive 2006/32/EC; European Parliament and the Council of the European Union, 2003)
- The 2007 lead market initiative for Europe where sustainable construction is one of the six selected markets (EU, 2007)

Standardisation works go with these initiatives such as the actions carried out in the ISO TC59 'Building construction' and SC17 'Sustainability in building construction' commissions. Many European research projects contributed (and will continue to contribute) to highlight political initiatives.

At a more international level, CIB Agenda 21 for building construction should also be mentioned as a specific sectorial response to the Bruntland agendas, developed through the worldwide construction community network (see http://www.cibworld.nl). The intention is to elaborate a conceptual framework with three objectives: create common language, establish a sound programme of collaborative activities and provide a source document.

The CIB commission W 108 (Climate change and the built environment) has been working since 2002 on the adaptation of the built environment to a new climatic context and works at collecting and updating information on national political initiatives.

5.4 Mitigation and Adaptation Policies

Though the link between climate change and GHG concentration is still in debate, the likelihood of a new climatic context during the twenty-first century has been considered important enough to trigger high-level political initiatives presented in the previous section. From these general frameworks, the question is how to elaborate and implement policies and regulations at a national level according to local contexts. A few cases are now presented mainly based on the French context and, whilst not exhaustive, do highlight questions raised in such a perspective.

Mitigation policies aim at both decreasing the energy consumption of buildings and limiting GHG emissions. They can be understood as a continuation of previous efforts made since the first oil shock to reduce building energy dependency and demand on fossil fuels. The associated policy measures are concentrated on buildings (existing and newly built) as such but also concern energy production patterns (e.g. the use of renewable energy and of local energy sources).

Adaptation policies refer to natural hazards consequences mitigation. They take root in prevention and protection concepts. The associated policy measures not only refer to isolated buildings but also consider buildings in their environment. Climate-related natural hazards such as flood, clay soil drying shrinking and rehydration swelling, wind and snow loads, urban heat islands and forest fires are considered.

These two types of policies are not independent and are both needed. To illustrate this, just consider the development of (individual) air conditioning as a spontaneous adaptation initiative of the population during a heat wave. People suffering from high day and night temperatures invest in this electrical equipment for some hundreds of euros. They feel temporarily more comfortable, but the sum of their individual decisions has a visible impact on energy demand as observed in the summer of 2003 (Salagnac, 2007). Due to the way electric energy is produced, this extra consumption may have a significant impact on GHG emissions. As far as this extra production of GHG impacts climate, the feedback of these individual adaptation initiatives on climate change process is negative. The role of policies is to avoid such situations by defining rules to design, exploit and maintain energy-efficient, low-carbon and climatic hazards-resilient buildings. Moreover, mitigation and adaptation policies must be consistent. They must not contain mitigation measures that would reduce the effects of adaptation measures, and vice versa.

5.4.1 An Example of Mitigation Policy

Similar to other European countries, France launched in 1974 a policy to improve the thermal performance of buildings. This first regulation concerned only new buildings and focussed on walls and roofs insulation. Through several steps in 1982, 1988, 2001 and 2006, the scope of thermal performance regulations for buildings was widened to partly take into account existing residential buildings as well as new non-residential buildings. These requirements were also reinforced in terms of performance, and more parameters were taken into account (e.g. the insulation of blind and transparent envelope elements, performance of equipment, contribution of solar energy and contribution of household equipment). The most recent regulations as well as the next (the latter being prepared for 2012) were elaborated according to the 2002 directive on energy performance of buildings (European Parliament and the Council of the European Union, 2003).

The effects of past regulations can be assessed by observing that the average energy consumption of residential buildings decreased from 372.5 kWh/m² to 245 kWh/m² (final energy) between 1973 and 2002 (ADEME, 2004). During the same period, the energy for heating decreased more (43%) but the electricity demand for specific purposes (e.g. domestic electrical equipment and computers) nearly doubled (from 13.3 kWh/m² to 26.4 kWh/m²).

The 2012 regulation will prescribe a maximum of 50 kWh/m² primary energy consumption for new buildings. This represents a huge gap that requires concomitant efforts in several directions:

- More efforts on building procurement to elaborate precise specifications even for ordinary buildings,
- More efforts on detailed building design and site implementation (e.g. control and quality management),
- Organisational and technical innovations,
- Training of constructions stakeholders (not only contractors) to shift from traditional organisation to a more cooperative organisation,
- Revision of liability of stakeholders and construction insurance schemes, and
- A better anticipation of exploitation and maintenance conditions.

In order to satisfy these constraints, the 'industrialisation' of building production is sometimes proposed as it may create better site conditions to answer organisational and technical questions. This has to be studied, but the business model of the resulting production organisation may be very different from that of the traditional business model of construction that has been used for centuries without major changes. The challenge is to reach the goal of constructing very energy-efficient buildings whilst keeping the cost affordable for clients. Mechanisms have been developed at national or regional levels to relieve clients of the financial burden (subventions, low-rate loans etc.). Further regulations will also have to deal in greater detail with the energy performance improvement of the building stock. This is a priority that came from the Grenelle de l'Environnement (the French Environment Round Table; see http://www.legrenelle-environnement.fr, English section) that allowed debates to outline future climate change-driven, environment-related policies.

5.4.2 Examples of Adaptation Policies

Climate-related natural hazards may hit any country. Buildings built according to local tradition and more recently built according to codified rules are all designed to withstand a certain degree of intensity of the considered hazard. For example, wind loads that are taken into account when designing a building result from local knowledge of windy events. In the case of extreme events, far beyond the design parameters' envelope, the so-designed building is likely to be damaged. Insurance can cover current situations when the consequences of natural hazards can be assessed, when the occurrence of the event is random and when the cost of insurance is affordable for the general population. This is for instance the case for storm, hail or snow. Events that do not meet these criteria cannot be covered by insurance mechanisms that rely on the mutualisation of risk consequences (a great number of insured pay premiums for randomly distributed events). The case of extreme events is out of this scheme because the occurrence is too rare (tornado in Europe, for instance) and/or the concerned community is not big enough (flood, for instance). In such cases, the traditional insurance scheme cannot be adopted, but specific regimes have been developed according to countries. The French CAT-NAT (natural catastrophe) system is an example (CCR, 2009). It was introduced by a 1982 law which creates a compensation fund that is triggered when two conditions are met:

- Declaration of the state of natural disaster by an interministerial decision; and
- The owner must have subscribed to a 'property damage' insurance policy that covers the damaged property (a mandatory contribution of 12% of the yearly premium is brought to the compensation fund).

This system relies on solidarity but demonstrates some drawbacks. It does not encourage the installation of prevention measures as those insured may consider they are covered in any circumstances. Besides such a solidarity-based system, a more market-driven system based on individual responsibility, as found in the United Kingdom for instance, provides (flood) insurance without government cover for householders and businesses (ABI, 2008). Between these two contrasted situations, there is room for many other systems. Insurance schemes do not stand alone but require information in order to assess risks. According to countries, this information is provided by private or public actors. In France, policies for the prevention of natural risks are based in particular on studies to identify and locate them, on land-use planning controls and on information issued with a view to increasing people's awareness of their vulnerability to natural risks and specifying what action they should take in the event of a natural disaster.

A 1995 law introduced the Foreseeable Natural Risk Prevention Plan (PPR). PPRs cover all natural risks, including floods (the most frequent), land movements, forest fires, avalanches, storms and cyclones. PPRs are drawn up by the state and approved in orders issued by the prefects. They are subsequently annexed to the land use plans, thus making compliance a statutory requirement for granting planning permission. The PPRs aim to ban all new construction work in the most exposed areas, and to regulate it and activities in less exposed areas (Institut français de l'environnement [IFEN], 2000). From the present context, based on regulation and on private actors' initiatives, the question has to be raised of the evolution of insurance cover in case of the significant evolution of climate-related natural hazards.

The white paper published in 2009 by the European Commission subtitled 'Adapting to climate change in Europe: towards a European framework for action' gives some indications of such evolution and states, 'A more strategic and long-term approach to spatial and urban planning will be necessary' (ECCP, 2009). Even if climate-related hazards may concern country areas, the consequences of climatic events do concern local built environments. Adaptation policies deal with local situations and should then be elaborated whilst taking the local context into consideration. National adaptation policy cannot ignore this situation. European member states will probably need to empower local authorities to implement adaptation measures at the local level.

As an example, a recent investigation on the use of national instructions concerning flood hazards in France confirmed the importance of the local decision making level which is in a better situation to appreciate the local context (e.g. knowledge of past events and capacity to raise awareness of the local population) than any remote actor at the national level (Marchand and Salagnac, 2009).

Defining adaptation measures is also a challenging issue of some urgency as it consists of decisions to be taken during the following years in order to cope with a near-future climatic situation which is difficult to precisely describe but may only slightly depend on mitigation measures' effectiveness. In spite of these difficulties, policy makers have begun to develop adaptation policies. The UK Climate Impacts Programme (UKCIP; http://www.ukcip.org.uk) carried out pioneering work that 'helps organisations to adapt to inevitable climate change. While it's essential to reduce future greenhouse gas emissions, the effects of past emissions will continue to be felt for decades'. An outcome of the 2004 French Climate Plan is the preparation of a national adaptation plan to climate change to be completed in 2011.

5.5 Solutions to Climate Change Challenges for the Built Environment

The elaboration of both mitigation and adaptation policies for the built environment is a challenging issue for the coming decades. We already experience many climate-related hazards in Europe, and the uncertainty is still high concerning the evolution of local climate. Each event demonstrates that the built environment is not always well adapted to hazards. Nevertheless, we have to be prepared for hazards that may become more frequent and maybe more intense with climate change, such as heat waves. This preparation requires policies addressing technical issues (to better withstand mechanical efforts on structures) as well as societal organization issues (such as the care of fragile persons during an urban heat island event).

Emerging political initiatives acknowledge the necessity to mitigate GHG emissions and to adapt to a new climatic context. This step is essential but nevertheless not sufficient to bring solutions to climate change challenges for the built environment. These solutions will emerge from the construction community as a whole when some conditions are met. These conditions include technical improvements through construction materials and processes innovations. Construction product manufacturers have proved to be very proactive in this domain in recent decades and will continue to be key players. Site production safety and productivity improvement will also lead contractors to use better daily practices during building construction.

In addition to these first conditions which are in line with traditional construction concerns, other conditions have to be considered which mainly address organisational issues. They represent crucial evolution for better cooperation between construction stakeholders during a construction project. Concepts such as *integration, global approach* or *systemic approach* will have to become daily realities for construction projects of any size and importance. By this way, construction details which contribute to the long-lasting quality of energy- and carbon-efficient, climate-adapted buildings are likely to be thoroughly considered.

The organisation which is necessary to implement an integrated approach requires tools and skills which are not yet widespread through the construction community. Presently, such an organisation is mainly adopted for big projects with important stakes. The adoption and generalisation of such cooperation among the multitude of construction professionals, the majority of which are (very) small and medium enterprises, comprise a major challenge. Accompanying actions have to be imagined, implemented and assessed to elaborate solutions and create the conditions for their diffusion among the construction community. Is it nevertheless realistic to imagine a fast (i.e. within some decades) transformation of a traditional organisation rooted in the past? A renewed interest for the 'industrialisation' of the building production has been observed during the 1990s and 2000s. An acceptable argument in favour of such a trend is that it is easier to take care of construction details in a workshop or in a factory than on a site. Nevertheless, the production cost decrease, which some claimed to be a consequence of industrialisation, never appeared to be fully convincing. A main reason for this result is the uniqueness of each construction project. This feature does not result from a posture of construction actors but is a reality. The development of more energy-efficient and better climate-adapted buildings will reinforce this uniqueness, as projects will carefully take into account the characteristics of the building location and the adaptation to the local context.

The flourishing development of IT tools since the 1980s will probably introduce unexpected evolutions. These tools may bring the underdeveloped capacity to cooperate without overturning the structure of the traditional organisation. The new generation of construction stakeholders (both professionals and clients) have been educated in such an IT tools environment, and an innovative way of organising building production may emerge.

Education and training are also crucial conditions to develop solutions to climate change challenges for the built environment. Even highly 'industrialised' construction will long encounter its '3D' reputation (dirty, dangerous and demeaning). Big efforts have to be made to change this situation. More aware and demanding clients, and more innovative building production organisation driving the construction of better quality buildings, are factors that will help stakeholders develop affordable solutions that emerging political initiatives tend to support.

References

ABI (2008) *ABI/government statement on flooding and insurance for England*. Association of British Insurers, London.

ADEME (2005) hes chiffres clís du Bâtiment – energie environnement. ADEME, Paris.

Agence de l'Environnement et de la Maîtrise de l'Energie (ADEME) (2004) Agence de l'Environnement et de la Maîtrise de l'Energie, Paris.

CCR (2009) Natural disasters in France. Available from http://www.ccr.fr/index.do?fid=1566517819374305701.

European Climate Change Programme (ECCP) (2009) *White Paper: Adapting to Climate Change: Towards a European Framework for Action*. Commission of the European Communities, Brussels.

European Parliament and the Council of the European Union (2003) Directive 2002/91/EC of the European Parliament and of the Council. *Official Journal of the European Communities*, 1 April, L1/65–71. Available from http://eur-lex.europa.eu/LexUriServ/LexUriServ.do?uri=OJ:L:2003:001:0065:0065:en:pdf.

European Parliament and the Council of the European Union (2006) Directive 2006/32/EC of the European Parliament and of the Council. *Official Journal of the European Communities*, 27 April, L 114/64. Available from http://eur-lex.europa.eu/LexUriServ/LexUriServ.do?uri=OJ:L:2006:114:0064:0064:en:pdf.

European Union (EU) (2007) Communication from the Commission to the Council, the European parliament, the European economic and social committee and the committee of the regions: A lead market initiative for Europe. Commission of the European Communities, Brussels.

Institut français de l'environnement (IFEN) (2000) *Spatial Planning and Environment: Policies and Indicators*, 2000 ed. Institut français de l'environnement, Paris.

Institut National Etudes Démographiques (INED) (2003) Biraben J-N. L'évolution du nombre des hommes. *Population et Sociétés*, **394**.

Intergovernmental Panel on Climate Change (IPCC) (2007) *Fourth Assessment Report: Climate Change 2007 (AR4), Synthesis Report*. Available from http://www.ipcc.ch

Marchand, R. & Salagnac, J-L. (2009) Capacity building and information quality. Paper presented at the Urban Flood Conference, Paris, 26–27 November.

Salagnac, J-L. (2007) Lessons from the 2003 heat wave: a French perspective. *Building Research & Information*, June, 450–457.

Taillefer, N. & Salagnac, J-L. (2010) Climate change: an opportunity for progress? Paper presented at the CIB World Congress, Salford, May.

United Nations (2007) *Urban Population, Development and the Environment, 2007*. United Nations, Department of Economic and Social Affairs, Population Division, New York.

United Nations Environment Programme (UNEP) (2007) *Buildings and Climate Change: Status, Challenges and Opportunities*. United Nations Environment Programme, Nairobi.

World Meteorological Organisation (WMO) (2009) *Building the Global Framework for Climate Services*. World Meteorological Organisation, London.

Green Economics Dialogue and the Built Environment

Miriam Kennet

6.1 Introduction

Much of current design, planning and new building in the United Kingdom has been focused on short-term investment gains, within a business cycle focused on the annual year end or during a political cycle of 3–5 years. However, the construction industry has the potential to unlock significant gains in the battle to arrest the onward march of catastrophic climate change and its effects.

For too long, our economy has in part been driven and led by the building of new offices and infrastructure and the increases it provides in GDP or economic growth indicators. So when the building industry got into trouble as a result of the subprime mortgage situation in the United States, much of the rest of the world caught a cold. This was particularly acute in Spain, where buildings were more than one third over-priced, and also in Ireland. People's individual wealth in many, but not all, western countries is still measured by the amount of property which they own. When property prices became unstable, as they did when biofuels began to compete for land use, commodity prices quickly followed suit and this compounded the crashing prices from oversupply of cheap mortgage-driven dwellings.

Similarly, short termism in construction planning and lack of clarity about the longer term effects of building projects have an actual direct effect in driving climate change. In developing countries such as Ethiopia, the desperate need for building individual makeshift dwellings has meant that millions continue to chop down any available tree for firewood and for building shelter. This has completely denuded the landscape of trees in the very heartland of where human civilisation began. The landscape and microclimate

---

Solutions to Climate Change Challenges in the Built Environment, First Edition.
Edited by Colin A. Booth, Felix N. Hammond, Jessica E. Lamond and David G. Proverbs.
© 2012 Blackwell Publishing Ltd. Published 2012 by Blackwell Publishing Ltd.

have meant lack of rain and caused soil erosion and soil poverty on an unimaginable scale reducing a once prosperous nation to a biblical-style economy where ploughing is being done by wooden ox ploughs, even pavements are few and far between, not much grows and the country is reduced to an income of $390 per person per year (IMF, 2009).

Women are found in many less developed countries working in the construction industry and usually earning less than half the men's wages on the supposition that they are unable to perform heavy lifting and, therefore, should not be paid the same. However, the evidence is that they are doing the same work as men but just not being paid for it equally. This is a part of the 'Green Economics' dialogue as it is regarded as part of the problem that economics assert, that equity is the price of survival and that a successful green economy is one which shares out its benefits fairly (Dey, 2009).

Elsewhere building materials, such as cement, in countries of South America have caused such awful health effects that an enormous lawsuit for deaths led to the establishment of the World Council of Business and Sustainable Development (WCBSD) to prevent the economy being adversely affected by environmental factors.

It is also especially important in green economics to be mindful of and prevent the effects of building on habitat. We are currently living in the sixth mass extinction of species since the earth began, and it is largely anthropogenic (Leakey and Lewin, 1995; Broswimmer, 2002). We require other species to regulate the climate and earth systems, and at present many of the tiny sea creatures and plankton are migrating northwards as the tropics and their former habitats become too warm and too acidic. As habitats get destroyed, and as remaining areas become too small for viable reproduction and species maintenance, we are losing species at an alarming rate with no possibility of saving them as their habitat is destroyed. For example, palm oil production in Asia and logging in Brazil destroy the rainforest and the important habitat it provides. The Three Gorges Dam project in China led directly to the extinction of the Chinese River *baiji* or white-fin dolphin (Lawrence, 2007).

Against this background, the *Stern Review* (Stern, 2006) has predicted that climate change will so affect the productive capacity of the earth that most areas will become uninhabitable for people. Stern (2006), supported by the *New Scientist* (2009), predicts only the polar regions and some parts of Scandinavia will be suitable for people to live in and the rest of the planet will turn into hot unproductive desert or go under water. This means people will need to work together in new kinds of building and construction planning for high-density, high-rise, low- or negative-carbon living. This new situation and challenge will occur within the next two centuries. It will also become much more difficult for agriculture and there is a coming crisis of global food production as a result. Far from land being used for biofuels as a substitute for fossil fuels (as required by the march of the logistics revolution of the 1990s, and those who believe technology will always overcome nature's problems), all land will be required for living, for subsistence and for producing food.

This chapter further examines some unsustainable building practices and the potential dangers they present in an era of climate change. Within this context the choices we face are presented from a green economics perspective. Finally some examples of green actions are presented.

6.2 Examples of Unsustainable Building Practices

6.2.1 Flood Risk – Unsustainable from an Insurance Perspective

In many areas, complete disregard for the future has allowed or even encouraged large numbers of new houses to be built on or next to the flood plains of rivers, even though such areas will eventually probably go under water as sea level rise accelerates. This was starkly brought to the world's attention after the tragedy of Hurricane Katrina. Many people of the world live in the major coastal cities, which will sink below the waves during the next century or at least be subject to regular inundation, for example London, New York and Tokyo. The cost of building on flood plains is rising all the time, and many homes in the United Kingdom are unable to get insurance as a result.

Building on the flood plain also prevents natural runoff and creates more flood hazard, and therefore green economics would strongly advise against increasing this kind of risky investment. The costs of dealing with floods can run into billions. According to the flood recovery minister at the time, John Healey, the total cost of the flood damage in 2007 was estimated at £2.7 billion. The Association of British Insurers and the Environment Agency claimed the cost of the flooding was as high as £3 billion (BBC, 2007). This will significantly increase premiums for everyone.

The Environment Agency (EA) estimates that 5 million people live and work in the 2.4 million properties in England that are at some risk of flooding and, at present, around £570 million is spent every year building and maintaining the defences required for them. Half a million of those properties are in the highest risk band, which means they are at risk of flooding due to extreme weather expected once every 75 years (Jha, 2010). Climate scientists predict that by the 2080s, sea levels could be around 70 cm higher around the southern parts of the United Kingdom, making serious storm surges and floods more frequent. UKCIP estimates that keeping all 2.4 million at-risk homes at the existing level of flood risk for the next 25 years will cost £1 billion per year by 2035. 'Assuming that no new properties are adding to that risk, then that investment is to maintain the existing infrastructure and to invest to make sure it isn't worsened, taking into account the uncertainties of climate change,' said Robert Runcie, the EA's director of flood and coastal risk management (quoted in Jha, 2010).

However, the last government pressed ahead with building new homes on flood plains. Planning for climate change must also include the avoidance of future problems and for homes to have at least a certain life span, perhaps 100 years.

6.2.2 Limited Lifespan of Buildings

An alternative visitors' site at Stonehenge was proposed by English Heritage (2000), designed to serve coach visitors who want a quick stop off, for say 3 hours, and some consumption of drinks, food and souvenirs, and would last a maximum of 30–60 years. However, my view is that the people who designed Stonehenge, whatever their plan, actually managed to build an amenity which has lasted 5000 years. Are we so poor of spirit and mind that we can't match even a tenth or hundredth of such an achievement? That we have to build poor-quality throwaway construction?

During the property boom there was a rush to convert every piece of possible, not necessarily available, land into profit-driven spaces, through such activities as 'garden grab.' Sheffield town centre, for example, has new buildings being pulled down to make way for new developments that make more money (see Figure 6.1). Many firms in construction became very profitable and very rich. As land became scarcer in the First World, many new projects were identified in developing countries and produced more profit for the construction companies and their home economies. Such projects were driven by the profit motive of the construction company rather than the need for the particular project in the host country, for example, large dam projects in Ethiopia and China, which have led to thousands if not millions of people actually being displaced and losing their homes, so construction and the built environment can be seen as a leading player in the contemporary economic downturn.

Figure 6.1 Demolition of relatively modern buildings in the town centre of Sheffield, UK (taken in 2010).

The Green Economics dialogue in the built environment, therefore, asks completely different questions and has completely different aims. It aims to use a foundation of social and environmental justice and to try to provide for all people everywhere, other species, nature, the planet and its systems (Kennet and Heinemann, 2006). It takes a very long-term view, for example, to recognise the reality over the rhetoric, so that climate change cycles are over 10 000 years and need to be considered in the long term in planning (Kennet and Heinemann, 2006; Kennet, 2007).

6.2.3 Lack of Urban and Spatial Planning

The very lack of sustainability of activities and lifestyles, for example building whole cities which are over their local carrying capacity, means that water has to be brought in from elsewhere, or bottled when the temperatures rise. These short-sighted planning activities have created an enormous folly, which is directly causing climate change. Schools, local shops and clinics have been encouraged to close and the properties sold off to make way for commercial property for sale. The local community is then forced to travel to reach the amenity, where the future hidden costs of transport to the local community are not factored into the planning economics calculation. This planning led by commercial building interests has led to the lack of local community amenities in most modern developments. This has meant a huge rise in the need for private cars even to go and buy, for example, a bottle of milk. With the exception of Section 106(S106) of the Town and Country Planning Act 1990 money, in the United Kingdom, only the short-term profit to the builder is considered.

The increase in transport, especially when powered by fossil fuels, increases carbon pumped into the atmosphere and accelerates climate change, and increases in aviation further exacerbate climate change. The use of peat for heating and other uses leads to methane release, and these greenhouse gasses are causing a climate change as rapid as anything for the last 55 million years since the PETM, which peaked about 55 million years ago (Stern, 2009). The use of oil for increasing transport has caused wars, and also pollution of the seas, already under pressure. Huge dead zones, or *hypotaxia*, are appearing in the sea, according to Jacqueline McGlade (head of the European Environment Agency) in 2010, and whole areas are unable to support any life at all. How long before we have a similar situation on land as already 30% of land habitat is fragmented due to urban sprawl, and 65% of the habitats and 52% of species covered by the EU Habitats Directive have unfavourable conservation status?

Rural areas are now characterised by a lack of complete amenities and the need to use more and more private transport, as community buses and facilities are removed to make way for commercial building. Cities are expanding to bursting point and unsustainability as they make way for the fact that more humans now live in towns and cities than in the country and the city has exceeded the natural carrying capacity of the surrounding countryside. Many cities are becoming unsustainable mega-metropolises to which people flee in a desperate search for work when countryside resources become exhausted (Seabrook, 2007). For instance, Dubai is perched on the edge of a desert.

What is fascinating in all this is that the reason that we have reached such an impasse is the drive for economics success as described by the mainstream. Economics has been described as the management of scarce resources, and recently it has been characterised by the very creation of scarcity, for example gold is scarce and therefore valuable, the rainforest is now scarce and is becoming valuable. However, we need to ask if scarcity of clean air, clean water and rainforest is what a sophisticated species should be creating, or rather should we be creating an 'Economics of Abundance' (Hoerschele, 2010)? Is it rather that we should be using our economy to provide for the needs, responsibilities and impacts of everyone and everything on the planet (Kennet and Heinemann, 2006)? For example, we have recently begun to realise that the planet is a self-regulating mechanism according to Gaia theory (Lovelock, 2000) with a temperature mean of about 14°C. We have noted that if the planet gets too hot, the glaciers melt rapidly – which they are doing almost everywhere at the moment. This then allows magma from volcanoes to rise up, as downward regulating pressure from the ice (which would prevent eruptions) is reduced. When the ice melts it allows the seismic activity to increase, releasing sulphuric aerosols and cooling the planet down. This incredible cycle is so simple, but we interfere with it at our peril. It was an eruption 80 000 years ago which wiped out half of humanity in the Tolba Eruption. It is not clever to go fiddling with such powerful systems. What is fascinating is that we have, for many hundreds of years, been assuming that technology will get us out of any predicament in which we find ourselves, and so it seems with the current climate dilemma. We have assumed that geo-engineering and other techno-fixes can cure all our climate problems and allow the onward march of progress and business as usual. Further, the philosophical aim and assumption have always been that humans can tame wild nature and harness it for their own needs.

Indeed, some of the larger construction projects, from the Tower of Babel to the contemporary Burj Khlifa Tower in Dubai through to the re-routing of large rivers in China, have been built on the assumption that nature can be overcome and that humans are in charge. The coming of the age of climate change has indicated not only that in fact humans are not in charge but also that now the economy is a tiny subset of nature, and nature is very firmly in charge and is visibly driving the agenda once again. This is also especially evident when we contemplate that Vesuvius is believed to be overdue for an eruption again but building continues all around it.

6.3 The Choices We Face

6.3.1 Limits to Growth

We will have to severely adjust our economy and with it our built environment to live within the earth's carrying capacity and within the realms and limits that nature imposes upon us. These will be technological constraints and also economic constraints. They have been termed 'the limits to growth' (Meadows *et al.*, 1972), where resources and population support have reached a limit and growth must now equate to flourishing and growth as in nature. No longer can rain forest destruction, where trees are cut down for firewood, be counted as growth – even if it leads to an increase in GDP – but now must be measured as destruction of natural assets or natural capital (Kennet, 2007). A country clearly is less well off with a renewable resource such as a forest being depleted. It is not better off no matter what the balance sheet or graphs are telling us.

6.3.2 Green Economics and What It Means

We need to recalibrate what our terms mean in economics. The very word *economics* is from the word *oikia* or household or estate management (Kennet and Heinemann, 2006). The integrity of the household today is the last thing considered in economics, and the realm of women is not included in *Homo economicus*, the basic unit in which economics modelling occurs. Green economics moves beyond this and can be considered 'post *Homo economicus*' and part of a new era where economics is practised by all people everywhere, and is not 'done to' or advised by one group on another group. Each person on the planet forms an equally valid economic unit. It is no longer led by white middle-class western-educated men, trained in Wall Street or Harvard (New Scientist 2008).

Green economics is about access for everyone, and provisioning for everything on the planet. The needs of the plankton regulating our climate are as important, if not more so, in the consideration of a new construction project as the men in suits. In reality this has a twofold aspect, firstly intrinsically (their own existence value, and their own right and needs of 'just simply being' part of life on earth) and secondly their use to us and our survival value, or need to continue as a species, as we need them there instrumentally in order for us to survive in a nonhostile climate. This instrumental approach is often called the 'ecosystem services approach', where the value of the ecosystem in providing human services is measured as a 'use value' to us.

An objection to the whole modelling and *Homo economicus* approach is that most of the planet is not *Homo economicus*. At least half is *Gyny oikonomika*. Many of the rest, one half of one fifth of people, are poor and hungry and have no ability to achieve their personal buying preferences as rational economic men. Indeed their decisions are based simply on meeting their needs from day to day. In fact the very word *civilisation* means living in towns. We have now become 'civilised'. We have tamed nature and we have achieved economic superiority to other species in harnessing much of the ecosystem at the moment to provide for our own purposes. However, as soon as we reached this position, it became clear that this is not a hospitable position to be in and neither is it sustainable or achievable for most people.

Women, for so long excluded from economics considerations, are now starting to run banks and whole economies in a more holistic way, redressing some of the worst imbalances. For example women have started an important trading floor in Ethiopia, are running the banks in Iceland and have taken over as heads of state in several countries. A green economy is characterised as a diverse and inclusive economy with special needs and all abilities recognised and valued. Learning is regarded as a lifelong activity, rather then something done just once at school. The key to getting out of poverty and recognising the 'Millenium Development Goals' (*New Scientist*, 2008) is regarded as ending maternal mortality and educating girls and women. This is one green way of reducing population and is regarded by Jacqueline McGlade (in 2010) as the single most important aspect of creating a green economy.

Two of the biggest questions currently are, what is the right solution to climate change, and what is the green solution? Is there a quick-fix solution to stop the runaway climate change predicted in most reports? Are we really going to get to 6 degrees of warming – as warm as the dinosaur period (Lynas, 2007)? Can humans really survive into such an era? As we bask in a huge heat wave again in the northern hemisphere with a major drought, what is the right approach?

6.3.3 Strategic Choices in a Green Economy

There are at least four strategic choices we can make.

6.3.3.1 *Market Mechanisms*

These are continuing to use the market to sort out the climate. In particular, the main well-known method is the Kyoto Protocol, where carbon is priced and then traded using the Clean Development Mechanism in less developed countries to trade and to allow money to be exchanged towards poorer countries. Those countries that have traditionally had higher carbon emissions trade and pay for their right to pollute. This method was invented by Chichilnisky (2009) and is now in another round; after the Copenhagen Conference COP 15, there was another round in Mexico (in December 2010). The latest idea is to provide geo-engineering and giant scrubbers to Africa to allow Africa to go carbon negative; in other words, in return for money income, they will clean up the carbon that richer countries use. However, although in theory (theoretical economics) this should work, in practice the *Stern Review* (Stern, 2006) showed that 'climate change is the single biggest market failure the world has ever seen'. In fact, more people are talking about the crisis of capitalism as a result and mainstream economics is facing some of its harshest criticism. Therefore, markets are undoubtedly part of the current mix.

6.3.3.2 *Geo-engineering*

A method that is being developed is geo-engineering where new technologies are being attempted in a rush to halt the rising tide. Some examples include creating artificial volcanoes, or carbon capture and storage where carbon is collected at a fossil fuel power station and stored underground for many years to remove it from the atmosphere. The problem is that few of these projects have been tested and no one knows if they will work; by the time we find out, it could be too late as the climate would have altered significantly and dangerously.

6.3.3.3 Regulation

There are many things we can do to halt the use of fossil fuels and methane production, which are causing melting of permafrost and tundra releasing very potent greenhouse causing climate agents.

It is known that the average user in the United States is consuming 25 tonnes per year of carbon dioxide equivalent (Kennet et al., 2009; Cologna, 2010). The average African is using less than 1 tonne and the average European about 10 tonnes, the average Chinese person now about 5 tonnes of carbon and the average Indian around 2 tonnes.

The first thing that could be done which would solve many of the issues is to educate women – as they tend to educate their children (Kennet, 2009) – and, secondly, to create a huge push towards 'Contraction and Convergence' (Kennet et al., 2009) would mean that the world's larger economies would contract so they would not consume as much and the world's poorest economies would be allowed and managed to increase. So there would be a convergence of levels both of economy and also of a climate inducing carbon footprint, eventually for equality. This was considered a radical idea when first proposed by Greens. However, today, the Stern team (Stern, 2009) is arguing that we need the 'fastest period of growth the world has ever seen in order to pay for the technical developments we need to meet the climate change imperative'. In fact, they argue that such growth will peak around 2030. This greatly enhanced growth looks increasingly unlikely to occur given the current economics downturn. Considering that with the last 50 years of 'high mass consumption of goods' (Rostow, 1960) and the artificially created demand through advertising, what has happened is an exhaustion of the world's resources and still one fifth of humanity is poor and hungry. A recipe of more of the same – and greater and accelerated growth and even more resource consumption and high markets – has led us to the state of massive debt that could take a generation to fix. Stern's (2009) high-market growth solution does not look attractive or viable, and this time we don't have time to experiment.

The strategy on the other hand of government regulation so that people can only use initially 2 tonnes of carbon and then negative carbon looks more promising, as if people know that they have to make changes and everyone is doing the same, they are far more likely to do it. This is effectively rationing, but it does provide for everyone equally and so removes the incentive for cheating. As climate change and polar ice melt become visible realities, the acceptability of such a scheme becomes much more likely and more desirable as a preferred option, a bit like a wartime adoption; everyone shares in it together equally and with pride.

6.3.3.4 Lifestyle Changes

The preferred green solution is lifestyle changes. By making lifestyle changes we argue that we are creating a new future for humanity, and it can be an exciting and high-tech future. We can choose to do different things differently. For example, we can all cut down our carbon footprint, measuring it with a carbon calculator, and cut it by 10% every 10 weeks. We can use slow travel when moving around and switch from plane to train even for business trips. I tend to try to use the train, this year alone having been to Montenegro, Italy, Norway and Spain using the overnight trains. The slow movement for food and for slow cities, the Citislow (Hoerschele, 2010) movement, is also starting to gather interest. We can choose to source locally from our community to remove embedded carbon and to recreate viable local communities and take an interest in planning, ensuring it meets the criteria outlined above for a longer term perspective. We can choose to cycle more and walk more.

In terms of our built environment, there has been much heralding of green technologies for new builds but a green view is that existing housing and buildings should be improved as they do not necessarily disturb the habitat of wildlife and also brownfield is a better option than disturbing wildlife on a greenfield site. ICLEI (Local Governments for Sustainability) works to combat climate change and build sustainable cities to improve whole life cycle costing such as the idea of cradle to cradle and also to implement green procurement, which is a major part of turning cities green. Much of the budget of a city will be bought-out or contracted items, and the role of green procurement, sustainable procurement, supply chains and procurement for the future and future generations is at the heart of practical implementation of green economics. It was the logistics revolution in 1990 which reduced the price of transport of goods – and masked the hidden external costs to the community as a whole, and this is in part one of the drivers of rapid climate change which in my opinion needs to be reversed and efforts should be made to turn procurement decisions into strategic decisions made with their impacts on the whole community in the years to come firmly in mind. Outsourcing such decisions led for example to the finding that children as young as 7 years old were working a 98-hour week for high-street shops in order to deliver low prices and young women were working very long hours for low wages in sweatshop conditions for the Microsoft supply chain (Hull and Sorrell, 2010). If we would never allow this here, then outsourcing and dumping social and environmental costs are hypocrisy of the worst kind and we need to *in source* these supply chains back – especially as other jobs disappear.

6.3.4 Shortage of Green Technology Raw Materials and Sourcing from China for Eco-tech

One of the biggest drawbacks of the green techno-fix approach is that much of the resources to make it work are from rare earth metals which are only found in China (Connelly *et al.*, 2005) and which are using child labour; also, China is increasingly reluctant to export these metals as it needs them for its own use. This will effectively halt some of the new green techno-fixes and technologies, leaving little choice but to opt for lifestyle changes in any case.

6.3.5 Facing Up to the Costs of Climate Change

Faced with a heat wave, and accelerating use of air conditioning which will make matters worse, the advice from governments is to cut the use of IT and switch it off when not in use, and to turn off ornamental fountains to reduce unnecessary energy and water use. Demand increases as temperatures increase, for example in the summer of 2010 on the east coast of the United States, where temperatures have exceeded 39°C. New York City hit record levels of electricity demand for three consecutive days, while a number of areas in New York State and Connecticut suffered power cuts.

Current economics models, I would argue, are beginning to tend towards the measurement of impacts to climate and to global environmental change, and this is very important and was a feature of environmental economics and welfare economics. For the first time the 'externalities' of a decision or project are factored in and costed so that the hidden costs to the community can be calculated. Similarly the willingness for a community to pay for the continued provision of a park or wilderness area or even rainforest is one of the most popular environmental methods for deciding on the value or suitability of a project, a method largely pioneered by David Pearce (Pearce et al., 1995).

Although today President Sarkosy of France and the European Environment Agency are making big strides in national green accounting processes, there is some suspicion that while measuring and alternative indicators are important parts of the programme and picture, they do not prescribe any actual change on the ground, but only measure it. Therefore, we need to act decisively if we want to avoid the worst effects of climate change. This action is designed to avert us from a possible or even likely future of 6 degrees of warming to hopefully a change of 2 or even 1.5 degrees of warming as requested at COP15 Copenhagen.

The taxation of unsustainable use of the built environment is also an area which has attracted interest, and there is a large movement for the implementation of a land value tax, which was advocated by Adam Smith, Henry George and Michael Gorbachev. This is regarded as one way of levelling the investment in land by rich investors and sharing it out among more of society.

Lifestyle choices really just mean taking a longer term view of our activities and in particular the effects of building and construction. In terms of building this can mean ensuring energy efficiency and greener choices in planning, and systems within those houses could enhance the development of a green economy. For example, in our green economics model (see Figure 6.2) we would consider reuse, recycle, reduce, relax and repair all important factors in creating this new way of working (Kennet and Heinemann, 2006). Some of these ideas are beginning slowly to percolate into the mainstream and into projects related to building.

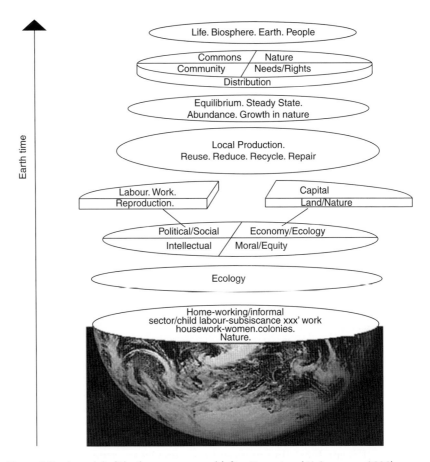

Figure 6.2 A model of the 'green economy' (after Kennet and Heinemann, 2006).

6.4 Green Projects in Action

6.4.1 Dongtan Green City Project

By 2030, 60% of the world's population will be urban dwellers, and so making our cities greener places to live is one way that business argues we can keep living in cities.

The plan was that in Dongtan, on Chongming island near Shanghai, half a million people would live in a car-free, zero-emission, recycling city with an ecological footprint one third that of people in Shanghai. This was to be the biggest single building project in the world, but it is taking place in a country which appears to accept the limitations of green to include only technical fixes rather than the complete scope of social and environmental justice. Urbanisation has led to 90 cities with more than a million residents, and 400 million people are expected to move from the countryside to cities in the next 30 years. Current plans for eco cities include the 'Sino-Singapore' Tianjin, Huangbaiyu, Nanjing and Rizhao.

Dongtan Green City Project in China, to be built by Ove Arup, is the world's first planned eco-city and will be near Shanghai. Details of technology changes include power that opens the door to an apartment coming from a solar cell on the roof, while the water used for the evening shower is recycled, as is all waste in this city of half a million residents. A dinner of boiled rice, spinach and spicy chicken would be locally produced using organic methods. Later, a person would stroll down to the car club, rent a battery-powered sports car and whizz through the tunnel back to Shanghai to cruise the Bund. This is one possible vision of China in 25 years' time, low carbon-footprint living in the eco-city of the future. This project concentrated on greener techno-fixes.

The green response is, however, why on earth do you need power to open a door? Such a response would also encourage lower tech and for people to replace gadgets, as labour would be more intensified in a green economy (Kennet and Heinemann, 2008). During the twentieth century, gadgets began to replace people or labour. In a green economy, this process is reversed as in for example organic farming where labour is intensified, labour and job taxes are reduced and work is shared out more equally with a basic income paid to all to prevent poverty and to keep everyone's skills up.

6.4.2 The Decent Homes Standard

One of the main ways green economics can be implemented is in securing 'Decent Jobs' (International Labour Organisation [ILO], 2008) under the ILO's Decent Jobs (ILO, 2008; Kennet and Heinemann, 2010) and Green Jobs Scheme (ILO, 2008, 2009) and also the Decent Homes Scheme (Kirklees Council 2009), which recommends thermal comfort for residents.

The Decent Homes Standard (2010) is a minimum standard that triggers action to improve social housing. It is a standard to which homes are improved. As constructed, the standard allows all landlords to determine, in consultation with their tenants, what works need to be completed, and in what order, to ensure the standard is met.

The Decent Homes Standard (2010) has four criteria:

1. It meets the current statutory minimum standard for housing (i.e. the dwelling should be free of Category-1 hazards under the Housing Health and Safety Rating System [HHSRS]).
2. It is in a reasonable state of repair.
3. It has reasonably modern facilities and services.
4. It provides a reasonable degree of thermal comfort.

The standard was revised in 2006 when the HHSRS was introduced to replace the unfitness standard as the statutory minimum standard.

By April 2009, 86% of homes in the social sector in the United Kingdom were classed as decent. The programme has also brought wider benefits such as improved housing management, tenant involvement and employment opportunities. As with many green areas of development, wider and deeper economics benefits occur simultaneously.

The original target was that all social sector homes would be decent by 2010, but by November 2009 the department was estimating that ~92% of social housing would meet the standard by 2010, leaving 305,000 properties 'nondecent'. 100% decency would not be achieved until 2018–2019.

An example of this is Kirklees Council (2009; and see http://www.kirklees.gov.uk/you-kmc/kmcbudget/2009/Capital2009-10to2013-14.pdf), which is using a number of schemes implementing government projects to contribute especially to criteria no. 4.

6.4.3   Warm Zone Project (Kirklees Council, 2009)

The Warm Zone Project has run from 2007 to 2010. Warm Zone contacted every householder, giving every Kirklees resident the opportunity to make their home warmer and more comfortable, contributing to reducing energy consumption and make a positive impact on the environment by providing free loft and cavity wall insulation and reducing CO_2 by 34 000 tonnes per year. The programme has support of over £20 million confirmed over a 3-year period. Also, as part of the Warm Zone Project, they offered every resident a free carbon monoxide detector.

It aims to help with fuel poverty, vulnerable residents with health issues, financial inclusion, and improving awareness and education. It provided a one-stop approach and delivering a low-carbon Kirklees. It was financed with innovative methods, and its total cost was around £20 million over the 3-year programme –£11 million in Carbon Emissions Reduction Target (CERT) funding from Scottish Power and £9 million from the Kirklees Council from its Capital Plan. The Capital Plan is funded by a complex mix of grants, revenue contributions, capital receipts and borrowing, but for most projects the type of funding is not hypothecated to tackling climate change.

Partners on this included Scottish Power as the main co-funder, National Grid as a co-funder and supplied resources to develop the business case and Yorkshire Energy Services (Y.E.S.) as the managing agent. They provide premises and equipment and integrated Warm Zone into their business. Warm Zones Ltd provided consultancy support, methodology, case studies, contact with other Warm Zones and networking opportunities with potential partners. Miller Pattison is the insulation contractor. The CAB, Kirklees Benefits Advice Service, the Pensions Service and Revenues and Benefits offered benefits and/or debt advice. West Yorkshire Fire Service provides fire safety checks and smoke alarms. Yorkshire Water provides water conservation advice. Carers Gateway offers support to people who care for friends or family. Safelincs are the contractor appointed to supply the carbon monoxide detectors. Energy Saving Trust provided advice packs.

6.4.4 The RE-Charge Scheme (Kirkless Council Capital Investment 2009–2014)

The scheme helps homeowners reduce their carbon footprint. RE-Charge offers interest-free loans to homeowners to install renewable energy and low-carbon technologies on their property. It was the first of its kind in the United Kingdom.

Renewable and low-carbon technologies are powered by abundant, free sources of clean energy such as wind, the Sun and even plant and animal matter. Generating energy from renewable sources can help reduce our dependence on resources like fossil fuels, producing much less carbon dioxide – a major climate change gas. It aims to generate a significant proportion of your home's heat or electricity requirements from renewable and low-carbon technologies and, with rising fossil fuel prices, significantly reduce people's energy bills. RE-Charge loans up to £10 000 interest free. Those costs above £10 000 would be paid in advance by the householder. A 'legal charge' (another name for a 'mortgage') will be taken as security over the property, and the sum of the loan will become repayable upon change of ownership of a property. Technologies available under the RE-Charge scheme are biomass heating systems, hydroelectricity, micro CHP (as this technology is in its infancy, it is anticipated it will be available in years 2 and 3), solar water heating, ground source heat pumps, air source heat pumps and solar photovoltaic (PV) systems. Wind energy will not be funded under RE-Charge.

6.4.5 Sustainable Cities

Sustainable cities are an important initiative often with transport hubs as core such as in Curitiba in Brazil and its system-wide sustainability benefits of integrated planning. This spatial strategy *maximizes* the efficiency and productivity of transportation, land-use planning and housing development by integrating them so they support one another to improve the quality of life in the city. Delivered through integrated urban planning and sustainability system-wide and decentralizing it. From the 1990s until today, the city's main planning focus has been on sustainable development and integration of Curitiba's metropolitan region.

By integrating traffic management, transportation and land use planning in the 1970s, the city was able to meet strategic objectives which sought to minimize downtown traffic, encourage social interaction by providing more leisure areas and pedestrian zones in the centre of the city and encourage the use of public transport and cycling in order to achieve an environmentally healthy city. Its eco-efficient, bus-only transportation system is a model for cities around the world. The 'speedy bus' runs along a direct line and stops only at tubular stations specially designed to move passengers quickly. An all-bus transit network with special bus-only avenues created along well-defined structural axes were pioneered by the city, plus the transit system is rapid and cheap.

Its efficiency encourages people to leave their cars at home. Curitiba has one of the highest rates of car ownership in Brazil, and high population growth. However, car traffic has dropped substantially. Curitiba has the highest public transport use of any Brazilian city (~2.14 million passengers a day), and it registers the country's lowest rates of ambient pollution and per capita gas consumption.

In addition, an inexpensive 'social fare' promotes equality, benefiting poorer residents settled on the city's periphery. A standard fare is charged for all trips, meaning shorter rides subsidize longer ones. One fare can take a passenger 70 km. Curitiba is referred to as the ecological capital of Brazil, with a network of 28 parks and wooded areas. In 1970, there was less than 1 square metre of green space per person; now there are 52 square meters for each person. Residents planted 1.5 million trees along city streets. Builders get tax breaks if their projects include green space. Flood waters diverted into new lakes in parks solved the problem of dangerous flooding, while also protecting valley floors and riverbanks, acting as a barrier to illegal occupation and providing aesthetic and recreational value to the thousands of people who use city parks.

The 'green exchange' employment program focuses on social inclusion, benefiting both those in need and the environment. Low-income families living in shantytowns unreachable by truck bring their rubbish bags to neighbourhood centres, where they exchange them for bus tickets and food. Under the 'garbage that's not garbage' program, 70% of the city's rubbish is recycled by its residents. Once a week, a lorry collects paper, cardboard, metal, plastic and glass that has been sorted in the city's homes. The city's paper recycling alone saves the equivalent of 1,200 trees per day (Sustainable Cities, 2009).

6.5 Conclusions

In a world where most people are now living an urban life, and where civilisation actually means urban dwelling, we have made surprisingly little progress in creating a sustainable future for ourselves as a 'civilised species.' We need to create sustainable urban living as well as rural communities which are part of their ecosystem and which don't destroy the very ecosystems on which they depend. Lamarck over 200 years ago thought we would destroy our habitat and then wipe ourselves out (Lamarck, 1802).

Our system of planning is very poor and short sighted. We need the built environment to lead in our economy and to enable other activities to become sustainable. The built environment has certainly created some of the problems we are experiencing today, but it also represents a very important opportunity and is an important key to solving them. It is therefore beginning to assume its rightful importance in the considerations about the green economy (Kennet and Heinemann, 2006). Ancient people managed to create buildings that have lasted for thousands of years. Surely we are capable of learning from their example?

References

BBC (2007) 2007 probably wettest UK summer. 31 August. Available from http://news.bbc.co.uk/1/hi/uk/6971370.stm.

Broswimmer, F. (2002) *Ecocide: A Short History of Mass Extinction of Species*. Pluto Press, London.

Chichilnisky, G. (2009) Surviving Kyoto's Do or Die Summit. *International Journal of Green Economics*, 3(3–4), 265–270.

Cologna, G. (2010) Speech at Bolzano University on climate change and green jobs, Bolzano, Italy, 17 May.

Connelly, N.G., Damhus, T., Hartshorn, R.M. & Hutton, A.T. (eds.) (2005). *Nomenclature of Inorganic Chemistry: IUPAC Recommendations 2005*. RSC Publishers, Cambridge.

Decent Homes Standard (2010) About decent homes. Available from http://www.decenthomesstandard.co.uk.

Dey, S. (2009) The socio-cultural impacts of economic changes to matrilineal Garo Society in Bangladesh. *International Journal of Green Economics*, 3(2) 184–198.

English Heritage (2000) *Stonehenge World Heritage Site: Management Plan*. English Heritage, London.

Hoerschele, W. (2010) *The Economics of Abundance*. Gower Publishing, London.

Hull, L. & Sorrell, L. (2010) The image Microsoft doesn't want you to see: too tired to stay awake, the Chinese workers earning just 34p an hour. Available from http://www.dailymail.co.uk/news/article-1266643/Microsofts-Chinese-workforce-tired-stay-awake.html#ixzz0z8xTMg99.

International Monetary Fund (IMF) (2009) *World Economic Outlook Database – April 2010*. Available from http://www.imf.org/external/pubs/ft/weo/2010/01/weodata/index.aspx.

International Labour Organisation (ILO) (2008) *Green Jobs: Towards Decent Work, in a Sustainable Low Carbon World*. UNEP, Nairobi.

International Labour Organisation (ILO) (2009) *Decent Work: Providing Safe and Healthy Workplaces for Both Men and Women*. International Labour Organisation, Geneva.

Jha, A. (2010) Cost of UK flood protection doubles to 1bn a year. *Guardian*, 8 September. Available from http://www.floatingconcepts.co.uk/news/cost-of-uk-flood-protection-doubles-to-1bn-a-year.html.

Kennet, M. (2007) Editorial progress in green economics: ontology, concepts and philosophy: civilisation and the lost factor of reality in social and environmental justice. *International Journal of Green Economics*, 1(3–4), 225–249.

Kennet, M. (2009) The costs of women's unequal pay and opportunity, transforming the unbalanced structure of our economy to meet the challenges of today, climate change, poverty, and the crisis of economy and economics, *International Journal of Green Economics*, 3(2), 107–129.

Kennet, M., Baster, N., Gale, M. & Tickell, M. (2009) Climate change: economics or science? The importance of the Copenhagen Summit, technological innovation, reduction in carbon use or market trading, and economic growth? The economics of the environment at a cross roads. *International Journal of Green Economics*, 3(3–4), 235–364.

Kennet, M. & Heinemann, V. (2006) Green economics setting the scene. *International Journal of Green Economics*, 1(1), 68–103.

Kennet, M. & Heinemann, V. (2008) *Managing the economy for lower growth: redefining prosperity*. Framework paper. Sustainable Development Commission, London. Available from http://www.sd-commission.org.uk/publications/…/Miriam_Kennet_thinkpiece.pdf.

Kennet, M. & Heinemann, V. (2010) Green jobs: a green reader. In: *Proceedings of the Bolzano Green Jobs Week May 2010 University of Bolzano*. Green Economics Institute, Reading, UK.

Kirklees Council (2009) *Capital Investment Plan 2009–2015*. Available from http://www.kirklees.gov.uk/you-kmc/kmcbudget/2009/Capital2009-10to2013-14.pdf.

Lamarck, J.B. (1802) *Hydrogeologie (Theory of the Earth)*. Paris.

Leakey, R. & Lewin, R. (1995) The sixth extinction. Available from http://www.well.com/~davidu/sixthextinction.html.

Lawrence, J. (2007) Extinct: the dolphin that could not live alongside man. *The Independent Newspaper*, 8 August.

Lovelock, J. (2000) *A New Look at Life on Earth*. Oxford University Press, Oxford.

Lynas, M. (2007) The hellish vision of life on a hotter planet. *The Independent Newspaper*, 3 February.

McGlade, J. (2010) Speech at Green Week Brussels European Commission June 2010. Available from http://ec.europa.eu/environment/greenweek/sites/default/files/speeches_presentations/mcglade_11_0.pdf.

Meadows, D.H., Meadows, D.L. & Randers, J. (1972) *The Limits to Growth*. Universe Books, New York.

New Scientist (2008) Top things to do to save the planet! 6 October.

New Scientist (2009) Future earth a world 4 degrees warmer: rapid climate change is making a dangerously hot world increasingly likely. What will it look like? 3 October.

Pearce, D., Markandya, A. & Barbier, E. (1995) *Blueprint for a Green Economy*. Earthscan, London.

Rostow, W.W. (1960) *The Stages of Economic Growth: A Non-Communist Manifesto*. Cambridge University Press, Cambridge.

Seabrook, J. (2007) *Small Guide to Big Issues*. Pluto Press, London.

Stern, N. (2006) *The Stern Review on the Economics of Climate Change*. H.M. Treasury, London.

Stern, N. (2009) Green growth. Speech to London School of Economics 21 September.

Sustainable Cities (2009) Curitiba Brazil. Available from http://sustainablecities.net/plusnetwork/plus-cities/curitiba-brazil.

Strategic Environmental Impact Assessment

Joseph Somevi

7.1 Introduction

Climate change has generated heated debates, necessitated solutions, occasioned agreements and divided opinion on its veracity. Division of opinion regarding climate change centres on its causes. While opponents on climate change actions will anchor their position to natural causes, others will blame anthropogenic actions as being the trigger for present climate change. Notable changes in climatic conditions have coincided with intensive human activity since the industrial revolution. Human activities manifested themselves in increased fossil fuel use, deforestation, methane release, nitrous oxide (N_2O) release and chlorofluorocarbons (CFCs) use (van den Hovea et al., 2002). The resulting global warming has increased the risks of hazards in the form of sea level rise, drought and changes in the frequency and severity of extreme events. Systems and people have become vulnerable to disasters, hurricanes, water stress, food insecurity, health risks, coastal erosion, inequalities, marginalisation and migration. How well systems can cope with the consequence of climate change depends on their exposure to hazards. This in turn depends on how vulnerable their underlying systems are; how they can absorb, cope with and address climatic hazards and how they can adapt to the changes brought about by climate change (OECD, 2008). Hitherto, climate change solutions seem to target products (i.e. fridges and appliances), elaborate policies (e.g. energy policy, economic policy or renewable policy) and address impacts from large projects (i.e. road infrastructure, housing and energy projects). Yet the environmental fallouts from strategic actions – policies, programmes, plans and strategies – which give rise to projects are only now receiving attention. Against this background, the paper argues that Strategic Environmental Impact Assessment (SEA) provides a coherent framework for reducing greenhouse gas emissions, reducing energy use,

---

Solutions to Climate Change Challenges in the Built Environment, First Edition.
Edited by Colin A. Booth, Felix N. Hammond, Jessica E. Lamond and David G. Proverbs.
© 2012 Blackwell Publishing Ltd. Published 2012 by Blackwell Publishing Ltd.

increasing energy efficiency, enabling renewable energy generation and increasing systems' resilience to climatic hazards. Following this introduction, the paper discusses strategic context and strategic environmental assessment. Next, the paper explains how SEA contributes to climate change solutions and explores the role of interest groups, before drawing conclusions.

7.2 Strategic Environmental Assessment

Sustainable development explores the tension between human progress and development on the one hand, and the limits of resources to perpetually fuel that progress and development on the other hand. Meeting the needs of current generation, future generations and other life forms lies at the very heart of that tension (Somevi and Hammond, 2008). Strategic environmental assessment (SEA) is a tool for promoting sustainable development presented in a written report. It does so by providing the framework for systematically assessing the environmental effects of policies, plans, programmes and strategies. It mitigates negative effects, enhances positive effects and sets out the context for monitoring future adverse effects. Through a step-by-step process of screening for significant effects, scoping for the level of details as well as assessing, mitigating, enhancing and monitoring environmental effects, the SEA process is aligned with policy making or planning processes (Somevi, 2002). SEA is not merely useful for promoting sustainable development. It can cope with indirect, secondary, time-crowded, additive, synergistic and cumulative effects beyond the reach of project environmental impact assessment. It can also consider a broader range of alternatives and mitigation measures (Therivel, 2004; Glasson et al., 2005; Fischer, 2007).

SEA evolved from the National Environmental Policy Act of 1969 (NEPA) of the United States. Today all major industrialised countries, including the United States, the United Kingdom, Europe, Canada, Australia, Japan, and Korea, and more than 50 transitional and developing countries in Asia, Latin America, Africa and the Pacific carry out SEA. This is supported by increased regulations (Thérivel, 1993; Dalal-Clayton & Sadler, 2005; Chaker et al., 2006). The advantage of SEA over other environmental management tools is its tiered process from policy to projects that allows environmental considerations to trickle down from high-level policies to low-level projects. As Glasson et al. (2005) have suggested, this allows sustainable development principles to trickle down from higher tier to lower tier strategic actions in such a way that parameters are set to promote sustainable development and to maintain carrying capacity of biophysical and socioeconomic systems. It also allows SEA to be carried out for all relevant tiers of strategic action and their alternatives; EIA to be developed for specific projects within constraints established by SEA and to ensure that monitoring and iterative feedback are carried out for earlier processes. As will be discussed in the next section, SEA contributes to climate change solutions not only because it addresses the shortcomings of project environmental assessment, but also because it addresses limitations of strategic actions, uses environmental receptors and their interrelationships to constrain its assessment, works within useful contexts, integrates the environment into planning processes, addresses significant environmental issues through the methods employed and has a scope for consultation.

7.3 Contributions of SEA to Climate Change Solutions

7.3.1 SEA Addresses Limitations of Strategic Actions

Policy makers, planners and strategy developers aim to address climate change problems through environmental policies. Yet the use of environmental policies, plans, programmes and strategies do not necessarily resolve climate change problems. For example, energy policies, plans and programmes can be diffused, fluid, conflicting, confused or subject to numerous interpretations. According to Thérivel *et al*. (1992), policy may not obviously be explicit but implicitly gleaned from different political statements. A transport ministry may have different policy outlook from ministries responsible for housing, energy, environment or finance. Energy policy may target specific energy efficient products yet overlook the effects of higher level policies that give rise to plans; followed by the plans that generate the programme; and the lower tier programme produced the projects. Even well-intended economic instruments like carbon taxes may adversely affect the more vulnerable in society (Thérivel *et al*., 1992). Again, energy policy making can have an imbalance in the allocation of resources. Policies may just be a mixed basket of ideas, aims and objectives lacking consistency and coherence (Thérivel *et al*., 1992). In this regard, un-assessed climate abatement policies or plans may fail to achieve its sustainability aims. Fortunately, SEA is a management tool that effectively renders policies, plans and programmes responsive to sustainable development. SEA looks at development actions systematically and brings co-ordination of decision making to the highest strategic level. It ensures that each policy is compatible and consistent with the overall climate change abatement goal under consideration (Somevi, 2002). Thus SEA enables strategic actions to reduce greenhouse emissions, tackle impact of transport greenhouse emissions, address energy use, increase energy efficiency, enable renewable energy generation, limit carbon sink depleting and increase the resilience of systems in a consistent and compatible manner through compatibility assessment. SEA differs from strategic actions in the sense that it is the crucible in which sustainability is really refined in the decision making process.

7.3.2 SEA Constraints Assessment through Receptors and their Interrelationships

Environmental receptors include water, soil, climatic factors, air, biodiversity, flora, fauna, landscape, cultural heritage, material asset, population and human health as well as their interrelationships (Weiland, 2010). The importance of water is seen in-house: it reacts with other substances to exacerbate greenhouse gases. In aquatic habitats where oxygen, nitrate, ferric iron and sulphate have been depleted as electron acceptors, the formation of the greenhouse gas methane through a terminal process of biomass degradation is facilitated. The construction of flood defence systems speeds up soil losses and greenhouse gas emissions from extensive concrete use. Changes to soil texture and structure through the application of nitrogen-based fertilizers in agricultural practices can increase greenhouse gas (i.e. nitrous oxide) emissions (Spokas & Reicosky, 2009). Wind farm development as a renewable option can cause soil losses, affect soil stability and facilitate methane emissions from high-carbon soils. In terms of population and human health, increased human numbers means increased use of natural resources and motor vehicles. More fossil fuel burning releases more greenhouse gases and leads to climate change. Population increases and corresponding housing, business and domestic growth puts pressure on water supplies and energy used in water production. With increased population is increased waste generation with greater consequence for greenhouse gas – methane – production. Ageing population could potentially lead to higher domestic emissions as older people stay home and turn on the heating. Creation of material assets such as roads, houses, industries and shops could use carbon intensive materials like steel and cement for their construction and maintenance. The use and maintenance of some historic features such as listed buildings could be counter-productive because of their poor energy efficiency. More energy will be consumed in their use and maintenance. Over-exploitation of biological resources such as forests could deplete the world's carbon sinks.

While pressure on environmental receptors exacerbates climate change, climate change in turn has grave impacts on the receptors by increasing their vulnerability to the effects of climate change. For example, climate change affects the abundance and quality of water resources. Lower summer precipitation is likely to lead to lower river flows. If lower summer precipitation is followed by higher winter precipitation, water quality and water quantity are likely to be adversely affected. When irrigation is used to cultivate biomass as a renewable energy option, hydrological regimes of soils can change so that soils become more vulnerable to the effects of climate change. Similarly hydro-renewable options such as dams could adversely affect terrestrial and aquatic receptors. In terms of biodiversity, droughts resulting from changing climates could stress biological resources, communities, habitats and species including wetlands, and change their pattern of distribution and lead to their migration or extinction. Extended winters mean more accidents from falls or hypothermia. Extended and hot summers will cause heat stresses and spread tropical-type diseases such as malaria (van Lieshout *et al.*, 2004). Climate change episodes like floods can disturb carbon-rich soils and erode coastal habitats and settlements. Furthermore, warmer temperatures and prolonged freezing temperatures could cause disruption to services and material assets. High insurance premiums and payouts could undermine the very value of material assets. Also, prolonged dry summers and wetter winters could negatively affect the historic environment. Subsidence from flooding or erosion of archaeological features along the coast could derogate from the enjoyment of the built features, their context and pattern of past historic use and associations of the historic environment. The environmental receptors shape and are shaped by climate change. Moreover, these environmental receptors are used in SEA as a framework for collecting baseline data, analysing plan context, identifying environmental problems, assessing impacts, devising mitigation and monitoring measures. Therefore, any assessment that considers how they affect climate change and are affected by it is likely to provide a credible solution to climate change. By using these environmental receptors to shape the form, content and process of SEA, the effects of climate change on the receptors as well as the effects of the receptors on climate changed are addressed.

7.3.3 SEA Works within Useful Contexts

Baseline environmental information, environmental problems and other relevant plans, programmes, strategies and environmental protection objectives do set the context for any meaningful SEA. SEA data management ensures that data is quantified with targets and trends. For instance, quantified data may look at the amount of electricity generated from renewable energy, sources and combined heat and power (CHP) located in the area, embodied energy in new buildings or average energy efficiency of buildings. It may look at energy use to include carbon emissions per person, total vehicle kilometres, total electricity and gas use or greenhouse gas emissions per region or per capita. It will measure the effects of climate change through sea levels, amount of rainfall and temperature (Donnellya *et al.*, 2007). It will include data relating to adaptation to include percentage developments with sustainable urban drainage systems (SUDS), percentage homes or roads in floodplain, ranges of habitats, number of heat or cold deaths, cases of subsidence, number of homes flooded, river flows and water quality, air quality in urban areas or cost of flooding to insurers or government agencies. By comparing trends in baseline of one agency against another at the same, higher or lower levels, agencies are able to gauge their relative performance compared with other agencies with similar constraints. In this context, SEA helps agencies or authorities to raise the environmental threshold and bar (João, 2007). OECD (2008) suggests that in a climate change adaptation-related SEA, data should cover climate variability, extreme events and related patterns of vulnerability in light of projected changes in precipitation and sea level rise and changes in the socioeconomic and environmental conditions that will affect vulnerability to climate-related hazards. Targets will invariably relate to international and national CO_2 reduction targets. Targets and trends inform changes needed to be made to plans being drafted. The nature or extent of the risks posed by climate change to development activities may only become evident if projected development trends are examined in parallel with projected changes in climate (OECD, 2008). Thus within the framework of SEA, planners, policy makers and developers are forced to consider problems and pressures affecting areas likely to be affected by the implementation of their proposals. SEA helps policy makers to take steps to avoid or remedy significant negative effects on fragile receptors.

Closely related to baseline data is how environmental problems are analysed in an area to which a particular plan relates. Analysing environmental problems and pressures in the area helps policy makers gauge the level of exposure of systems and areas to climate-related hazards and their underlying vulnerability. This vulnerability is often interlaced with economic, social, political, environmental and cultural factors that affect their capacity to resist, absorb, cope and respond to climate change related stresses and hazards. In this way SEA helps to analyse the adaptive capacity of systems future climate variability (OECD, 2008). Thus SEA forces the negotiation of actions to avoid damage to the ecosystem. Another context setting issues relates to broader policy context. SEA under the European Union Directive 2001/42/EC, for instance, requires member states to consider the relationship of other relevant plans, programmes and environmental protection objectives in the SEA process. This stems from the fact that policies, strategies or plans may fail to set ambitious climate abatement targets when they are focusing on limited department goals. By allowing other relevant plans and programmes outside their departmental bubble, particularly those related to climate change, to set contexts for their policies, strategies and plans, they can raise climate change mitigation and adaptation targets. None of these context setting issues are mutually exclusive. Compilation and analysis of baseline data will help identify problems in an area. Environmental problems throw more light on the baseline information. Other relevant plans, programmes and environmental protection objectives help establish targets for receptors in baseline information and also throw more light on environmental problems.

7.3.4 SEA Integrates the Environment into Planning Processes

From the evolution of SEA, it has become an integral part of the planning covering very broad sectors like agriculture, fisheries, forestry, energy, industry, telecommunications, transport, tourism, town and country planning, waste management and water management. As well as serving as the steps in the SEA process, screening, scoping, environmental assessment and post-adoption statement also aligns neatly with the planning processes. The aim of screening in relation to climate change is to determine whether the strategic action is likely to significantly increase or reduce greenhouse gas emissions, either directly or indirectly. Screening helps to identify whether the plan is likely to significantly affect the ability to adapt to the effects of climate change, in the area, in the future. Thus from the very outset SEA helps to identify and address significant climatic factors. The level of detail is the subject matter of a scoping report. Scoping within the context of climate change looks at whether a strategic action and adaptation measures are likely to have a significant effect on climatic factors through appropriately thought out methodology. In an environmental report stage in the process, the significant effects on climatic factors are predicted and evaluated to pave the way for mitigation measures. In this respect, SEA helps to prevent, reduce or offset those parts of the plan that may lead to adverse climate change effects. SEA enables climatic factors to be taken into account when finalising the plan so that monitoring is aligned to the potential climate impacts.

A major strength of strategic planning is its capacity to set the framework for development consents through spatial strategies, policies, conditions and supplementary guides (Glasson, 1995). That is the ability of higher tier policies to constrain lower tier plans, programmes, strategies and projects (Buurena & Nooteboom, 2010). Climate change conditions will aim to improve energy efficiency, and reduce demand for energy. Through options appraisal process in the planning and SEA processes, SEA and planning will encourage no foreclosure of options like carbon capture and storage in national and regional plans (Kørnøv, 2009). The SEAs of transport strategies and plans will encourage emissions reduction from the transport sector. Measures aimed at effective and integrated public transport system, efficient use of existing infrastructure and reducing the need to travel will be supported. Furthermore, in land use or transportation planning a development goal will be shaped to promote patterns that reduce the need to travel achieved through mixed use development, the allocation of sites close to existing public transport routes and facilitate modal shift (Kørnøv, 2009). In housing, SEA will promote high energy efficiency standards, support the use of solar gain through layout and design, will promote smaller housing development at higher density and encourage the need to increase tree cover. In housing development, the priority will also be to reuse brownfield land, support car free developments and provide transport choices that encourage modal shift, walking, cycling and car sharing. In waste management, a more sustainable modal shift in transportation of waste will be the goal of SEA in planning (Environmental Assessment Team, 2010).

Reducing energy use, increasing energy efficiency and promoting renewable energy generation are other priorities for SEA. Facilitating biomass as a substitute for fossil fuels in planning is as useful as providing a policy framework to support renewable energy development. Through SEA, the spatial framework for renewable energy development in appropriate locations is essential (Josimović & Pucara, 2010). In this context the use of CHP, energy from waste, supporting microrenewables on buildings or in developments can be accommodated within planning. If methane production is to be limited, it is essential to reduce farm wastes and other waste sent to landfills. Thus within the planning framework, new waste facilities would have to be provided for recycling, composting and thermal treatment. It also means reducing waste from new buildings and encouraging the reuse and management of construction and demolition wastes. If soil carbon sinks are to be protected, SEA has to set conditions for agricultural and forestry sector plans to protect areas contributing to net sink for carbon. Peatland should be protected from forestry or agricultural uses. Peat soil loss and sealing should also be protected from energy, housing, infrastructure or commercial developments (Environmental Assessment Team, 2010).

7.3.5 SEA Addresses Significant Environmental Issues through Methods

Thérivel (2004) provides a comprehensive list of methods of assessing effects (i.e. matrices, checklists, overlays, networks, quantitative or index methods) and more complex techniques (i.e. photomontage, remote sensing, GIS, compatibility matrices, scenario analysis, modelling, exclusion zoning, life cycle analysis, ecological footprint, expert opinion and cost–benefit analysis among others). More simply, effects can be assessed against SEA objectives and indicators or against environmental receptors. In terms of climate change solutions these tools have in-built climate-related criteria for describing, analysing and comparing environmental effects against which the performance of plans are gauged (Finnvedena et al., 2003). Climate change effects are predicted to determine whether the strategic action has negative, positive, uncertain or neutral effects on environmental receptors. Thus, prediction addresses the extent to which a particular plan or programme increases or decreases energy use, improves or worsens energy efficiency, uses lower carbon fuels or uses renewable energy. It also looks at a strategy's effect in relation to the production of waste, emissions (carbon or CO_2), mode of travel and disturbance of peat soil, among others. When predicted effects are further evaluated, SEA helps to determine the significance of the predicted effects on the receptors. The extent to which the predicted effects can, through some remedial actions, be reversed, pose significant risks, persist over duration of time (i.e. permanent, temporary, short term, medium term and long term) or measure their significance (Glasson et al., 2005; Fischer, 2007). Sometimes effects of separate development or strategic actions have minimal climatic effects when they act alone. When they combine and interact with other effects their effects tends to be significant. This is cumulative (direct, indirect, secondary and synergistic) measured within the context of SEA. Despite the contribution of SEA, through its methodology, to address climate change solutions, Somevi (2009b) argues that common assessment methods, from a wide repertoire of methods, tend to be very qualitative in nature. Consequently, predicting environmental effects can be subjective, evaluating significance problematic, choosing between alternatives challenging, assessing cumulative effects indeterminate, communicating results debatable and decision making based on the SEA outcome inconclusive. To maximise the role of SEA, he employed an ecological footprint (EF) method to quantitatively predict effects, evaluate significance, appraise options and assess cumulative effects across all environmental receptors. Because the method can also access carbon footprints without extra efforts, climatic effect of any strategic action can be assessed and quantified. Thus using an appropriate SEA methodology, not only will climatic effects resulting from a strategic action be quantified, but solutions can be diagnosed as well.

7.3.6 SEA Has a Scope for Climate Change Mitigation and Adaptation

Through mitigation hierarchy, effects likely to exacerbate climate change are avoided, abated, repaired or compensated. Positive effects are also enhanced. For instance, using a mitigation hierarchy in transportation, the emphasis in SEA will be to reduce need to convey forest resources like timber across long distances; and only if transportation of these resources is necessary will SEA encourage a more sustainable mode of transport to be chosen. In terms of climate change, the key focus of mitigation measures includes maintaining soil carbon stocks, maximising sequestration potentials and minimising flood risks. It also focuses on reducing greenhouse gas emissions, reducing transport impacts, reducing energy use, increasing energy efficiency and enabling renewable energy generation. The goals of climate change adaptation include increasing systems' resilience to the effects of climate change (Somevi, 2002). It means improving systems' adaptive capacity, developing infrastructure and approaching strategic planning so as to withstand increased precipitation and flooding; increased winds, unpredictable storminess and severe hurricanes; prolonged warmer climate, droughts and heat waves; as well as soil erosion and landslides (OECD, 2008; Mirza, 2003). In this regard, SEA mitigation measures ensure that strategic actions reduce greenhouse gas emissions. In the agriculture sector plans, SEA will reduce methane emissions from livestock; in economic sectors, SEA will emphasise low-carbon technologies and promote decreased energy demands from business.

SEA also provides the scope to increase the adaptive capacity of systems to cope with climate change hazards (Dessaia & Hulmea, 2007). This is achieved through conditions imposed on strategic actions and development actions through SEAs. Such conditions inevitably include the need to avoid constructing buildings and infrastructure in flood risk areas or coastal areas at risk, using sustainable drainage systems, increasing permeable surfaces and greenspace in new developments, protecting transport infrastructure from future flood risk; protecting supply of resources, goods, services and energy from disruption from floods (Dessaia & Hulmea, 2007; Environmental Assessment Team, 2010). Being a proactive environmental assessment tool, SEA enables planners to consider resilience of developments and structures to increased winds and storminess. Thus the effects of climate change will be taken into account through the design and specification of building, transport, drainage and sewerage infrastructure (Koutseris *et al.*, 2010). Any robust SEA will also consider the resilience of systems to warmer climate, droughts and heat waves. SEA will encourage building design for environmental performance that reduces the need for cooling and help to address urban heat island effect. Through SEAs greater use of green roofs and greenspace and tree cover will be encouraged. Consideration will also be given to future water needs and availability in planning for new developments. Measures such as the use of rainwater and grey water will be encouraged (Environmental Assessment Team, 2010). Through the SEA process, sites vulnerable to erosion and landslides will be protected to increase their resilience (Ford *et al.*, 2006; Koutseris *et al.*, 2010). Promoting increased forest cover in vulnerable areas will be a focus of SEAs. While SEA will specify the need to protect existing energy, minerals and infrastructure assets from the risk of erosion, SEAs will recommend that vulnerable areas be avoided in the development of new resources. In this context, SEA has become a powerful climate change mitigation and adaptation tool at the strategic level (OECD, 2008).

7.3.7 SEA Has a Scope for Consultation of Stakeholders

Interest groups and stakeholders are those who initiate the strategic actions and are responsible of their implementation, such as government ministries, departments, agencies, local authorities, public bodies and climate change experts. Others involved in and affected by the implementation of policies, plans and programmes will include individuals, communities corporate entities, insurance companies and banks – concerned about the effects of development actions. Other socioeconomic interests groups will be most vulnerable to climate change and will be affected by the implementation of the strategic action (OECD, 2008). The legal requirement to involve agencies and the public in consultation throughout the SEA process has helped the engagement of some stakeholders. However, the level of technical details in environmental reports makes them inaccessible to the average person. The requirement that environmental reports should contain a nontechnical summary tends to become an executive summary. If SEA is to become a powerful tool for consultation, the report needs to be accessible to the average reader (Vicentea & Partidario, 2006). A conscious effort must be made to bring interest groups together to participate in climate change solutions. Part of the process of SEA should also include capacity development (Somevi, 2009a).

7.4 Concluding Remarks

From the above discussions, SEA addresses direct, indirect, secondary, cumulative and synergistic impacts from climate change. SEA provides solutions for climate impacts generated by strategic actions and in turn deals with the effects of climate change on strategic actions. It also renders energy policies, plans and programmes consistent, compatible and sustainable. SEA allows impacts to be prevented, mitigated, ameliorated and avoided by communicating them to policy makers and making sure that SEA will inform and influence decision making at policy, plan, programme and project levels. It forces the introduction of systematic practices in the identification of relevant environmental issues, establishing the appropriate context and assessment of environmental impacts during planning processes. SEA helps to determine the need and feasibility of government initiatives and proposals, avoiding the foreclosure of options. SEA helps to integrate environmental issues into the development of policies, planning and programme decisions.

Through engagement with stakeholders, it makes planning processes open and transparent. Environmental consequences of energy production and use will be tackled. Greenhouse gas emissions from deforestation, carbon store depletion and methane production will be curtailed. SEA therefore provides a coherent framework for reducing energy use, increasing energy efficiency, enabling renewable energy generation and increasing systems' resilience to climatic hazards. In this sense, SEA will substantially contribute to climate change solutions.

References

Buurena, A.V. & Nooteboom, S. (2010) The success of SEA in the Dutch planning practice: how formal assessments can contribute to collaborative governance *Environmental Impact Assessment Review*, 30(2), 127–135.

Chaker, A., El-Fadl, K., Chamas, L. & Hatjian, B. (2006) A review of strategic environmental assessment in 12 selected countries. *Environmental Impact Assessment Review*, 26(1), 15–56.

Dalal-Clayton, B. & Sadler, B. (2005) *Strategic Environmental Assessment: A Sourcebook and Reference Guide to International Experience*. Earthscan, London.

Dessaia, S. & Hulmea, M. (2007) Assessing the robustness of adaptation decisions to climate change uncertainties: A case study on water resources management in the East of England. *Global Environmental Change*, 17(1), 59–72.

Donnellya, A., Jonesa, M., O'Mahony, T. & Byrne, G. (2007) Selecting environmental indicator for use in strategic environmental assessment. *Environmental Impact Assessment Review*, 27(2), 161–175.

Environmental Assessment Team (2010) *Consideration of Climatic Factors within Strategic Environmental Assessment (SEA)*. Scottish Government, Edinburgh.

Finnvedena, G., Nilsson, M., Johansson, J., Persson, A., Moberga, A. & Carlsson, T. (2003) Strategic environmental assessment methodologies – applications within the energy sector. *Environmental Impact Assessment Review*, 23(1), 91–123.

Fischer, T.B. (2007) *The Theory and Practice of Strategic Environmental Assessment: Towards a More Systematic Approach*. Earthscan, London.

Ford, J.D., Smit, B. & Wandel, J. (2006) Vulnerability to climate change in the Arctic: A case study from Arctic Bay, Canada. *Global Environmental Change*, 16(2), 145–160.

Glasson, J. (1995) Regional planning and the environment: time for a strategic environmental assessment change. *Urban Studies*, 32, 4–5.

Glasson, J., Thérivel, R. & Chadwick, A. (2005) *Introduction to Environmental Impact Assessment*, 3rd ed. Routledge, London.

João, E. (2007) The importance of data and scale issues for strategic environmental assessment (SEA). *Environmental Impact Assessment Review*, 27(5), 361–364.

Josimović, B. & Pucara, M. (2010) The strategic environmental impact assessment of electric wind energy plants: case study 'Bavanište' (Serbia). *Renewable Energy*, 35(7), 1509–1519.

Kørnøv, L. (2009) Strategic environmental assessment as catalyst of healthier spatial planning: the Danish guidance and practice. *Environmental Impact Assessment Review*, 29(1), 60–65.

Koutseris, E., Filintas, A. & Dioudis, P. (2010) Antiflooding prevention, protection, strategic environmental planning of aquatic resources and water purification: the case of Thessalian basin, in Greece. *Desalination*, 250(1), 316–322.

Mirza, M.M.Q. (2003) Climate change and extreme weather events: can developing countries adapt? *Climate Policy*, 3(3), 233–248.

OECD (2008) Strategic environmental assessment and adaptation to climate change. OECD/DAC Advisory Note on SEA and Climate Change Adaptability. Available from http://www.unece.org/env/eia/sea_manual/links/climate_change.html

Somevi, J. (2002) *Potential Role of Strategic Environmental Assessment in delivering Sustainable Energy Policies, Plans and Programmes for Ghana*. Unpublished PhD thesis. Oxford Brookes University, Oxford.

Somevi, J. (2009a) Partnership approach to delivering SEA. In: *International Association for Impact Assessment (IAIA). Impact Assessment and Human Well-Being. Accra, 11–22 May 2009*. Fargo, ND, IAIA.

Somevi, J. (2009b) Application of ecological footprint to SEA. In: *International Association for Impact Assessment (IAIA). Impact Assessment and Human Well-Being. Accra, 11–22 May 2009*. Fargo, ND, IAIA.

Somevi, J. & Hammond, F.N. (2008) Poverty, urban land and Africa's sustainable development controversy. In: *Environmental Management, Sustainable Development and Human Health*, ed. Goosen, M.F.A. *et al*. Taylor and Francis Group, London, 95–109.

Spokas, K.A. & Reicosky, D.C. (2009) Impacts of sixteen different biochars on soil greenhouse gas production. *Annals of Environmental Science*, 3, 179–193.

Thérivel, R. (1993) Systems of strategic environmental assessment. *Environmental Impact Assessment Review*, 13(3), 145–168.

Therivel, R. (2004) *Strategic Environmental Assessment in Action*. Earthscan, London.

van den Hovea, S., Le Menestrel, M. & de Bettignies, R. (2002) The oil industry and climate change: strategies and ethical dilemmas. *Climate Policy*, 2(1), 3–18.

Thérivel, R.E., Heaney, D. & Thompson, S. (1992) *The Practice of Strategic Environmental Assessment*. Earthscan, London.

van Lieshout, M., Kovats, R.S., Livermorec, M.T.J. & Martensa, P. (2004) Climate change and malaria: analysis of the SRES climate and socio-economic scenarios. *Global Environmental Change*, 14(1) 87–99.

Vicentea, G. & Partidario, M. (2006) SEA – enhancing communication for better environmental decisions. *Environmental Impact Assessment Review*, 26(8), 696–706.

Weiland, U. (2010) Strategic environmental assessment in Germany: practice and open questions. *Environmental Impact Assessment Review*, 30(3), 211–217.

Methods for Valuing Preferences for Environmental and Natural Resources: An Overview

Jessica E. Lamond and Ian Bateman

8.1 Introduction

If, as so many believe, climate change is the defining issue of the age, then it may be argued that concerns about greenhouse gases should outweigh any economic concerns in the design of the built environment. A fundamental question arises as to whether the models and methods used to evaluate the costs and benefits of construction can still be relevant in this new era. However, as demonstrated by Stern (2006), economic arguments cannot be ignored and may be helpful in understanding the incentives that can increase climate-sensitive behaviour. In addition, the belief that the actions of humans are having an impact on the climate and therefore contributing to the alteration of natural environments in a negative sense leads to an imperative need to be able to quantify the value of the environments so changed.

It can also be argued that if we understand the change in value caused by the actions of humans, then we can properly weigh environmental issues with financial ones in directing future policy and thereby help to prevent negative environmental change (Stern, 2008). Whilst incorporating global climate concerns into current thinking, it becomes increasingly important to counter-balance what may be changing policies on carbon emissions with local concerns such as air quality (Nernet et al., 2010) and local uses of natural resources such as forests (Karky and Skutsch, 2010). These issues cause debate at international levels and can be informed by knowledge of the value of natural resources over and above their carbon sink status.

While this is true on national and global scales, the valuation of environmental and natural resources is also important for built environment

---

Solutions to Climate Change Challenges in the Built Environment, First Edition.
Edited by Colin A. Booth, Felix N. Hammond, Jessica E. Lamond and David G. Proverbs.
© 2012 Blackwell Publishing Ltd. Published 2012 by Blackwell Publishing Ltd.

professionals on a more local scale. Climate change mitigation and adaptation technologies may have a large role in shaping the design of the future built environment. Future visions of urban spaces that are robust to climate change include such things as sustainable urban drainage, renewable energy and climate-resilient materials (Building Research Establishment, 2009; Building Futures/ICE, 2010).

Mitigation targets have also led to the call for carbon-neutral construction for new buildings. This is a demanding standard involving the provision of renewable energy technology on development sites. Developers and investors may wish to know the impact of the selection of renewable energy sources on the value of these sites and whether features chosen for their mitigation and adaptation properties may also have costs or benefits due to the preferences of potential future residents.

Choices in adaptation also derive from flooding of towns and cities, which is predicted to increase due to greater intense rainfall events, sea level rise and increased storminess. The response to these risks could be to use hard-engineered flood defences, but an alternative choice is to allow the return of floodplain areas to more natural states and thereby increase wetland provision, green spaces and biodiversity within urban areas. Investment in flood protection schemes falls far short of the demand for such schemes, and therefore prioritisation of projects requires extensive cost–benefit analysis (CBA), and the inclusion of values for the protection or creation of natural habitats could significantly affect the selection of schemes.

The valuation of environmental and natural resources is therefore important for both designing new buildings and developments and adapting our existing urban environments towards anticipated climate change. However, the question of whether environmental goods can be valued in any sense useful to decision makers is open to debate.

Critics of the concept itself argue that it leads to a degradation of human and spiritual values (O'Riordan and Turner, 1983). Critics of current practice in the area note that the difficulties in arriving at a number when techniques are applied poorly or in inappropriate circumstances can mislead decision makers to the extent that it would be better not to make the attempt (Diamond and Hausman, 1994). Countering these views is the assertion that when the right conditions prevail, these methods, although not a panacea for all ills, can provide a useful input to project appraisals (Barde and Pearce, 1991; Hidano, 2002; Day et al., 2007; Laird et al., 2009; Nickel et al., 2009).

This chapter considers several of the issues surrounding the valuation of the environment in turn. Firstly, it addresses the fundamental question of how far monetary evaluation can assist in arriving at a true assessment of the value of the environment. Secondly, having recognised the limitations within which monetary evaluation is set, it examines the methods by which monetary evaluation of the environment can be effected and compares their strengths and weaknesses. The chapter concludes by summarising the strengths and weaknesses of the available practical approaches in the light of the climate change debate. Possible research avenues are suggested.

8.2 Monetary Evaluation of Environmental Preferences: Theory

The first and most important clarification is that the techniques discussed in this chapter cannot be said to value the environment per se. What these methods truly do is to quantify human preferences for environmental goods and services. This means that if humans do not value a given entity, nor does that entity contribute in some indirect way to things that humans care about, then, from an economic perspective, that entity lies outside the remit of economic valuation. This also means that the values that humans hold for environmental goods, just as for manufactured goods, are not immutable or set in stone. Rather, if preferences change, then so will those values.

A second clarification lies in the difference between price and value encapsulated starkly in the observation made by Adam Smith 200 years ago that the very high value of water was in no measure reflected in its very low price. A more modern example of this is the very low price of waste assimilation services of the North Sea. The value of this service is becoming apparent now that alternative sewage disposal methods are being enforced (Ministry of Transport and Public Works, 1990).

The third clarification lies in the question of from whose perspective we wish to examine value. What is the extent of the moral reference class (Turner et al., 1994)? Are humans of the future to receive equal weighting with humans of the present? Should the interests of other species or ecosystems feature within decision making, as some authors suggest (Goodpaster, 1978; Watson, 1979)? While the future value to humans can be determined by appropriate weighting of future costs and benefits, as shall be discussed later in this chapter, the rights and values of other species are beyond the remit of economics, and such nonhuman intrinsic values cannot sensibly be monetised. Making these judgments requires a different philosophy beyond the economic perspective.

Therefore, in returning to the human valuation of natural or environmental goods the expression of value is often reduced to a monetised form by the concept of willingness to pay. This concept relates to individuals or societies being willing to pay for more of a good or willing to accept payment for less of it. This concept is not restricted to environmental goods but applies to all economic goods including frequently traded consumer goods.

Willingness to pay in turn can be broken down into the actual price paid plus the consumer surplus, that is, the amount the consumer would be willing to pay over and above the actual price paid. This is summed up by the following equation:

$$\text{Value} = \text{Willingness to pay (WTP)} = \text{Price paid} + \text{Consumer surplus}$$

For a privately supplied consumer good, the action of supply and demand will dictate that the consumer surplus is likely to be moderate and the value and market price will tend to be quite close. Usually in these cases, market price is used as a proxy for value and the two terms are used interchangeably. However, for an environmental good such as the waste assimilation example above, the market price may be close to zero and the consumer surplus potentially large as companies would be willing to pay much more to dispose of their waste than they have been asked to pay. Other examples of natural and environmental goods may not be traded at all. Many public goods are freely available, such as clean air or walks in public parks, but individuals would be willing to pay for these goods. Under these circumstances, maximum willingness to pay cannot be measured directly from transactions as the surplus exists in the mind of the consumer only. Therefore, for these goods the concept of consumer surplus becomes central to the valuation and different approaches are required to measure the willingness to pay for natural and environmental goods.

Notwithstanding, the use of willingness to pay has its drawbacks, not least the fact that equating value with willingness to pay has serious equity implications. Willingness to pay for a good is affected by an individual's ability to pay for it and therefore their income bracket. The preferences of the better off may therefore have a higher valuation than the preferences of those on lower incomes, and the former may dictate the agenda. For example, if the willingness to pay to avoid the risk of flooding (evaluated either through stated preference or by, say, house price depreciation) is used to determine the allocation of flood defence funding, the projects with highest value may well be determined by the wealth of the inhabitants. In terms of efficient allocation of resources, or the benefit per monetary unit invested, willingness to pay estimates can predict the highest yielding option because these areas may contain the greatest asset values to protect. However, the question of whether it is fair to protect the better off at the expense of the lower income areas is not addressed by this concept of value. Income distribution weightings can be employed to offset this problem if it is recognised and under certain circumstances are recommended in flood and coastal appraisal techniques (Penning-Rowsell et al., 2005). Often, for clarity, the question of equity is dealt with separately from the economic valuation.

Another way of disaggregating the value of a good is to consider utility. The total economic value of a good can be disaggregated into use and non-use value. Use value is a concept that reflects the usual ideas of utility. Non-use value embraces the idea of sustainability and encompasses an existence value, which may be based on moral or ethical beliefs that the fact that something exists is a benefit to all regardless of any plans to actually use the benefit oneself. An illustration of this might be the conservation of rare species or world heritage sites.

Figure 8.1 illustrates the categorisation of human values in the context of valuing environmental goods. This complex array of values need not always be in harmony; a project may simultaneously improve the utilitarian value while destroying the ecological function. As an example, the paving of front gardens to provide parking and improve the market value of property in towns is said to be destroying the function of gardens as a player in flood control.

The notion of sustainability so central to current debate within the built environment profession has led to a tendency to give higher weight to the needs and preferences of future generations. In economic valuation theory, future costs and benefits are often evaluated in monetary terms. Future cash

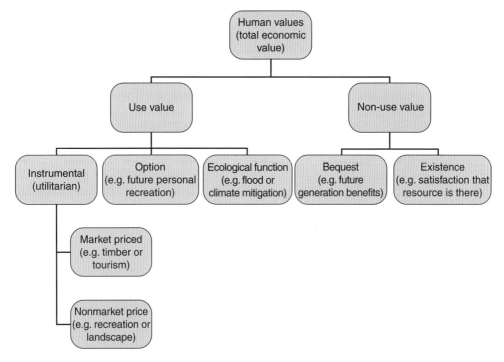

Figure 8.1 The categorization of human values (after Bateman, 1993).

flows are then assigned weights, which decline as the flows move further into the future. This process is known as discounting and results in a discounted cash flow estimate. Inherent in this process is the choice of weighting factor, which can make a large difference to the feasibility or desirability of proposed projects. It is in the nature of built environment projects to be designed to last decades at the least; therefore costs and benefits will need to be evaluated far into the future. In the context of climate change, one of the new challenges for valuing future costs and benefits is the large amount of uncertainty associated with future climate and also with future human preferences in the light of changing priorities. Climate science is introducing the complexities of probability theory into all decision making processes, and this presents a stumbling block for some evaluation methods. However, it seems likely that many of the methods described below will be amenable to the inclusion of uncertainty (Graham, 1981).

8.3 Methods for Monetary Evaluation of Environmental Preferences

A rich history of valuation of environmental goods stretches back to the 1940s (Bateman, 1993). However, there has been a recent proliferation of methods, which are divided, for pragmatic purposes here, into 'valuation' and 'pricing' approaches. Valuation approaches are those methods which are applied to estimate theoretically consistent values, and pricing approaches produce estimates which are based on the cost of attaining a good but do not necessarily reflect its theoretical value. These methods are discussed below.

8.3.1 Pricing Methods

The range of pricing methods encompasses opportunity costing, cost of alternatives, mitigation behaviour, shadow project costs, government payments and the dose–response method. In essence, these approaches examine the costs of protecting or obtaining a good or various alternatives seen to be equivalents. They are analogous to prices in that they measure the cost of goods but do not assess maximum willingness to pay, therefore neglecting the consumer surplus.

8.3.1.1 Opportunity Costing

When resources are limited, then their use for one purpose precludes their deployment elsewhere and represents a lost opportunity to pursue other projects. The opportunity cost approach examines what opportunities have to be forgone in order to realise an environmental goal. For example, the creation of wetland habitats from areas of coastal land represents the loss of the opportunity to develop the same area as a profitable farming resource.

The advantage of this method is that its value can be readily assessed as the expected net value of the products of the land as farmland. Disadvantages include the perhaps arbitrary choice of lost opportunity and, in some cases, the distortion caused by politically motivated subsidies which reflect the wishes of major interest groups and which will bias project appraisal.

8.3.1.2 Cost of Alternatives Method

The value of something that is freely available can be equated to the cost of an alternative means of achieving the same goal. One example is the value of the dumping of sewage in the North Sea, which could be equated to the cost of treating the sewage by the construction and operation of sewage works. The value of the use of a flood meadow as flood storage could be equated to the cost of building a flood storage tank of the same capacity.

8.3.1.3 Mitigation Behaviour

The willingness to make expenditure in order to mitigate environmental nuisance can be used to obtain a monetary assessment of the amenity value. One example is the cost of installing flood mitigation measures or the purchase of insurance against flooding (Lamond et al., 2009). Another might be the purchase of double glazing to mitigate noise pollution. However, mitigation actions are only partial solutions to the nuisance and are likely to undervalue environmental amenities, such as ignoring the facts that the double glazing installed to cut noise pollution must remain closed or the noise will still be heard and that external areas of the property will be unaffected by the installation of glazing.

8.3.1.4 Shadow Project Cost

If an environmental good is to be destroyed, then another pricing option is to look at the cost of providing an equal alternative environmental good elsewhere. An example of this is given in Buckley (1989), where the value of a threatened wildlife habitat was estimated by three alternative shadow projects for providing that wildlife habitat elsewhere. The cost of the alternative options can be seen as the 'price' of the threatened habitat, but great care is needed to ensure the adequacy of the proposed shadow project.

8.3.1.5 Government Payments

Where governments have already determined the value of environmental goods and services, for example by providing subsidies for environmentally benign production methods, such value can be incorporated into future project appraisals as they were in the appraisal of the Aldeburgh sea wall (Turner, 1992). Government valuations related to climate variables such as emissions targets are likely to drive such schemes and are subject to change in the long term.

8.3.1.6 Dose–Response Method

Where direct impact of the loss of environmental goods can be measured on a sector, the monetary evaluation of the damage can be used as the price of the environmental good. For example, the cost of air pollution may be measured through its effect on reduced crop production.

All of the above methods can be a useful starting point in valuing environmental goods and services which might otherwise be regarded as free. They are, on the whole, better than having no number, but all have limitations. Most provide benchmarks against a limited range of alternatives, which may over- or under-value the resource. In particular where government subsidies or direct government valuation is involved in the market, the political drivers may distort CBA to achieve political goals.

8.3.2 Valuation Approaches

The valuation approach centres on the estimation of a demand curve for the environmental good in question. Theoretically, therefore, the impact of incremental changes in provision of the good can be predicted with consistency and conceptual rigour, a 'true' valuation. These approaches are categorised further into stated and revealed preference methods. As suggested by the name, stated preference methods ask people to state the value they attribute to environmental goods. Conversely, revealed preference methods infer the value of the environmental goods from expenditure on other goods which are necessary to enjoy it, such as the purchase of petrol to travel to the amenity.

8.3.2.1 Stated Preference Methods

The contingent valuation (CV) method, contingent ranking (CR) and choice experiments (CE) are examples of stated preference methods. These methods ask participants to state directly what they would be prepared to pay or which options they prefer out of a suite of choices. Contingent valuation uses surveys to ask people directly what they would be willing to pay (WTP) or willing to accept (WTA) for a given gain or loss of a specified good. It is the most commonly used method in environmental economics representing a vast body of literature (Carson and Hanemann, 2005). Contingent methods are highly flexible and can be used to address preferences directly, to rank different preferences to compare consumer versus citizen preferences and to assess options which are outside the range of current choices. As such, they are useful in understanding complex motivations cross-sectioning society, for example the tendency to be WTP a higher price for exclusive rights to enjoy amenities than for community amenities.

However, there are also well-documented shortcomings for the stated preference suite of methods (Bishop and Heberlein, 1979). Along with the flexibility of the method comes the complication of framing the options to answer the desired question and the danger that a poorly specified survey may mislead. Further, in asking individuals to express preference rather than assess actual behaviour the methods are open to deliberate or accidental falsehood. Particularly in face-to-face encounters a subject may express socially acceptable views or employ gamesmanship (Bishop and Heberlein, 1979). In addition, there is the possibility of lack of experience or lack of engagement with the topic resulting in nonresponse or vague estimates of choices which are unlikely ever to be experienced, as in the study of floodplain discount where nonresidents anticipated a larger discount than residents of the floodplain (Brookshire *et al.*, 1982).

These problems aside, in a climate change context, stated preference methods can be used to envisage future scenarios which cannot easily be tested with revealed preference methods, such as those described below.

8.3.2.2 Travel Cost Method

The travel cost method uses the cost incurred by individuals travelling to reach a site as proxy for its recreation value. Travel costs are composed of two elements: the actual expenditure and the value of travel time. On the whole, the greater the travel cost to a destination the less frequent the visits are likely to be. Thus, a demand curve can be constructed for different travel costs and for the frequency of visits or consumption of the good. The value of the good can then be calculated.

8.3.2.3 Hedonic Pricing

Another revealed preference method, the hedonic pricing (HP) method, first described by Rosen (1974), has often been applied to amenities such as landscape, noise and air quality. The analysis is in two stages: firstly, the price paid for baskets of environmental goods is evaluated from changes in the price of houses subject to varying baskets of environmental goods. Secondly, the specific focus environmental good is separated from the rest of the basket by mapping out the demand curve for that variable holding all other environmental goods steady.

Hedonic pricing studies have been carried out for a large variety of valuation purposes, for example the value of open spaces (Bolitzer and Netusil, 2000; Luttik, 2000; Kopits et al., 2007; Cho et al., 2008), woodland (Garrod and Willis, 1992; Powe et al., 1997) and flood defences (Chao et al., 1998). However, because the range of potential variables for inclusion in hedonic models is extremely large (Sirmans et al., 2005) and the potential for misspecification great (Freeman, 1979; Butler, 1982), special care must be taken that omitted variables are not correlated with the environmental amenity of interest. In particular, neighbourhoods, which are often assumed to be exogenous to hedonic models, may in reality be long-term manifestations of a combination of amenity values (Cheshire and Sheppard, 1998). Recent extensions of the hedonic model include the time varying model and the repeat sales and repeat sales-hedonic hybrids which can offset some of these shortcomings (Palmquist, 1982; Case et al., 2006; Lamond et al., 2010).

The advantage of revealed preference methods is that they rely on actual actions of economic entities; however, they can only assess variations and combinations of factors that actually exist, and therefore projections beyond the range of historic data should be treated with extreme caution. They also reveal very little about the motivations of actors.

8.4 Solutions to Valuation of Environmental and Natural Resources

This chapter has considered two things, the conceptual and philosophical approaches to valuing environmental goods in the era of climate change and the theoretical strengths and weaknesses of methods available to quantify the monetary equivalents of environmental goods.

Firstly, it is recognised that the sum of human evaluation of the environment is eternally hampered by the human perspective. None of the evaluation methods above attempts to address the nonhuman valuation of the environment as some critics have suggested that it should. We conclude that evaluations of the preferences of other species or of the environment itself cannot be contemplated within a structured economic analysis.

Secondly, we have seen that 'valuation' methods have wider applicability and usefulness than the simpler and more easily applied 'pricing' methods but that neither is without its weaknesses and limitations. The use of 'valuation' methods, however, enhances our ability to incorporate environmental goods and services under a common monetary unit with a sound theoretical basis.

Finally, it has been identified that climate change increases the need to value the environment but simultaneously adds to the difficulty of such valuation. Further research which will address methods to explicitly incorporate climate change issues into project appraisal could strengthen the decision making process in built environment projects. Adaptation to climate change and mitigation of the effects of construction on the climate will both require consideration. This chapter has identified two major challenges, but there are no doubt many more. The first challenge will involve the temporal aspects of appraisal techniques such as alternative discount factors for world shepherding scenarios in CBA. The second challenge is the incorporation of vastly increased uncertainty inherent in models of future climate impacts on the built environment into estimates of future costs and benefits. It seems likely that the theoretically robust nature of valuation methods will lend itself most naturally to these potential developments.

References

Barde, J-P. & Pearce, D.W. (eds.) (1991) *Valuing the Environment: Six Case Studies*. Earthscan, London.

Bateman, I.J. (1993) Valuation of the environment, methods and techniques: revealed preference methods. In: *Sustainable Environmental Economics and Management Principles and Practice* (ed. R.K. Turner). Belhaven Press, London.

Bishop, R.C. & Heberlein, T.A. (1979) Measuring values of extramarket goods: are indirect measures biased? *American Journal of Agricultural Economics*, **61**(5), 926–930.

Bolitzer, B. & Netusil, N.R. (2000) The impact of open spaces on property values in Portland, Oregon. *Journal of Environmental Management*, **59**, 185–193.

Brookshire, D.S., Thayer, M.A., Schulze, W.D. & D'Arge, R.C. (1982) Valuing public goods: a comparison of survey and hedonic approaches. *The American Economic Review*, **72**(1), 165–177.

Buckley, G.P. (1989) *Biological Habitat Reconstruction*. Belhaven Press, London.

Building Futures/ICE (2010) *Facing Up to Rising Sea-Levels: Retreat? Defend? Attack?* RIBA, London.

Building Research Establishment (2009) *Life Handbook: Long-Term Initiatives for Flood-Risk Environments*. BRE, Bracknell.

Butler, R.V. (1982) The specification of hedonic values for urban housing. *Land Economics*, **58**(1), 96–108.

Carson, R.T. & Hanemann, W.M. (2005) Contingent valuation. In: *Handbook of Environmental Economics*. Elsevier, Amsterdam.

Case, B., Colwell, P.F., Leishman, C. & Watkins, C. (2006) The impact of environmental contamination on condo prices: a hybrid repeat-sales/hedonic approach. *Real Estate Economics*, **34**(1), 77–107.

Chao, P.T., Floyd, J.L. & Holliday, W. (1998) *Empirical studies of the effect of flood risk on housing prices*. IWR report. US Army Corps of Engineers, Washington, DC.

Cheshire, P. & Sheppard, S. (1998) Estimating the demand for housing land and neighbourhood characteristics. *Oxford Bulletin of Economics and Statistics*, **60**(3), 357–382.

Cho, S-H., Poudyal, N.C. & Roberts, R.K. (2008) Spatial analysis of the amenity value of green open space. *Ecological Economics*, **66**, 403–416.

Day, B., Bateman, I.J. & Lake, I.R. (2007) Beyond implicit price: recovering theoretically consistent and transferable values for noise avoidance from a hedonic property price model. *Environmental Resources Economics*, **37**, 211–232.

Diamond, P.A. & Hausman, J.A. (1994) Contingent valuation: is some number better than no number? *Journal of Economic Perspectives*, **8**(4), 45–64.

Freeman, A.M. (1979) Hedonic prices, property values and measuring environmental benefits: a survey of the issues. *Scandinavian Journal of Economics*, **81**(2), 154–173.

Garrod, G. & Willis, K. (1992) The environmental economic impact of woodland: a two stage hedonic price model of the amenity value of forestry in Britain. *Applied Economics*, **24**, 715–728.

Goodpaster, K.E. (1978) On being morally considerable. *The Journal of Philosophy*, **75**, 308–325.

Graham, D.A. (1981) Cost-benefit analysis under uncertainty. *The American Economic Review*, **71**(4), 715–725.

Hidano, N. (2002) *The Economic Valuation of the Environment and Public Policy: A Hedonic Approach*. Edward Elgar Publishing, Aldershot.

Karky, B.S. & Skutsch, M. (2010) The cost of carbon abatement through community forest management in Nepal Himalaya. *Ecological Economics*, **69**(3), 666–672.

Kopits, E., Virginia, M. & Walls, M. (2007) *The trade-off between private lots and public open space in subdivisions at the urban-rural fringe*. Discussion paper. Resources for the Future, Washington, DC.

Laird, J., Geurs, K. & Nash, C. (2009) Option and non-use values and rail project appraisal. *Transport Policy*, **16**(4), 173, 182.

Lamond, J., Proverbs, D. & Hammond, F. (2009) Accessibility of flood risk insurance in the UK: confusion, competition and complacency. *Journal of Risk Research*, **12**(5), 825–840.

Lamond, J., Proverbs, D. & Hammond, F. (2010) The impact of flooding on the price of residential property: a transactional analysis for the UK. *Housing Studies*, **25**(3), 335–356.

Luttik, J. (2000) The value of trees, water and open space as reflected by house prices in the Netherlands. *Landscape and Urban Planning*, **48**(3–4), 161–167.

Ministry of Transport and Public Works (1990) *Formal Declaration of the Third International Conference for the Protection of the North Sea: The Hague 1990*. MTPW, The Hague.

Nernet, G.F., Holloway, T. & Meier, P. (2010) Implications of incorporating air-quality co-benefits into climate change policymaking. *Environmental Research Letters*, **5**.

Nickel, J., Ross, A.M. & Rhodes, D.H. (2009) Comparison of project evaluation using cost-benefit analysis and multi-attribute tradespace exploration in the transportation domain. In: *Second International Symposium on Engineering Systems*. MIT Press, Cambridge, MA.

O'Riordan, T. & Turner, R.K. (1983) *An Annotated Reader in Environmental Planning and management*. Pergamon Press, Oxford.

Palmquist, R.B. (1982) Measuring environmental effects on property values without hedonic regressions. *Journal of Urban Economics*, **11**, 333–347.

Penning-Rowsell, E.C., Chatterton, J. & Wilson, T. (2005) *The Benefits of Flood and Coastal Risk Management: A Handbook of Assessment Techniques*. Middlesex University Press, London.

Powe, N.A., Garrod, G.D., Brunsdon, C.F. & Willis, K.G. (1997) Using a geographic information system to estimate an hedonic price model of the benefits of woodland access. *Forestry*, **70**(2), 139–149.

Rosen, S. (1974) Hedonic prices and implicit markets: product differentiation in pure competition. *Journal of Political Economy*, **82**(1), 34–55.

Sirmans, G.S., Macpherson, D.A. & Zietz, E.N. (2005) The composition of hedonic pricing models. *Journal of Real Estate Literature*, **13**(1), 3–43.

Stern, N. (2006) *Stern Review: The Economics of Climate Change*. HM Treasury, London.

Stern, N. (2008) The economics of climate change. *American Economic Review: Papers and Proceedings*, **98**(2), 1–37.

Turner, R.K. (1992) *Speculations on weak and strong sustainability*. CSERGE Global Environmental Change Working Paper. Centre for Social and Economic Research on the Global Environment, London.

Turner, R.K., Pearce, D.W. & Bateman, I.J. (1994) *Environmental Economics: An Elementary Introduction*. Harvester Wheatsheaf, Hemel Hempstead.

Watson, R.A. (1979) Self-consciousness and the rights of non-human animals. *Environmental Ethics*, **1**(2), 99.

Ecological Value of Urban Environments
Ian C. Trueman and Christopher H. Young

9.1 Introduction

Urban environments are an increasing proportion of the Earth's habitats and will undoubtedly have greater influence on future ecosystems. Ecologists working within these areas agree that urban environments are important not only because they may hold populations of interesting and sometimes characteristic flora and fauna but also because they are inherently distinctive (Harrison *et al.*, 1987; Duguay *et al.*, 2007). Urban environments exhibit combinations of environmental characteristics unusual elsewhere and often present environmental conditions analogous to those anticipated for climate change scenarios (Wildby and Perry, 2006) resulting in peculiar and unique species, habitat associations and communities (Goode, 1989). Specifically, these are:

- A characteristic climate that buffers the environment from extremes (Catterall *et al.*, 1998);
- Direct or indirect disturbances resulting from human activity (Rebele, 1994);
- Extreme subdivision and isolation of habitats (Dickman, 1987) leading to an intimate mosaic of differing site sizes (Young *et al.*, 2008);
- Many transient habitats (Hansson and Angelstam, 1991);
- Habitats exhibiting extreme environmental conditions (e.g. devoid of vegetation) (Rebele, 1994);
- Availability of a varied or particularly rich food supply (Catterall *et al.*, 1998); and
- Extreme interventionist management (Young *et al.*, 2008).

Solutions to Climate Change Challenges in the Built Environment, First Edition.
Edited by Colin A. Booth, Felix N. Hammond, Jessica E. Lamond and David G. Proverbs.
© 2012 Blackwell Publishing Ltd. Published 2012 by Blackwell Publishing Ltd.

9.2 Ecological Value

The natural environment provides people with goods and services that are fundamental to human wellbeing (Department of the Environment, Food and Rural Area [DEFRA], 2007). Probably the most direct benefits of nature for people concern health. Exposure to natural scenes reduces stress (Ulrich et al., 1991), while contact with even very small green spaces in cities leads to improvement in children's abilities to pay attention, delay gratification and manage impulses (Faber Taylor et al., 1998; Kuo, 2001). In addition to these 'cultural services', there are also 'regulating services' such as climate regulation, water purification and flood protection which are provided by seminatural vegetation (Wentworth, 2006).

The principal organisms in the urban environment are people, and the study of urban ecology is largely the study of anthropogenic influences. However, botanical recording in the urban area readily demonstrates high plant species richness per unit area, quite comparable with countryside levels (Table 9.1). Some of this is undoubtedly due to the presence of large numbers of relatively small anthropogenic biotopes, each with its own set of characteristic species. A vast rural forest might include more species than a tiny urban copse, but alongside the latter there might well be canals, lawns, car parks, rubbish dumps, road verges, graveyards and umpteen idiosyncratic gardens all in the equivalent area. Apart from the opportunities given for plant colonisation and survival, the structural complexity also increases the habitat niches available for all kinds of organisms.

Many animals are adapted to urban environments exploiting the full range of biotopes available and utilising the structural complexity they offer. Few are solely reliant upon urban areas, though some have strong associations, for example Black Redstart in the United Kingdom (Birmingham and Black Country BAP, 2000), but many exploit urban niches

Table 9.1 The ten most species-rich tetrads in Montgomeryshire.

Tetrad grid reference	No mapped species	Locality	Habitats
SJ22Q	275	Llanymynech Hill	Rural (post-industrial)
SJ22K	246	Llanymynech Hill	Rural (post-industrial)
SO29X	224	Cornden Hill	Rural
SJ31C	203	Breidden	Rural
SJ20I	201	Welshpool	Mostly urban
SJ11S	189	Near Meifod	Rural
SN98M	198	Llanidloes	Mostly urban
SJ20D	188	Near Welshpool	Urban fringe, mostly rural
SO08J	188	Llandinam	Rural
SO29I	187	Montgomery	Includes entire town, but mostly rural

Note: The figures relate to the number of mapped species per tetrad (i.e. 2 km × 2 km square) recorded for the Flora of Montgomeryshire (Trueman et al., 1995). The figure excludes the 184 commonest species in the vice county, which were not mapped.

leading to important populations of some species. Due to the variability in the nature of resources in urban environments, the most successful species tend to be generalists who are flexible in their life history requirements and are able to adapt to the unusual spatial and temporal distributions of food resources. The ability of urban areas to support these species leads to some being found in greater densities than would otherwise be possible (e.g. foxes and badgers) (Dickman, 1987). However, for some species their urban success is down to the fact there are direct analogues of their natural habitats present (Szacki *et al.*, 1994; Rossi and Kuitenen, 1996). For example, brownfield sites provide ideal conditions for many invertebrates that rely on open ground and early successional stages in their life cycles (Eversham *et al.*, 1996). Some of the more charismatic species that exploit these opportunities are raptors, such as peregrines and kestrels (Pomarol, 1996), which nest on buildings using them in the same way as they would use natural cliff faces.

Modern urban improvement schemes tend to sweep away this complexity and replace it with vast warehouse cities or neat neo-suburbias. Such is the intricacy of city life that the pattern might re-establish itself eventually, but we should try to make landscape architects (and maybe architects and designers of roads) consider how to engineer a slightly more environmental complexity into their new worlds. Wildlife can live successfully alongside people in those instances, where their needs do not cause conflict, can be accommodated or can be managed. Not only do these situations ensure the wider viability of wildlife, but it is these kinds of direct encounters with animals that give an extra dimension to urban living and often provide an introduction to the natural world to urban residents.

9.3 Urban Habitats

Urban environments, globally, contain the full range of ecological habitats (with the possible exception of true wildland) and, as such, the generic problems of climate change such as sea level rise and changing weather patterns may well have profound impacts on many of these (Wildby and Perry, 2006). There are a range of habitats that are, however, widely found or else have particular resonance in an urban context, and we deal explicitly with these below.

9.3.1 Seminatural Habitats

Individual species-rich habitats do occur in urban landscapes. There are many urban examples of classical seminatural vegetation (fondly known as *trapped countryside*), and they are at least as difficult to conserve as rural examples. The advantages in protecting these sites include the fact that well-tested criteria for designation can be used. Once designated, urban planners are usually quite good at giving protection to such sites, even second- and third-tier conservation sites. The main problems, as in the countryside, are the usual twin evils of intensification and neglect. Although agricultural intensification is often less of a problem than in the countryside, pressures for development are more likely, and even excessive public interest (nature for *too many* people, most of whom are more interested in cycling, dog walking and jogging, amongst other activities) can be very damaging. On the other hand neglect, leading to succession – loss of fences, no grazing, desultory cutting, emotive issues with cutting down any trees whatsoever and fly tipping amongst others – can be even commoner than in the countryside.

9.3.2 Created Habitats

Creative conservation sites are probably special cases of the above. Some believe imitation hay meadows and ancient woodlands are inappropriate in the urban context, but they are not really any more artificial than flowerbeds or mown lawns and they do (sometimes) have the advantage of supporting a wider range of species and (occasionally) looking quite attractive (Trueman and Millett, 2003). Whatever the merits, the issue is usually again how to provide long-term management, although there is, potentially, a very large issue about 'officious' habitat creation in inappropriate sites. Like every other kind of development, habitat creation requires detailed prior site assessment for existing nature conservation value. For instance, the hybrid-created habitats, with mixed natives and ornamentals, of the Dutch Heemparks and what are possibly their descendants from the Sheffield Landscape School (Dunnett, 1999), must surely present particular problems of management despite their often striking beauty. It is reported the Heemparks are very expensive to manage, especially in terms of labour.

9.3.3 Brownfield and Early-Successional Habitats

At the other extreme to the created and seminatural classical habitats are those habitats resulting from the natural processes of invasion and succession after site demolition and clearance, which can be such a feature of the urban landscape. Although these are commonly termed *brownfield sites*, they often have complex origins and interesting and diverse communities as a result. Gilbert (1989) made the case for these *spontaneous successional communities* to be allowed to go through the early stages without hindrance. Beyond the obvious need to prevent premature landscaping of such sites and to protect them from abuse, conservation is probably a landscape issue. Long-term survival of the species of these early successional communities probably does not depend on the conservation of the individual site, which would be futile anyway without programmes of repeated disturbance. The crucial factor must be the presence at any one time of sufficient examples of such sites within the conurbation, which can together regenerate the landscape seed bank for the next generation of examples. Ruderals with their abundant seed production, nonspecific seed dispersal and long-lived seed banks would seem to be well adapted for survival under these conditions, but how important is connectivity between sites in the urban landscape? It stands to reason it is important (Dawson, 1994), but the urgent investigations (Austin and Angold, 2000) could not demonstrate any positive effect of the proximity of key linear features – canals, railways and rivers – on site plant assemblages. One problem may actually be recognising a corridor when you see one. Private gardens are perhaps part of connectivity and also examples of seed-generating foci. Also, allotments and other examples of frequent cultivation need to be included in the assessment.

The conservation of woodland, which arises through these processes of succession, also needs to be addressed. Such sites require an unusual length of time to elapse before they mature and they are, therefore, uncommon. They usually include an array of native and alien species and, therefore, cannot be easily matched with the National Vegetation Classification. They are often unscheduled and unprotected, and require further study.

9.3.4 Post-industrial Sites

Somewhere in between the classical sites and the successional sites are what are possibly the most characteristic urban sites. Anyone who has undertaken urban botanical recording, in particular, would recognise them. Often rich in species, with regionally or even nationally scarce species often present, the vegetation is typically unusual and possibly recombinant (Barker, 2000) without any exact model in the countryside. The conditions are often such that succession is proceeding sufficiently slowly for 'quality' assemblages to develop and, therefore, also for the feasibility of conserving the individual site to be contemplated. In the UK West Midlands, these develop particularly on industrial waste, but also in quarries and on railway lines. Some old post-industrial sites have long been celebrated for their floristic richness (Davis, 1976; Greenwood and Gemmell, 1978) and a few are managed as nature reserves, but all conurbations include both large and small examples (Trueman *et al.*, 2001).

Urban-specific site-context evaluations of such sites directly compare them using identical criteria to adjacent urban patches have helped to confirm their relative merits. Using a contextual landscape evaluation method that takes a suite of criteria and applies them equally across the landscape (Young and Jarvis, 2001a), post-industrial habitats within sites, such as Ladymoor Pools in Bilston, West Midlands, in the United Kingdom show up as being of equivalent or superior 'value' to adjacent more characteristically 'worthy' grassland, woodland and open water areas (Figures 9.1 and 9.2). A West Midlands example of an 'ancient' post-industrial site is the nineteenth-century blast furnace slagheap and associated wetland in Bilston which carries a rich vascular plant, lichen and bryophyte flora, including winter annuals such as *Cerastium semidecandrum* and *Aira caryophyllea*, mire species such as *Triglochin palustris* and four species of sphagnum moss.

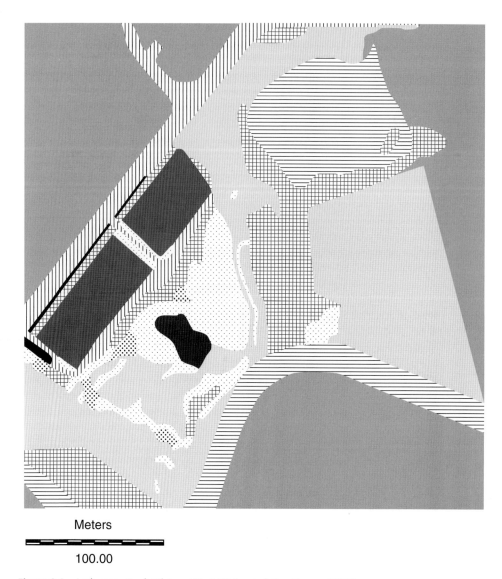

Figure 9.1 Ladymoor Pool, Bilston, West Midlands (after Young, 1999).
Note: Light grey = grassland, dots = scrub, dashes = woodland, horizontal stripes = open water, medium grey = residential, dark grey = industrial waste (slag), dotted = tall/short herb communities, angled stripes and black = miscellaneous and crosshatch = transport.

What facilitates the development of interest? Repeated or even continuous disturbance may play a part. Stress such as nutrient poverty, drought, phytotoxicity and mineral imbalance is clearly crucial in allowing large numbers of species to coexist indefinitely. Most of these controlling factors are subject to amelioration by succession, but the process is often very slow because conditions are extreme. This impedance of succession is probably at the root of their ecological interest.

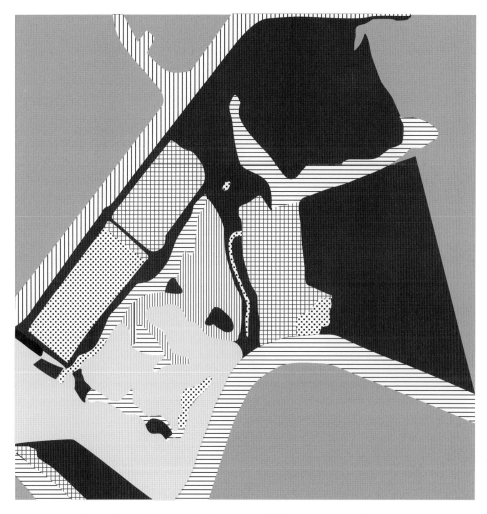

Figure 9.2 Habitat Value evaluation of Ladymoor Pool (after Young, 1999).
Note: The brighter areas have higher 'value' as calculated by the Habitat Value Index. Note the relatively high values in the centre-west associated with the area of most significant remaining industrial slag waste. Black = no data.

9.3.5 Garden Habitats

Garden evaluation is another important developing topic particularly relevant to the maintenance of a rich urban environment. Gardens are a significant proportion of the urban fabric (19–27%) and constitute a significant area of extensive interconnected greenspace (Smith *et al.*, 2005; Mathieu *et al.*, 2007). Possibly a nationally or internationally accepted classification of garden vegetation and garden types needs to be developed to aid this assessment and to give a fuller picture of the diversity and, therefore, greater insight into their true ecological value. A recent attempt at gaining an understanding of this diversity has been undertaken by the

Sheffield-based Biodiversity of Urban Gardens project (Thompson et al., 2003). Their findings corroborate the single-garden study by the Owens who monitored the ecology of their Leicester garden over an extended period and found significant numbers of many different groups represented (Owen, 1991). As an illustration, on the basis of the ichneumon wasps alone, their garden should have been a Site of Special Scientific Interest (SSSI).

On a small scale what happens in gardens reflects the kinds of changes happening in the wider urban landscape, with constant turnover of species, extreme management, successional changes and all-year-round food availability, amongst others. As such, gardens can be a resource in and of themselves for some groups, such as birds (Chamberlain et al., 2004), but they also provide linking greenspace permitting movement for other wildlife across the urban fabric, for example butterflies (Young, 2008). Significantly, the totality of urban gardens provides added value where they abut or are near to conventional conservation-worthy sites and, therefore, extends the influence of these sites beyond the constraints of their physical and administrative boundaries (Goddard et al., 2010).

Indeed, urban areas are often thought of as ecological deserts, yet this is far from the case as the intricate mix of gardens, conventional greenspace, seminatural habitats and successional communities on brownfield sites contributes significant ecological variability even before the more hostile bits in between these areas are addressed. From a climate change perspective the role of gardens as refugia for species, with the added potential to buffer threats and provide ecosystem services is important and adds significantly to the wider value of urban greenspaces.

9.4 Landscape Scales and Urban Areas

From a landscape perspective, urban areas have a particular character that has significant impacts on the ecology of individual habitats, sites and areas that makes them distinctive (Young et al., 2008):

- Patch sizes tend to be smaller on average than non-urban areas;
- Patches are highly fragmented with obvious (and often characteristic) linear 'corridors'; and
- Boundaries are sharp with contrasting patch types next to each other.

Current trends to conserve wildlife in urban areas are developing very much towards the landscape-scale appreciation of the importance of areas as opposed to the site-by-site approaches traditionally adopted (Young and Jarvis, 2001b; Freeman and Buck, 2003; Thompson et al., 2003; Smith et al., 2005; Daniels and Kirkpatrick, 2006). Such a strategy recognises the importance of the successional and temporally variable nature of many urban ecological variables and the importance of functional and spatial linkages both within the built environment and between the built environment and the rural peripheries.

Individual sites may not be large enough to support viable populations or may not be large enough in themselves to attract support from conservation organisations for protection. Strategies that can account for the totality of urban green infrastructure its configuration, and the associated matrix variability in all its guises as well as provide scope for the movement through the urban matrix are going to have increasing significance since they provide one solution to help in buffering both flora and fauna from the effects of climate change (Rudd *et al.*, 2002; Melles *et al.*, 2003). The importance of such developments has been highlighted in a recent meta-study of the last 22 years of recommendations for adapting to climate change, which identified improving landscape connectivity, so that species can move, as the most widely cited recommendation (Heller and Zavaleta, 2009).

9.5 Ecological Implications of Climate Change

Britain is in a good position to understand the implications of climate change on its flora. The Botanical Society of the British Isles (BSBI) undertook intensive surveys of flora in the middle and end of the twentieth century, which aimed to map the distribution of all vascular plant species on a 10 km grid. These gave an excellent opportunity to study changes in the flora, and these are explored both in the 2002 *Atlas* (Preston *et al.*, 2002a) and in a specific publication (Preston *et al.*, 2002b). These investigations demonstrated massive change in the flora, mostly attributable to habitat loss and degradation. Plant species with different types of distribution in the northern hemisphere were also compared, the flora being divided into four types: (1) widespread, (2) northern (arctic and boreal), (3) temperate and (4) Mediterranean. A marked decline in 'northern species' in southern Britain was attributed to habitat loss, but there was also a relative increase in all three other types. The authors conclude it is possible that at least some of the Mediterranean species have been the beneficiaries of recent climatic trends. The study also showed that when the flora is divided into natives, archaeophytes (introduced before 1500) and neophytes (introduced after 1500), there has been a relative increase in neophytes. The topic was further explored in Braithewaite *et al.* (2006), who compared 811 systematically distributed 2 km × 2 km samples across Britain surveyed by BSBI recorders in 1987–1988 and 2003–2004 and concluded that climate change has led to an increase in some species, particularly southerly species in neutral and calcareous grassland and ruderal species found in urban habitats and the transport network.

Although it is the general lack of sensitivity of plants to short-term, weather-related phenomena, which is particularly compelling, other groups have shown similar patterns and responses. For example, butterfly and moth distributions ebb and flow in response to weather and site management and are, therefore, somewhat easier to dismiss by sceptics. Yet this same sensitivity and responsiveness makes them good indicators of climate change, as they have quite particular climatic restrictions and respond to them rapidly.

Changes in distribution patterns of these groups, over the last 15 years, indicate northward movement of warm-adapted species and declines in species typical of northern and upland areas (Morecroft et al., 2009). These trends are noted as much in urban areas as elsewhere, with the advantage of significantly larger numbers of interested observers noting changes in these kinds of fauna on their local 'patch'. Similarly, Hickling et al. (2005) noted northward shifts in 37 species of nonmigratory British odonata over the past 40 years, seemingly as a result of climate change. Although such species are largely benign, some problematic invasive species, for example the Chinese mitten crab *Eriocheir sinensis*, may also take advantage of warmer conditions (Wilby and Perry, 2006).

9.6 Implications of Climate Change for Urban Ecology

Implications for the value of urban environments probably focuses mainly on the interactions of these changes. There is much in the literature to suggest neophytes are the group most closely associated with human activity (Pyšek, 1988) and urban areas (Hill et al., 2002). Human activity constantly introduces species from all over the world and creates disturbed conditions suitable for their spread. Very few neophytes are a problem, but some, such as Japanese knotweed *Fallopia japonica* and Himalayan balsam *Impatiens glandulifera*, invade natural habitats and suppress biodiversity very effectively. Many recent invasive species in the United Kingdom, such as the water fern *Azolla filiculoides* and floating pennywort *Hydrocotyle ranunculoides*, are from milder climates than our own. It is possible the results of climate change will be gradual replacement of valued habitats not with different but still biodiverse ones, but with monotonous stands of invasive species. An alternative and more optimistic scenario would be for climate change to give another dimension to the species assemblages and habitats by adding elements of the diverse continental flora and fauna to the largely retained existing individualistic communities. Whichever is the predominant outcome, further and sustained study of urban biodiversity will be necessary both to help determine the balance between gains and losses, and to act as a measure of the long-term and general impact of climate change on nature.

9.7 Solutions to Climate Change Challenges for the Built Environment

If we therefore acknowledge that climate change is affecting our flora and fauna whether we like it or not, there are a number of ways in which we can adapt to the challenges this offers.

- We should maintain and develop networks of open spaces of different sizes, shapes and configurations within urban areas to ensure that they are ecologically permeable. The built environment is part of the wider landscape dynamics, rather than separate from it.

- At the same time we must be vigilant, since ecological permeability is particularly helpful to invasive alien species, which themselves may well be encouraged by climate change.
- It also needs to be recognised that examples of species-rich habitats, whether primary and post-industrial, are, in addition to their intrinsic value, reservoirs of biodiversity. These may need to be utilised in the re-creation of habitats lost in climate change.
- The pressure to convert gardens into multiple dwellings should be resisted, as the garden habitat is unique and provides both significant urban greenspace and helps to deliver ecosystem services.
- We should resist the urge to 'tidy' as much as we can. Maintaining successional and transient communities alongside more conventional greenspaces will give the existing species, and those moving through, the resources they need.
- Change should be acknowledged as part of the built environment for wildlife as much as for people. Excessive effort to maintain things as they are will be money down the drain in the light of wider climate changes.

Changes may be inevitable, but we can adapt how we manage the built environment to work with the changing patterns of biodiversity as long as we take a longer term view. In many cases, this can be for the benefit of both people and wildlife – surely the best all-around outcome.

References

Austin, K.C. & Angold, P.G. (2000) Influence of landscape components on species recruitment in cities. *Aspects of Applied Biology*, **58**, 115–122.

Barker, G. (ed.) (2000) *Ecological Recombination in Urban Areas: Implications for Nature Conservation*. CEH, Monks Wood, UK.

Birmingham and Black Country BAP (2000) Black Redstart (*Phoenicurus ochruros*) Species Action Plan. Wildlife Trust for Birmingham and the Black Country. Available from http://www.wildlifetrust.org.uk/urbanwt/ecorecord/bap/html/redstart.htm.

Braithewaite, M.E., Ellis, R.W. & Preston, C.D. (2006) *Changes in the British Flora 1987–2006*. BSBI, London.

Catterall, C.P., Kingston, M.B., Park, K. & Sewell, S. (1998) Deforestation, urbanisation and seasonality: interacting effects on a regional bird assemblage. *Biological Conservation*, **84**, 65–81.

Chamberlain, D.E., Cannon, A.R. & Toms, M.P. (2004) Associations of garden birds with gradients in garden habitat and local habitat. *Ecography*, **27**, 589–600.

Daniels, G.D. & Kirkpatrick, J.B. (2006) Does variation in garden characteristics influence the conservation of birds in suburbia? *Biological Conservation*, **133**, 326–335.

Dawson, D. (1994) *Are Habitat Corridors Conduits for Animals and Plants in a Fragmented Landscape?* English Nature Research Report 94. English Nature, London.

Davis, B.K.N. (1976) Wildlife, urbanisation and industry. *Biological Conservation*, **10**, 249–291.

Department of the Environment, Food and Rural Areas (DEFRA) (2007) *An Introductory Guide to Valuing Ecosystem Services*. HMSO, London.

Dickman, C.R. (1987) Habitat fragmentation and vertebrate species richness in an urban environment. *Journal of Applied Ecology*, 24, 337–351.

Duguay, S., Eigenbrod, F. & Fahrig, L. (2007) Effects of surrounding urbanization on non-native flora in small forest patches. *Landscape Ecology*, 22(5), 89–99.

Dunnett, N. (1999) Annual meadows. *Horticulture Week*, February.

Eversham, B.C., Roy, D.B. & Telfer, M.G. (1996) Urban, industrial and other man-made sites as analogues of natural habitats for Carabidae. *Annales Zoologici Fennici*, 33, 149–156.

Faber Taylor, A., Wiley, A., Kuo, F.E. & Sullivan, W.C. (1998) Growing up in the inner city – green spaces as places to grow. *Environment and Behaviour*, 30(1), 3–27.

Freeman, C. & Buck, O. (2003) Development of an ecological mapping methodology for urban areas in New Zealand. *Landscape and Urban Planning*, 63, 161–173.

Gilbert, O.L. (1989) *The ecology of urban habitats*. Chapman and Hall, London.

Goddard, M.A., Dougill, A.J. & Benton, T.G. (2010) Scaling up from gardens: biodiversity conservation in urban environments. *Trends in Ecology and Evolution*, 25(2), 90–98.

Goode, D.A. (1989) Urban nature conservation in Britain. *Journal of Applied Ecology*, 26, 859–873.

Greenwood, E.F. & Gemmell, R.P. (1978) Derelict industrial land as a habitat for rare plants in S.Lancs and W.Lancs. *Watsonia*, 12(1), 33–40

Hansson, L. & Angelstam, P. (1991) Landscape ecology as a theoretical basis for nature conservation. *Landscape Ecology*, 5(4), 191–201.

Harrison, C., Limb, M. & Burgess, J. (1987) Nature in the city – popular values for a living world. *Journal of Environmental Management*, 25, 347–362.

Heller, N.E. & Zavaleta, E.S. (2009) Biodiversity management in the face of climate change: A review of 22 years of recommendations. *Biological Conservation*, 14(2), 14–32.

Hickling, R., Roy, D.B., Hill, J.K. & Thomas, C.D. (2005) A northward shift of range margins in British Odonata. *Global Change Biology*, 11(3), 502–506.

Hill, M.O., Roy, D.B. & Thompson, K. (2002) Hemeroby, urbanity and ruderality: bioindicators of disturbance and human impact. *Journal of Applied Ecology*, 39, 708–720.

Kuo, F.E. (2001) Coping with poverty: impacts of environment and attention in the inner city. *Environment and Behaviour*, 33(1), 5–34.

Mathieu, R., Freeman, C. & Aryal, J. (2007) Mapping private gardens in urban areas using object-oriented techniques and very high resolution satellite imagery. *Landscape and Urban Planning*, 81, 179–192.

Melles, S., Glenn, S. & Martin, K.O. (2003) Urban bird diversity and landscape complexity: species–environment associations along a multi-scale habitat gradient. *Conservation Ecology*, 7(5). Available from http://www.consecol.org/vol7/iss1/art5.

Morecroft, M.D., Bealey, C.E., Beaumont, D.A., Benhamd, S., Brooks, D.R., Burt, T.P., Critchley, C.N.R., Dick, J., Littlewood, N.A., Monteith, D.T., Scott, W.A., Smith, R.I., Walmsley C. & Watson, H. (2009) The UK Environmental Change Network: emerging trends in the composition of plant and animal communities and the physical environment. *Biological Conservation*, 142, 2814–2832.

Owen, J. (1991) *The Ecology of a Garden: The First Fifteen Years*. Cambridge University Press, Cambridge.

Pomarol, M. (1996) Artificial nest structure design and management implications for the lesser kestrel *Falco naumanni*. *Journal of Raptor Research*, 30, 169–172.

Preston, C.D., Pearman, D.A. & Dines, T.D. (eds.) (2002a) *New Atlas of the British and Irish Flora*. Oxford University Press, Oxford.

Preston, C.D., Telfer, M.G., Arnold, H.R., Carey, P.D., Cooper, J.M., Dines, J.M., Hill, M.O., Pearman, D.A., Roy, D.B. & Smart, S.M. (2002b) *The Changing Flora of the UK*. DEFRA, London.

Pyšek, P. (1998) Alien and native species in central European urban floras: a quantitative comparison. *Journal of Biogeography*, 25, 155–163.

Rebele, F. (1994) Urban ecology and special features of urban ecosystems. *Global Ecology and Biogeography Letters*, 4, 173–187.

Rossi, E. & Kuitunen, M. (1996) Ranking of habitats for the assessment of ecological impact in land-use planning. *Biological Conservation*, 77, 227–234

Rudd, H., Vala, J. & Scaefer, V. (2002) Importance of backyard habitat in a comprehensive biodiversity conservation strategy: a connectivity analysis of urban green spaces. *Restoration Ecology*, 10, 368–375.

Smith, R.M., Gaston, K.J., Warren, P.H. & Thompson, K. (2005) Urban domestic gardens (V): relationships between landcover composition, housing and landscape. *Landscape Ecology*, 20, 235–253.

Szacki, J., Glowacka, I., Liro, A. & Matuszkiewicz, A. (1994) The role of connectivity in the urban landscape: some results of research. *Memorabilia Zoologica*, 49, 49–56.

Thompson, K., Austin, K.C., Smith, R.M., Warren, P.H., Angold, P.G. & Gaston, K.J. (2003) Urban domestic gardens (I): putting small-scale plant diversity in context. *Journal of Vegetation Science*, 14, 71–78.

Trueman, I.C., Cohn, E.V.J.C., Tokarska-Guzik, B., Rostański, A. & Woźniak, G. (2001) Calcareous waste slurry as wildlife habitat in England and Poland. In: *Proceedings of GREEN 3: The Third International Symposium on Geotechnics Related to the European Environment, Berlin 2000*, ed. R.W. Sarsby and T. Meggyes. Thomas Telford, London.

Trueman, I.C. & Millet, P. (2003) Creating wild-flower meadows by strewing green hay. *British Wildlife*, 15(1), 37–44.

Trueman, I.C., Morton, A. & Wainwright, M. (1995) *The Flora of Montgomeryshire*. Montgomeryshire Field Society and Montgomeryshire Wildlife Trust, Welshpool, UK.

Ulrich, R., Simons, R.F., Losito, B.D., Fiorito, E., Miles, M.A. & Zelson, M. (1991) Stress recovery during exposure to natural and urban environments. *Journal of Environmental Psychology*, 11(3), 201–230.

Wentworth, J. (2006) Ecosystem services. Postnote 281. UK Parliamentary Office of Science and Technology, London.

Wilby, R.L. & Perry, G.L.W. (2006) Climate change, biodiversity and the urban environment: a critical review based on London, UK. *Progress in Physical Geography*, 30(1), 73–98.

Young, C. (1999) *Greening the city: habitat evaluation in Wolverhampton*. Unpublished PhD thesis, University of Wolverhampton.

Young, C.H. (2008) Butterfly activity in a residential garden. *Urban Habitats*, 5(1), 84–102. Available from http://www.urbanhabitats.org/v05n01/butterfly_full.html.

Young, C. & Jarvis, P. (2001a) A simple method for predicting the consequences of land management in urban habitats. *Environmental Management*, 28(3), 375–387.

Young, C.H. & Jarvis, P.J. (2001b) Measuring urban habitat fragmentation: an example from the Black Country, UK. *Landscape Ecology*, **16**, 643–658.

Young, C., Jarvis, P., Hooper, I. & Trueman, I. (2008) Urban landscape ecology and its evaluation: a review. In: *Landscape Ecology Research Trends*, ed. Dupont, A. and Jacobs, H. Nova Science Publishers, New York, 45–69.

10

The Pedological Value of Urban Landscapes

Jim Webb, Michael A. Fullen and Winfried E.H. Blum

10.1 Introduction

The increasing world population is placing ever-greater demands on soils for the production of sufficient food. This poses special problems in a climatic environment that is changing at an increasing pace but perhaps ultimately in an uncertain direction. The growing population will also lead inevitably to expansion of urban areas, which cover and consequently sterilize some soil resources or through pollution render others less suitable for crop production. Currently, the net world population is increasing by about 10 000 people per hour. Scrutiny of the 'world population counter website' gives a startling view of the rate of population growth (see http://www.ibiblio.org/lunarbin/worldpop).

10.2 Urban Soils: The 'Grey Areas' on Soil Maps

Sealing of urban soil has been identified as one of the eight main threats to the soil resources of Europe (European Community [EC], 2002; Vřšcaj et al., 2008). Other identified threats include soil erosion; decreasing soil organic matter (SOM) content; loss of biodiversity; contamination; compaction and salinization, floods and landslides. Of these threats, sealing and contamination predominate in urban and adjacent areas.

Urbanization has effectively sterilized extensive areas of soil, much of it very fertile. This is a major issue in Western Europe and North America, where rapid urban expansion started in the nineteenth century. In recent years about 120 hectares per day have been converted to urban use in Germany, 15 in Austria and The Netherlands and 10 in Switzerland

---

Solutions to Climate Change Challenges in the Built Environment, First Edition.
Edited by Colin A. Booth, Felix N. Hammond, Jessica E. Lamond and David G. Proverbs.
© 2012 Blackwell Publishing Ltd. Published 2012 by Blackwell Publishing Ltd.

(EC, 2001). Urban expansion is now especially problematic along the north Mediterranean coast. In 1996, 43% of the coastal zone in Italy was completely occupied by urban areas and only 29% was completely free from buildings and roads (EC, 2002). The proportion of sealed surfaces varies considerably among EU member states and regions, from 0.3% to 10% (EC, 2006).

Airports are generally located on flat, well-drained peri-urban land, often with high soil fertility and agricultural potential. For instance, the soils derived from loess over Thames terrace gravels beneath Heathrow Airport were very fertile. Ronaldsway Airport on the Isle of Man (British Isles) covers much of the very restricted area of Grade 1/2 agricultural land on the Island; thus, effectively sterilizing fertile calcareous brown earths (Harris et al., 2001). In addition, urban areas may have great impacts on soils, changing water flow and drainage patterns and fragmenting areas of valuable biodiversity.

Currently, urban expansion is particularly rapid in the developing world. Increasing urban slums pose particular problems. For instance, the expanding urban slums or *favelas* of Brazil are encroaching on steep land at the margins of urban areas. This puts them at risk of flash floods and gully erosion, which is a serious threat to life, health and property (Guerra, 1995; Guerra and Favis-Mortlock, 1998). The tragic consequences of this were highlighted in the devastating mudflows in Rio de Janeiro in April 2010. Construction sites in particular are at risk of high erosion rates (Wolman, 1967).

A further and most critical aspect of the expansion of urban soils is the loss of fertile land for food production and for the production of biomass in general (e.g. for biofuels). All traded food worldwide is produced on 12% of the land surface (Blum and Eswaran, 2004). Most of the best soils are near urban agglomerations, because our ancestors usually settled and cultivated the best available soils. It is estimated that worldwide we are losing 280–300 km^2 of fertile land and soil per day through urbanization. This is alarming and may lead to food security problems in the near to medium future.

Urban soils are not usually mapped in detail and many scientists have argued they deserve more attention (e.g. Blume, 1989; Bullock and Gregory, 1991; Effland and Pouyat, 1997). For instance, on the Soil Survey of England and Wales (1983) 1:250,000 map, they are represented as 'unsurveyed, mainly urban and industrial areas' and shaded grey. However, urban soils are very diverse and include gardens, parks, cemeteries, allotments, grass verges along roads, playing fields and sometimes much derelict and industrial land. The industrial land includes disposal sites, demolition and building sites, waste and derelict land, rubbish tips, spoil heaps, canal and railway land, collieries, docklands, power station land, shipbuilding land, scrapyards, dried-out industrial lagoons, sewage works and land associated with mining, smelting and manufacture. Such diversity, fragmentation and complexity make predictions of the responses of urban soils to climate change extremely problematic.

As a consequence of urbanization soils, are moved, mixed and compacted, and may receive mineral and chemical additives that change soil characteristics (Schleu et al., 1998; Lorenz and Kandeler, 2005; Blume and

Felix-Henningsen, 2009). Building causes considerable soil disturbance, and in some areas soils may be, in effect, destroyed (Randrup and Dralle, 1997).

Craul (1985) identified eight environmentally important characteristics of urban soils:

1. Great vertical and spatial variability, reflecting their often complex history of construction, management and modification.
2. Compaction by trafficking, walking and vibration. This is reflected in increased bulk density and low porosity. The low organic content of many urban soils increases their susceptibility to compaction. Craul (1985) quoted bulk density values as high as 2.18 g cm^{-3}. High bulk densities (>1.7 g cm^{-3}) severely restrict root development and limit nitrogen (N) fixation by legumes, which places urban plants under high stress (Gilbert, 1989).
3. The presence of a surface crust on bare soil, which is usually water repellent.
4. Modified pH. Incorporation of building rubble containing mortar rich in calcium carbonate ($CaCO_3$) often increases soil pH (Gilbert, 1989). Use of calcium or sodium chloride for de-icing roads or use of Ca-rich irrigation water can have the same effect. Craul (1985) reported pH values of 8–9 in some US urban soils. Brick rubble also contains high concentrations of phosphorus (P), potassium (K) and magnesium (Mg), which may be released as the brick rubble weathers. Plaster is a major source of both calcium and sulphur, derived from the weathering of gypsum ($CaSO_4$).
5. Restricted aeration and drainage.
6. Interrupted nutrient cycling and decreased activity of soil organisms. The lack of annual organic matter inputs (e.g. autumn leaf fall) decreases the organic content.
7. The presence of anthropogenic materials and contaminants, such as glass, wood, metal, plastic, asphalt, masonry, concrete, cloth, polythene, cardboard, tiles, cinders, ash, soot, bricks and rubble (Bridges, 1991; Blum, 1998).
8. Generally increased temperatures, because of the urban heat island effect.

Probably the most significant urban soil characteristic is that it is polluted, especially by heavy metals, and can thus act as a source of pollution (Kelly and Thornton, 1996; Blum, 1998; Biasioli et al., 2006). Due to emissions arising from industrial and other activities within urban areas, soils may be contaminated with toxic heavy metals in both urban and peri-urban areas. This effect can be apparent for distances of 15–20 km, depending on the prevailing wind direction (Blum, 1998). These metals are retained by soils and subsequent dispersion following soil cultivations may lead to ingestion by the human population long after the industrial activity that led to the original emissions has ceased. The exposure of children to trace metals can increase greatly through their ingestion of metal-laden soil particles and dust via frequent hand-to-mouth activities (Wong et al., 2006).

Purves and Mackenzie (1970) showed evidence of elevated concentrations of heavy metals in vegetables grown in urban environments. Li *et al.* (2001) compared the concentrations of copper (Cu), lead (Pb) and zinc (Zn) in soils in the urban and country parks of Hong Kong. The mean concentrations of Cu and Zn in urban soils (24.8 and 168 mg/kg, respectively) were at least four and two times greater than those of rural soils (5.17 and 76.6 mg/kg, respectively), while the mean Pb concentration of urban soils (89.9 mg/kg) was one order of magnitude greater than that of rural soils (8.66 mg/kg).

Large concentrations of heavy metals in urban soils need not be due to contamination. Davis *et al.* (2009) reported that, in a study of the metal contents of soils in both urban and rural areas, in general, barium (Ba) and manganese (Mn) were consistently grouped and may represent naturally occurring metals in all soils. In contrast, they concluded that both Pb and mercury (Hg) were derived primarily from anthropogenic sources. Hg was equally distributed regardless of land use and numbers of industrial facilities, suggesting ubiquitous long-range distribution from atmospheric releases as wet or dry deposition from manufacturing, coal-fired power plants and other combustion processes associated with industrial facilities. Madrid *et al.* (2009) reported evidence that the behaviour of elements predominantly of anthropogenic origin (e.g. Cu, Pb and Zn) is different from that of elements primarily of geochemical origin (e.g. Fe and Mn). UK government policy on defining contaminated land takes a risk-based approach. Specifically, land is assessed in terms of whether contaminants pose a 'significant possibility of significant harm' (SPOSH) (Defra, 2008).

Industrial sites often have pollution problems, with very varied pollutants, such as heavy metals (Alloway, 1995) and hydrocarbons (Bridges, 1991). Industrial sites are also major sources of eroded sediment, as these soils are often poorly vegetated and steep. Old mine and colliery mounds are major sediment sources, as in South Wales (Bridges and Harding, 1971; Haigh 1979, 1992; Higgitt *et al.*, 1994), South Yorkshire (Haigh and Sansom, 1999) and the United States (Haigh, 1988). However, they can be visually striking and it has been argued that such disturbed sites should be preserved for their historical and landscape heritage interest (Quinn, 1988, 1992).

The properties of urban soils may nevertheless display inherent characteristics of the original soil. Madrid *et al.* (2009) reported different soil properties among the cities of Seville, Turin and Glasgow. Soils from Seville were more calcareous than those from Turin, while those from Glasgow had no free $CaCO_3$. This accorded with the pH values, with neutral to slightly alkaline values in all soils from Seville and most of those from Turin, but acidic soils in Glasgow. Soil organic carbon (SOC) was less abundant in Seville and Turin than in Glasgow, whereas soil N contents were comparable. Consequently, C:N ratios were markedly greater in Glasgow, as would be expected from the climate of that city, which favours SOC accumulation.

He and Zhang (2009) proposed that black carbon (BC), being a recalcitrant material derived primarily from human activity, could be used to study the impact of human activity on urban soils. They reported a study to estimate BC concentrations in urban soils and its change during the history of

Nanjing, China in order to illustrate the relationship between BC and heavy metal (Cu, Pb and Zn) contents in cultural layers, to improve the understanding of factors influencing urban soil quality.

The BC:OC ratio showed a distribution pattern similar to that of BC: the average in the port area where industrial activities had long been practised was 0.44, much greater than the 0.15 in the administrative and residential area. The ratios were considered to indicate the composition of SOC, and a large value was taken to mean that soil OC in urban soils is more inert than that in agricultural soils. He and Zhang (2009) suggested that a small BC/OC ratio (·0.1) represented biomass burning, while large values (~0.5) arose from fossil fuel combustion. Previous studies had found that the average ratio of BC/OC in roadside soils was >0.45, evidence that the main source of BC was from fossil fuel combustion, especially emissions from vehicles (He and Zhang, 2006).

The change of land use from nature or agriculture to urban significantly changes soil C pools and fluxes (Groffman *et al.*, 1995; Pouyat *et al.*, 1995). Urban land conversions, however, often result in poor conditions for plant growth (Pouyat *et al.*, 2002). As urban areas expand, both direct and indirect factors can affect soil C pools. Direct effects include physical disturbances, burial or coverage of soil by fill material and impervious surfaces (Pouyat *et al.*, 2002). Pouyat and Effland (1999) considered that direct effects often lead to 'new' soil parent material on which soil development then proceeds. Indirect effects involve changes in the abiotic and biotic environment that can influence soil development as areas are urbanized. Pouyet *et al.* (2002) listed indirect effects as: the urban heat island effect; soil hydrophobicity; introductions of exotic species and atmospheric deposition. Moreover, soil C fluxes can be significantly affected by toxic or sublethal effects of the urban environment on soil decomposers and primary producers (Pouyat *et al.*, 1994, 1997).

Pouyat *et al.* (2002) found from a 10-year study of oak stands located along an urban–rural land use gradient that abiotic and biotic environmental factors can substantially change as the adjacent land use becomes more urbanized. In turn, these changes can affect soil chemistry, temperature regimes, soil community composition, and N and C fluxes (Groffman *et al.*, 1995; Pouyat *et al.*, 1995). These indirect effects suggest that urbanization and the resultant environmental changes that occur can influence soil C pools even in woodland ecosystems that are not directly or physically disturbed by urban development.

In their analysis of soils not affected by earthworms, Pouyat *et al.* (2002) reported that woodland floor mass was greater in the urban core than in stands >40 km away. A possible explanation for the increase in woodland floor mass in non-earthworm affected stands in or near the urban core may, in part, be due to the input of lower quality leaf litter in these stands. Pouyat *et al.* (2002) cited Carreiro *et al.* (1999), who found red oak (*Quercus rubra* L.) litter collected from rural stands decomposed more rapidly than litter from suburban and urban stands, in both field and laboratory incubations. Pouyat *et al.* (2002) concluded that while urban conditions (i.e. the presence of non-native earthworms and elevated soil temperatures) tend to accelerate decay, urban litter quality may tend to decrease decay rates. Thus, in the absence of earthworms, woodlands exposed to urban environments may have the potential to sequester and store more C than rural stands of the same canopy composition. This pattern can depend on consistent annual reductions in litter quality in urban areas, but also on SOC accumulation, caused by the increasing input of N from all kinds of incineration processes, including traffic, decreasing microbial decomposition of organic matter; thus, stimulating carbon sequestration (Janssens *et al.*, 2010).

Disturbed and made soils examined by Pouyat *et al.* (2002) exhibited a wide range in SOC contents. Soil C densities ranged from 1.4 kg m^{-2} in clean fill materials deposited almost 100 years ago in central Washington DC, to 28.5 kg m^{-2} in loamy fill material underlying a golf course in New York City. Similar results were reported for surface soils in Baltimore, where mineral soil (0–15 cm) SOC concentrations and densities varied widely among land use types. Conversely, for any particular soil disturbance or fill type, Pouyat *et al.* (2002) found the C densities were surprisingly similar. For example, residential sites in Chicago, United States and Moscow, Russia, had very similar SOC densities. Similarly, C densities for old dredge materials across four different sites varied by only 1 kg C m^{-2}. Pouyat *et al.* (2002) concluded that if these consistencies in soil C densities apply to other urban ecosystems, it may be possible to use SOC data as a criterion in developing soil series concepts for highly disturbed and 'made' soils.

Comparisons reported by Pouyat *et al.* (2002) of SOC density between 'made' and forest soils suggest that residential areas have nearly the same C density as US Northeastern woods and greater density than Mid-Atlantic woods. This finding may seem surprising, since many residential lawns are cut frequently and the cuttings removed. However, lawns often receive large nutrient inputs and much water, which would be expected to increase both above- and below-ground productivity. Pouyat *et al.* (2002) showed that the relatively large amounts of SOC in residential areas are primarily due to increases in belowground productivity. Lawns also have a much longer growing season than woods. While residential soils were found to be similar to wooded and steppe ecosystems with respect to SOC, the disturbed and 'made' soils had similar C densities to US Northeastern and Mid-Atlantic croplands (Pouyat *et al.*, 2002). This was not unexpected, since cropland soils are highly disturbed with a large proportion of the biomass produced being harvested. However, like highly maintained lawns, croplands often receive large inputs of nutrients, especially N and water. The main difference between cropland and lawns was considered to be the physical disturbance from soil cultivation reducing soil C inputs.

Pouyat *et al.* (2002) noted that reductions in litter quality, which increase passive C and decrease labile C pools, can affect various microbial processes important to ecosystem functioning. For example, Groffman *et al.* (1995) observed less microbial biomass and Goldman *et al.* (1995) observed reduced rates of methane uptake in urban compared with suburban and rural woods.

Pouyat *et al.* (2002) investigated woodland soils along an urban–rural land use gradient in the New York City metropolitan area. The research included the quantification of land use characteristics, soil physico-chemical properties, soil C and N dynamics, plant community structure and populations of soil biota. In the upper 10 cm of urban soils, temperatures were greater and there were up to fivefold differences in heavy metal and total salt concentrations compared with soils in suburban and rural areas (Pouyat et al., 1995). In contrast, soil physical properties, such as bulk density and texture, did not differ appreciably among areas (Pouyat et al., 2002). Groffman et al. (1995) measured smaller pools of labile C and larger pools of passive C in urban relative to rural woodlands.

From a parallel study in Baltimore, soils under low-density residential and institutional land use had 44 and 38% greater SOC densities, respectively, than those under commercial land use (Pouyat et al., 2002). Soils under woodland, medium-density and high-density residential, and transportation rights of way had intermediate organic C densities.

10.3 Policy Responses for Urban Soils

The quality of urban life can be improved by the careful management of urban soils to create habitats that are more favourable to wildlife, provide amenity sites for recreation or limit pollution. We are increasingly recognizing urban soils may play important roles in both adapting to and mitigating climatic change.

To limit pollution, we must understand the severity of the problem and define mitigation strategies. Wong et al. (2006) pointed out that assessments of urban soils concentrated on trace metals that were significant for the environment and health, particularly cadmium (Cd), Cu, Pb and Zn, while the distribution of other trace metals in the urban environment has received less attention. Much of the earlier monitoring activity concentrated on Pb, since Pb pollution was a serious health problem in the urban environment. These concerns lead to the cessation of Pb additions to petrol. Wong et al. (2006) posed the question of what the next priority should be for heavy metal monitoring. They concluded that road traffic continues to emit metal pollutants to the urban environment.

Rare earth elements (REEs), platinum (Pt) group elements (PGEs), and Mn have been recommended as environmental indicators of traffic and other urban activities, especially where leaded petrol is no longer used. Wong et al. (2006) cited UK results between 1982 and 1998, which demonstrated that there had been an increase in PGEs in road dust. Further indications that traffic was the source of Pt were obtained by comparing Pt with Au (gold) in soils and dust sampled in the London Borough of Richmond in 1994 (Farago et al., 1995, 1996). Concentrations of Pt, like those of Pb, which originate from traffic, were greater in road dust than they are in soil samples. For Au, which does not originate from traffic, concentrations were greater in soils than in road dust. Measurements of the mineral magnetic properties of urban dusts and road deposited sediments have the potential to act as surrogate proxies of pollution (Booth et al., 2007; Crosby et al., 2009).

Wong et al. (2006) considered the fundamental factors governing the selection of an appropriate soil remediation treatment to include the level of cleanup desired, length of time allowed, chemical forms and amounts of contaminants, site characteristics, and the cost involved. In urban environments, phytoremediation (the use of plants, sometimes in conjunction with micro-organisms and chemical reagents) can be a preferable soil remedial technique for the removal of trace metals. Since the upper layer of urban soils tends to be contaminated, hyper-accumulators and high-biomass plants with an active root zone <2 m can be used to treat contaminated urban soils. While phytoremediation can be used to restore contaminated soils over any area, it is especially advantageous for soils that are fragmented and small in size, where applications of traditional remedial methods can be impractical and cost-prohibitive. This is frequently the case with contaminated urban soils (Wong et al., 2006).

The amenity value of urban soils is an important consideration. A possible avenue of improving urban landscapes is by habitat creation. Changes in farming practises since the 1940s have led to catastrophic losses of species-rich meadows in the United Kingdom and elsewhere. The decline in floristically rich grasslands has altered the visual impression of the British countryside, particularly in the spring and summer months. The remaining species-rich hay meadows are of rich botanical and ecological interest and are now the focus of conservation initiatives (Jones et al., 1995; Atkinson et al., 1995). Only a fraction of this once extensive environment remains, existing as isolated pockets. Although habitat creation cannot hope to fully compensate for these losses, attempts at reconstruction provide a means of reversing trends, which have continued for several decades.

In urban areas, creation of hay meadows can improve recreational and amenity sites by increasing diversity of structure and species, which makes landscaping schemes more interesting, visually attractive and results in management economies. These hay meadows make established multi-functional contributions to biodiversity, landscape aesthetics and education and particularly benefit inner-city residents. This not only improves the aesthetic value of the area, but also provides environmental and educational resources. Many of these sites are developed on mineral soils deficient in nutrients, which might encourage wildflower assemblages that require low soil fertility. Since these soils are also often deficient in organic matter, they have considerable potential to store SOC and so act as C sinks. This should help promote urban hay meadows as part of national environmental development policies. However, new data indicate that through N input from urban activities, especially incineration processes by traffic and other activities, soil respiration and C losses will decrease, leading to soils richer in SOC (Janssens et al., 2010).

Between 1983 and 1989, 17 species-rich meadows were established in various locations within and around the Wolverhampton and Dudley conurbations of the West Midlands of England. These studies have formed the basis of a growing corpus of knowledge on habitat creation and management (Besenyei and Trueman, 2001; Trueman and Millett, 2003). These include studies on site establishment, soil fertility and monitoring and

Table 10.1 Soil fertility conditions suitable for species-rich hay meadows.

Soil property	Optimum
Extractable phosphorus (mg 100 g^{-1})	<7.0
Extractable potassium (mg 100 g^{-1})	10–30
Soil pH	>5.2

Source: McCrea et al. (2001a).

management. Management records at the different sites have been used to investigate the causes of change in the vegetation (Besenyei et al., 2002) and the minimum management input needed to retain or increase species diversity.

The habitat creation technique in the Wolverhampton area is to strew hay cut in old floristically diverse rural meadows, such as along the border areas between England and Wales (the Welsh Marches) (Trueman and Millett, 2003). After several years, vegetation comparable to the donor site establishes itself in the new urban-industrial environment. Desirable species include key indicators of old grassland, such as crested dog's tail (*Cynosurus cristatus*) and lesser knapweed (*Centaurea nigra*) and attractive meadow species, such as oxeye daisy (*Leucanthum vulgaris*), yellow rattle (*Rhinanthus minor*) and green-winged orchid (*Orchis morio*). Soil conditions affect the success of habitat creation schemes. Grime (1973) proposed the 'hump-back model' for the relationship between soil fertility and species diversity. If soil fertility is too low, introduced species cannot establish themselves. If fertility is too high, then a few species dominate and exclude others from the site, as found in the Rothamsted Park Grass Hay Experiment (Catt and Henderson, 1993). Tall grasses are particularly effective in dominating such sites. This means that creation of species-rich grassland can only be successful in areas of moderate to low soil fertility (Marrs, 1993).

Attempts are in progress to define 'envelopes of soil fertility.' If soil conditions fall within the envelope, then habitat creation is likely to be successful. If conditions fall outside, then the scheme is likely to fail. Table 10.1 shows the range of fertility conditions in which habitat creation for hay meadows is likely to be successful (McCrea et al., 2001a). Fertility depletion studies have shown that arable crops can deplete available nutrients and allow rapid establishment of a diverse sward. Barley is particularly efficient in depleting nutrients compared with potatoes, maize or tobacco (McCrea et al., 2001b).

In recent years, the importance of 'open areas' in promoting infiltration into soils has been increasingly recognized. These open areas include parks, allotments and gardens. Sealing these areas with impermeable surfaces increases surface runoff volume and velocity and contributes to urban flooding. Given that most general circulation models (GCMs) predict wetter winters in much of northern Europe, the combination of climate change and surface sealing by urban construction will exacerbate flooding problems. The spate of devastating winter floods in northern Europe during the early twenty-first century may well be testimony of this interaction. Thus, policies that prevent or decrease the urban sealing of new land should be encouraged. Local councils are now discouraging house owners from sealing their gardens for car parking. Parking areas can also be designed to facilitate drainage. A plausible instrument to promote the participation of house owners is through local taxation (Council Tax in the United Kingdom). Participating householders could be rewarded with lower local tax charges. Encouraging drainage could be integrated into the planning process, on both brownfield and greenfield sites. The retention and/or installation of even moderate areas of vegetated soil within urban development schemes would encourage infiltration. A challenge for the future is to develop intelligent drainage schemes to promote infiltration rather than runoff. Urban planning is also encouraging re-use of city centre brownfield sites in urban development schemes. Not only does this mean the re-use of often contaminated soils, but impedes urban sterilization of precious greenbelts and encourages revitalization of inner-city life.

These trends suggest a broad recognition of the multifunctionality of urban environments may well be a positive direction for management policy. Green areas play crucial roles in the drainage and filtering of water, thus decreasing flood magnitudes and often encouraging groundwater recharge. Only recently have we recognized there is potential to sequester C (as atmospheric CO_2) into urban soils through the input of N (Janssens et al., 2010) and as part of a broader series of 'wedges' (progressive cumulative actions) to try to ameliorate global warming (Pacala and Socolow, 2004). The importance of green space for the social and psychological well-being of urban communities has long been known (Trueman and Young; this volume). The integration of these multifunctional roles may well be a positive direction as we plan strategies to manage our urban soils.

References

Alloway, B.J. (1995) *Heavy Metals in Soil*, 2nd ed. Blackie Academic and Professional, Glasgow.

Atkinson, M.D., Trueman, I.C., Millett, P., Jones, G.H. & Besenyei, L. (1995) The use of hay strewing to create species rich grasslands (ii) monitoring the vegetation and the seed bank. *Land Contamination and Reclamation* 3, 108–110.

Besenyei, L. & Trueman, I.C. (2001) Creating species-rich grasslands using traditional hay cutting and removal techniques. *(England) Ecological Restoration* 19, 114–115.

Besenyei, L., Trueman, I.C., Atkinson, M.D., Jones, G.H. & Millett, P. (2002) Retaining diversity in hay meadows by traditional management. *Grassland Science in Europe* 7, 762–763.

Biasioli, M., Barberis, R. & Ajmone-Marsan, F. (2006) The influence of a large city on some soil properties and metals content. *Science of the Total Environment* 356, 154–164.

Blum, W.E.H. (1998) Soil degradation caused by industrialization and urbanization. In: *Towards Sustainable Land Use, Vol. I, Advances in Geoecology 31*, ed. Blume, H-P., Eger, H., Fleischhauer, E., Hebel, A., Reij, C. & Steiner, K.G. Catena Verlag, Reiskirchen, 755–766.

Blum, W.E.H. & Eswaran, H. (2004) Soils for sustaining global food production. *Journal of Food Science* 69(2), 37–42.

Blume, H-P. (1989) Classification of soils in urban agglomerations. *Catena* 16, 269–275.

Blume, H-P. & Felix-Henningsen, P. (2009) Reductosols: natural soils and technosols under reducing conditions without an aquic moisture regime. *Journal of Plant Nutrition and Soil Science* 172, 808–820.

Booth, C.A., Winspear, C.M., Fullen, M.A., Worsley, A.T., Power, A.L. & Holden, V.J. (2007) A pilot investigation into the potential of mineral magnetic measurements as a proxy for urban roadside particulate pollution. In: *Air Pollution XV*, ed. Borrego, C.A. & Brebbia, C.A. Wessex Institute of Technology Press, Southampton, UK, 391–400.

Bridges, E.M. (1991) Waste materials in urban soils. In: *Soils in the Urban Environment*, ed. Bullock, P. & Gregory, P.J. Blackwell Scientific Publications, Oxford, 28–46.

Bridges, E.M. & Harding, D.M. (1971) Micro-erosion processes and factors affecting slope development in the Lower Swansea Valley. In: *Slopes: Form and Process*, ed. Brunsden, D. I.B.G. Special Publication No. 3. Alden & Mowbray, Oxford, 65–79.

Bullock, P. & Gregory, P.J. (1991) Soils: a neglected resource in urban areas. In: *Soils in the Urban Environment*, ed. Bullock, P. & Gregory, P.J. Blackwell Scientific Publications, Oxford, 1–4.

Carreiro, M.M., Howe, K., Parkhurst, D.F. & Pouyat, R.V. (1999) Variation in quality and decomposability of red oak leaf litter along an urban-rural gradient. *Biology and Fertility of Soils* 30, 258–268.

Catt, J.A. & Henderson, I.F. (1993) Rothamsted Experimental Station – 150 years of agricultural research. The longest continuous scientific experiment? *Interdisciplinary Science Reviews* 18, 365–378.

Craul, P.J. (1985) A description of urban soils and their desired characteristics. *Journal of Arboriculture* 11, 330–339.

Crosby, C.J., Booth, C.A., Worsley, A.T., Fullen, M.A., Searle, D.E., Khatib, J.M. & Winspear, C.M. (2009) Application of mineral magnetic concentration measurements as a particle-size proxy for urban road deposited sediments. In: *Air Pollution XVII*, ed. Brebbia, C.A. & Popov, V. Wessex Institute of Technology Press, Southampton, UK, 153–162.

Department for Environment, Food and Rural Affairs (DEFRA) (2008) Guidance on the Legal Definition of Contaminated Land. HMSO Publication PB 13149. Available from http://www.defra.gov.uk/environment/quality/land/contaminated/documents/legal-definition.pdf.

European Community (EC) (2001) The Soil Protection Communication. DG ENV Draft. Commission of the European Communities, Brussels.

European Community (EC) (2002) Towards a Thematic Strategy for Soil Protection. Communication from the Commission to the Council, The European Parliament, The Economic and Social Committee and the Committee of the Regions COM(2002), 179 final. Commission of the European Communities, Brussels.

European Community (EC) (2006) Thematic Strategy for Soil Protection. Communication from the Commission to the Council, the European Parliament, the European Economic and Social Committee and the Committee of the Regions. COM(2006)231 final, Commission of the European Communities, Brussels.

Davis, H.T., Aelion, C.M., McDermott, S. & Lawson, A.B. (2009) Identifying natural and anthropogenic sources of metals in urban and rural soils using GIS-based data, PCA, and spatial interpolation. *Environmental Pollution* 157, 2378–2385.

Effland, W.R. & Pouyat, R.V. (1997) The genesis, classification and mapping of soils in urban areas. *Urban Ecosystems* 1, 217–218.

Farago, M.E., Kavanagh, P., Blanks, R., Simpson, P., Kazantzis, G. & Thornton, I. (1995) Platinum group metals in the environment: their use in vehicle exhaust catalysts and implications for human health in the UK. UK Department of the Environment, London.

Farago, M.E., Kavanagh, P., Blanks, R., Kelly, J., Kazantzis, G., Thornton, I., Simpson, P.R., Cook, J.M., Parry, S. & Hall, G.M. (1996) Platinum metal concentrations in urban road dust and soil in the United Kingdom. *Fresenius Journal of Analytical Chemistry* 354, 660–663.

Gilbert, O.L. (1989) Soils in urban areas. In: *The Ecology of Urban Habitats*, ed. Gilbert, O.L. Chapman and Hall, London, 41–54.

Goldman, M.B., Groffman, P.M., Pouyat, R.V., McDonnell, M.J. & Pickett, S.T.A. (1995) CH_4 uptake and N availability in forest soils along an urban to rural gradient. *Soil Biology and Biochemistry* 27, 281–286.

Grime, J.P. (1973) Competitive exclusion in herbaceous vegetation. *Nature* 242, 344–347.

Groffman, P.M., Pouyat, R.V., McDonnell, M.J., Pickett, S.T.A. & Zipperer, W.C. (1995) Carbon pools and trace gas fluxes in urban forest soils. In: *Soil Management and the Greenhouse Effect*, ed. Lal, R., Kimble, J., Levine, E. & Stewart, B.A. CRC Press, Boca Raton, FL, 147–157.

Guerra, A.J.T. (1995) Catastrophic events in Petrópolis City (Rio de Janeiro State), between 1940 and 1990. *GeoJournal* 37, 349–354.

Guerra, A.J.T. & Favis-Mortlock, D. (1998) Land degradation in Brazil. *Geography Review* 12, 18–23.

Haigh, M.J. (1979) Ground retreat and slope evolution on plateau-type colliery spoil mauls at Bleenavon, Gwent. *Transactions of the Institute of British Geographers* 4, 321–328.

Haigh, M.J. (1988) Slope evolution on coal-mine disturbed land. In: *Environmental Geotechnics and Problematic Soils and Rocks*, ed. Balasubramanian, A.S., Chanda, S., Bergado, D.T. & Prinya Natalaya, A.A. Balkema, Rotterdam, 3–13.

Haigh, M.J. (1992) Degradation of 'reclaimed' lands previously disturbed by coal mining in Wales: causes and remedies. *Land Degradation and Rehabilitation* 3, 169–180.

Haigh, M.J. & Sansom, B. (1999) Soil compaction, runoff and erosion on reclaimed coal-lands (UK) International *Journal of Surface Mining, Reclamation and Environment* 13, 135–146.

Harris, J., Fullen, M.A. & Hallett, M.D. (2001) *Agricultural soils of the Isle of Man*. Centre for Manx Studies Research Report 9. Centre for Manx Studies, Liverpool.

He, Y. & Zhang, G.L. (2006) Concentration and source of organic carbon and black carbon of urban soils in Nanjing (in Chinese with English abstract). *Acta Pedologica Sinica* 43(2), 177–182.

He, Y. & Zhang, G.L. (2009) Historical record of black carbon in urban soils and its environmental implications. *Environmental Pollution* 157, 2684–2688.

Higgitt, D.L., Walling, D.E. & Haigh, M.J. (1994) Estimating rates of ground retreat on mining spoils using caesium-137. *Applied Geography* 14, 294–307.

Janssens, I.A., Dieleman, W., Luyssaert, S., Subke, J-A., Reichstein, M., Ceulemans, R., Ciais, P., Dolman, A.J., Grace, J., Matteucci, G., Papale, D., Piao, S.L., Schulze E-D., Tang, J. & Law, B.E. (2010) Reduction of forest soil respiration in response to nitrogen deposition. *Nature Geoscience* 3, 315–322.

Jones, G.H., Trueman, I.C. & Millett, P. (1995) The use of hay strewing to create species rich grasslands (i) general principles and hay strewing versus seed mixes. *Land Contamination and Reclamation* 3, 104–107.

Kelly, J. & Thornton, I. (1996) Urban geochemistry: a study of the influence of anthropogenic activity on the heavy metal content of soils in traditionally industrial and non-industrial areas of Britain. *Applied Geochemistry* 11, 363–370.

Lorenz, K. & Kandeler, E. (2005) Biochemical characterization of urban soil profiles from Stuttgart, Germany. *Soil Biology and Biochemistry* 37, 1373–1385.

Li, X.D., Poon, C.S. & Liu, P.S. (2001) Heavy metal contamination of urban soils and street dusts in Hong Kong. *Applied Geochemistry* 16, 1361–1368.

Madrid, F., Reinoso, R., Florido, M.C., Díaz Barrientos, E., Ajmone-Marsan, F., Davidson, C.M. & Madrid, L. (2009) Estimating the extractability of potentially toxic metals in urban soils: A comparison of several extracting solutions. *Environmental Pollution* 147, 713–722.

Marrs, R.H. (1993) Soil fertility and nature conservation in Europe: theoretical considerations and practical management solutions. *Advances in Ecological Research* 24, 241–300.

McCrea, A.R., Trueman, I.C., Fullen, M.A., Atkinson, M.D. & Besenyei, L. (2001a) Relationships between soil characteristics and species richness in two botanically heterogeneous created meadows in the English West Midlands. *Biological Conservation* 97, 171–180.

McCrea, A.R., Trueman, I.C. & Fullen, M.A. (2001b) A comparison of the effects of four arable crops on the fertility depletion of a sandy silt loam for grassland habitat creation. *Biological Conservation* 97, 181–187.

Pacala, S. & Socolow, R. (2004) Stabilization wedges: solving the climate problems for the next 50 years with current technologies. *Science* 305, 968–972.

Pouyat, R.V. & Effland, W.R. (1999) The investigation and classification of humanly modified soils in the Baltimore ecosystem study. In: *Classification, Correlation, and Management of Anthropogenic Soils*, ed. Kimble, J.M., Ahrens, R.J. & Bryant, R.B. USDA Natural Resource Conservation Service, Lincoln, NE.

Pouyat, R., Groffman, P., Yesilonis, I. & Hernandez, L. (2002) Soil carbon pools and fluxes in urban ecosystems. *Environmental Pollution* 116, S107–S118.

Pouyat, R.V., McDonnell, M.J. & Pickett, S.T.A. (1995) Soil characteristics of oak stands along an urban-rural land use gradient. *Journal of Environmental Quality* 24, 516–526.

Pouyat, R.V., McDonnell, M.J. & Pickett, S.T.A. (1997) Litter decomposition and nitrogen mineralization in oak stands along an urban/rural land-use gradient. *Urban Ecosystems* 1, 117–131.

Pouyat, R.V., Parmelee, R.W. & Carreiro, M.M. (1994) Environmental effects of forest soil-invertebrate and fungal densities in oak stands along an urban-rural land use gradient. *Pedobiologia* 38, 385–399.

Purves, D. & Mackenzie, E.J. (1970) Enhancement of trace-element content of cabbages grown in urban areas. *Plant and Soil* 33, 483–485.

Quinn, M-L. (1988) Tennessee's Copper Basin: a case for preserving an abused landscape. *Journal of Soil and Water Conservation* 43, 140–144.

Quinn, M-L. (1992) Should all degraded land be restored? A look at the Appalachian Copper Basin. *Land Degradation and Rehabilitation* 3, 115–134.

Randrup, T.B. & Dralle, K. (1997) Influence of planning and design on soil compaction in construction sites. *Landscape and Urban Planning* 38, 87–92.

Schleu, U., Wu, Q. & Blume, H-P. (1998) Variability of soils in urban and periurban areas in Northern Germany. *Catena* 33, 255–270.

Soil Survey of England and Wales (1983) *1:250,000 Soil Association Map of England and Wales*. Lawes Agricultural Trust, Harpenden.

Trueman, I.C. & Millett, P. (2003) Creating wild-flower meadows by strewing green hay. *British Wildlife* 15, 37–44.

Vřšcaj, B., Poggioa, L. & Marsana, F.A. (2008) A method for soil environmental quality evaluation for management and planning in urban areas. *Landscape and Urban Planning* 88, 81–94.

Wolman, M.G. (1967) A cycle of erosion and sedimentation in urban river channels. *Geografiska Annaler* 49A, 385–395.

Wong, C.S.C., Li, X. & Thornton, I. (2006) Urban environmental geochemistry of trace metals. *Environmental Pollution* 142, 1–16.

11 Insights and Perceptions of Sustainable Design and Construction

David W. Beddoes and Colin A. Booth

11.1 Introduction

Shelter for humans is a primary need. Early humans made use of natural shelters, such as caves, and then progressed to the use of locally available materials to construct a form of building. The requirements then, as now, of the building or environmental envelope is to provide an internal environment that is safe and comfortable for the occupants. Some buildings serve to provide aesthetic attraction (e.g. places of worship), but the fundamental reason for buildings existence still remains as a provider of a suitable internal environment.

Due to the limited resources of the planet and increasing effects of climate change, this now has to be achieved with sustainability in mind (Intergovernmental Panel on Climate Change [IPCC], 2007). The day-to-day climate is beyond our immediate control, but the construction industry is able to reduce its own impact on long-term climate change by utilising materials and enhanced design to significantly influence the climatic behaviour and sustainability of our buildings. Today, construction involves a complicated erection process that requires technical knowledge and organisational control of many different skilled trades, with design having a direct effect on production and use of resources.

Nowadays, the built environment can benefit from new lighter materials that exceed the functional requirements of traditional materials but necessitate a more scientific approach in their application. These new materials could open the way for designers to explore novel and more complex structural forms. However, a fundamental weakness in building construction can be the artistic nature of the designer, who often seems

Solutions to Climate Change Challenges in the Built Environment, First Edition.
Edited by Colin A. Booth, Felix N. Hammond, Jessica E. Lamond and David G. Proverbs.
© 2012 Blackwell Publishing Ltd. Published 2012 by Blackwell Publishing Ltd.

unaware of issues associated with production and economic use of resources that are latent within designs. Attempts to resolve this weakness have ranged from design and build, to negotiation and collaboration within contracts, but the imbalance between design and construction sometimes still remains a fundamental issue (Foster and Harrington, 2000).

However, having never fully addressed the issues of the designer being divorced from production and resource selection, the industry is not best equipped to face the additional burdens of new structural form, combined with the recent re-emergence of the importance of sustainability in construction.

This chapter provides insights and perceptions into sustainable construction. It proffers the idea that the construction industry cannot continue to build as it has done in the past, it should now design and build to address the changes that lie ahead. Although the initiatives presented in this chapter are set within the UK and European context, the approaches and principles are applicable elsewhere.

11.2 Sustainable Construction

Sustainable construction (Halliday, 2007; Pitt *et al.*, 2009) relates to how the activities, products and services used in construction work contribute to the continued maintenance of ecosystem components and functions for future generations, where the components are plants, animals and humans, as well as the physical environment. That said, humans must demonstrate a balance of three primary aspects, economic, environmental and social/cultural, whilst also satisfying the functional performance of construction work (International Standards Organisation, 2008).

The contribution that sustainable construction can make to combat climate change and adapting to the changing climate is relatively clear, since buildings, during both construction and operation, consume large amounts of resources. That said, buildings account for the largest proportion of UK energy consumption by providing necessary materials and operation of buildings (Figures 11.1 and 11.2) (Pout *et al.*, 2002). Therefore, building technology must continue to provide an acceptable internal environments but using less energy and, furthermore, not relying on a limitless supply of raw materials.

Too often, 'new build' has previously been a wasteful enterprise of a throwaway society, subsidised by cheap energy and creating waste disposal problems. Construction technology must now move away from being a linear process, which does not endure in nature, and transform into a sustainable cyclic process (Edwards, 2010). This must encompass pre-cradle to deconstruction allowing for recycling materials and reducing energy needs throughout. In other words, construction must be 'designed to be deconstructed' (Addis and Schouten, 2004) and follow the 'cradle-to-cradle' philosophy (McDonough and Braungart, 2002) (Figure 11.3). A truly 'green' building can emerge only by adhering to the principles of not generating

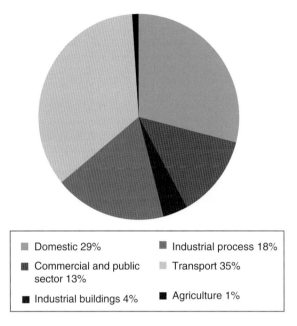

Figure 11.1 Total UK delivered energy consumption by sector (after Pout *et al.*, 2002).

- Domestic 29%
- Commercial and public sector 13%
- Industrial buildings 4%
- Industrial process 18%
- Transport 35%
- Agriculture 1%

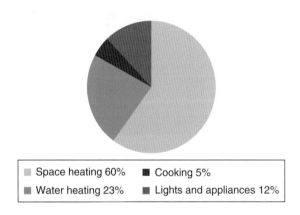

- Space heating 60%
- Water heating 23%
- Cooking 5%
- Lights and appliances 12%

Figure 11.2 Carbon emissions from buildings by source (after Pout *et al.*, 2002).

waste (cyclic materials loop) (Sassi, 2008), of benefiting from renewable energy systems and of concomitantly celebrating diversity by integrating with the local landscape and social/cultural requirements, plus the embracement of added value.

Alongside 'new build', there is also the imperative need to address similar issues for existing buildings. For instance, 66% of the existing UK property portfolio is anticipated to still be with us in 2050 (Stern, 2006). Therefore,

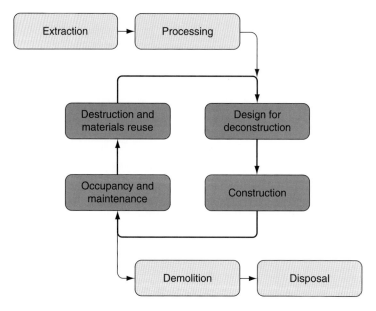

Figure 11.3 An example of the closed-loop 'designing for deconstruction' stepwise process.

the upgrade or renovation of these buildings rather than demolition and 'new build' is desirable because it conserves initial embodied energy and produces less waste in shorter contract times (King and Weeks, 2010). To meet the required reduction in emissions, renovation of these buildings requires innovative materials and technologies that conserve energy, whilst also addressing ecological issues (Lomas, 2010).

11.3 Drivers for Sustainable Construction

The four principle drivers for sustainable construction are presented beneath: (1) government, (2) energy and materials, (3) consumers demanding green living and (4) innovation in technology and materials. The following sections provide the reader with information about these drivers.

11.3.1 Government

In the United Kingdom (UK), adoption and implementation of sustainable development strategies at national and local government level paved the way for the production of similar strategies for individual industries. For the construction industry, this was implemented through the publication of *Building a better quality of life: a strategy for sustainable construction* (Department of the Environment, Transport and the Regions [DETR], 2000), which address key themes: (1) design for minimum waste, (2) lean construction and minimise waste, (3) minimise energy in construction and use, (4) do not pollute, (5) preserve and enhance biodiversity, (6) conserve water resources, (7) respect people and local environments and (8) monitor and report.

More recently, a shared recognition between government and the industry produced a joint initiative, Strategy for Sustainable Construction (Department of Business, Enterprise and Regulator Reform [BERR], 2008), to clarify the existing policy framework. The five key priorities are (1) more investment in people and equipment for a competitive economy, (2) achieving higher growth whilst reducing pollution and use of resources, (3) sharing the benefits of growth more widely and more fairly, (4) improving towns and cities and protecting the quality of the countryside and (5) contributing to sustainable development internationally. These are delivered by a set of overarching targets, relating to procurement, design, innovation, people, better regulation, climate change mitigation, climate change adaption, water biodiversity, waste and materials.

The threat posed by climate change to our common future has led to action by governments worldwide. For instance, within the European Union (EU), the Energy Performance of Buildings Directive (EPBD) came into force in 2006 to promote awareness and encourage energy savings in buildings. It is the main legislation affecting the EU building sector and applies to new and refurbishment projects. The objectives of this European Directive are (1) to set a common basis for calculating energy performance, (2) to set minimum energy performance requirements and (3) to introduce the requirement of Energy Performance Certificates (EPCs). In the UK, the EPBD covers central and local government buildings, health trust and public buildings; plus, domestic properties also require an EPC when sold, leased or rented.

The UK government, in an attempt to mitigate the effects of climate change, has introduced the Climate Change Act (2008) that sets a target of 80% reduction in carbon emissions by 2050 (Department of the Environment, Food and Rural Area [DEFRA], 2008). At present, ~27% of the United Kingdom's total carbon emissions comes from the domestic sector with space heating being responsible for the majority of emissions (DEFRA, 2008; Department for Communities and Local Government [DCLG], 2010). Therefore, part of the government target for reduction in emissions is designed to be achieved by stepwise changes to UK Building Regulations, which apply to new and existing housing stock. This will involve the imposition of zero carbon dioxide emission standards for 'new build' by 2016, so that new homes will not generate further emissions, and also by the promotion of energy saving methods for the existing portfolio to reduce their contribution to emissions.

To promote sustainability, and guide towards even lower emissions targets in the future, the DCLG has adopted the Code for Sustainable Homes (CfSH), which is an environmental assessment used to rate and certify the performance of new homes (DCLG, 2009). The CfSH is a national standard adopted in England to set targets for energy use, carbon dioxide emissions, water use and site ecology (DCLG, 2010). In doing so, the CfSH rates the sustainability of homes from level 1 to 6 on a points-based system. In this rating, level 1 indicates a small improvement over current 'Building Regulations' and would typically cost less than 1% of the base build cost to implement (simply building to current building regulations equates to level 0 rating); whereas level 6 includes zero carbon and typically involves a 30–40% cost increase. Building to the code remains voluntary but does provide certification displaying a rating against 'the Code'. However, it is already mandatory that all government funded social housing must achieve level 3 (Planning Advisory Service, 2009). Similarly, proposed changes to UK Building Regulations Part L 'Conservation of Fuel and Power' (Planning Portal, 2010) will update energy efficiency standards of dwelling emission rates to a mandatory CfSH level 3.

During 2007 the EU also agreed to adopt the European Renewable Energy Directive, which is a commitment to produce 20% of its energy from renewable resources by 2020. In the UK, this has influenced the Climate Change Act (2008) to allow government backing and promotion of small-scale generation using renewable energy (such as photovoltaic, wind, hydro and biomass sources). The current UK process of planning permission being separate from building regulations will become difficult to operate because, as the building regulations are progressively tightened to promote energy efficiency then the measures necessary to achieve the requirements will need planning permission. This has been addressed to some extent by changes to permitted development rights in 2008 that allows limited use of some renewable technologies but the amended rights do not extend to building fabric or wind power. It may soon be necessary to apply with full planning applications (planning permission and building regulations together), which takes more time and results in more expense for the applicant.

Who will now produce the innovative design for building fabric to address these mandatory requirements? The government invests in Standard Assessment Procedures (SAP, or ratings for energy use and CO_2 emissions) and CfSH training and software for its own use, it then promotes new jobs via training for individuals to become accredited energy assessors and produce assessments with software they can purchase. Many surveyors have invested heavily to train, only to find that the local building control department is now carrying out SAP calculations and undercutting market prices. Combined with a building regulations application, many local building control departments will now take on the SAP calculations and provide an EPC for the applicant at reduced fees. The department that has to inspect and ratify the calculations is now the one that is paid to produce them and, furthermore, they will recommend measures to satisfy SAP calculations if necessary. There is a serious conflict of interest with this procedure. As far as the applicant is concerned, building control is cheaper and, ultimately, they have to pass the calculations, so why would they use an independent assessor? Unfortunately, the implications are that as the buildings regulations tighten, the only body that will be working to satisfy the government's aims will be the local building inspectors as they produce, modify and ratify their own SAP calculations. There will be no innovation in design and use of materials and no progress towards the government's aims.

The UK government has acted like this before when it implemented home inspection plans (HIPS) and convinced many professionals to invest in and train for surveying requirements that were subsequently omitted from HIPS. With effect from May 2010 HIPS have been suspended by the government, the EPC is still mandatory but local building control, using government resources, can supply at reduced rates. The government has also suspended the requirement for sellers to provide CfSH certificates to buyers of newly constructed homes. This undermines the work by some developers, who encouraged by the government, have gone to the expense of building above basic level 0 to higher CfSH levels in an attempt to build in a responsible and sustainable way. It is perhaps not surprising there is a lack of confidence in the UK that prevents a company from committing money to research and development needed into technologies, materials and aspects of sustainable construction. The government has been shown to change the parameters on its policy without any concern for industry investments.

11.3.2 Energy and Materials

Energy is not only used in the operation of a building but is also needed to make the components and to construct the building. The energy used to gain raw materials, process, manufacture, transport and build is known as the embodied energy of a building (Treloar *et al.*, 2001). Many standard building materials use large amounts of energy for their production. The assessment of embodied energy from basic data supplied by manufacturers can be difficult. Manufacturers, of course, have a vested interest in presenting favourable data and therefore some may base their data on primary energy, which is the total energy used; whereas others base their data on delivered energy, which is the energy used at the point of use. It can be very misleading and, in some cases, the embodied energy calculated from primary energy values can be double that given by the delivered energy data (Edwards, 2010). Therefore, there is an increasing ownership on designers to understand and to use materials efficiently and specify those with minimum environmental impact that can also be readily reused or recycled (Harrison, 2006).

11.3.3 Consumers Demanding Green Living

Green living is no longer something that dreadlocked tree-huggers worry about because people all over the country are beginning to consider their own carbon footprints. Nowadays, most consumers understand that collective energy savings will result in lower national carbon emissions. This is because they are well informed and, typically, concerned about their ecological footprint (personal impact on the environment). Furthermore, they also recognise that energy efficient homes can serve to alleviate climate change (more than expensive hybrid cars). However, for many consumers, reductions in energy demand simply translate to a direct reduction in household fuel bills (Rowlands *et al.*, 2002; Eves and Kippes, 2010).

As energy prices rise and the cost of heating a home rises too, it is likely there will be a knock-on increase in demand for ways to lower energy costs through more efficient designs and products as well as small-scale energy generation from renewables, which may engender a shift in consumer attitudes towards greener technologies (Rowlands *et al.*, 2002). That said, consumers can be influenced through various avenues (e.g. media, retail, work, friends and family). For instance, when large retail companies are seen to embrace sustainable practices or create new greener stores, society enjoys the association of being 'perceived to be green' through their patronage. It is recognised that this can provide some companies with a competitive edge in a market and, thus, encourages others to develop their own green corporate image (United Nations Environment Programme, 2003). Whether society is demanding or companies are leading environmental awareness is not entirely clear. However, it is apparent that most people, nowadays, want to live and work in a greener and more sustainable environment (Dye and McEvoy, 2008).

Despite many people embracing green living, not everyone is trusting of government involvement in green issues. For instance, the Institute for Public Policy Research (IPPR), in 2009, conducted a survey on how the public thinks low-carbon behaviour could be made mainstream. Participants had good awareness of the effects of climate change; particularly weather and all expressed a dislike of waste and pollution. However, participants also showed a weariness and fatigue about the actual mention of climate change. Some were only motivated to act because of cost and considered this to be more important than the environment. Moreover, there was a general cynicism about the way the UK government takes action with the suggestion that it was just another way to increase taxation. It is, perhaps, somewhat ironic that the IPPR survey also revealed that any cost benefits enjoyed by participants were likely to be spent on potentially carbon-intensive behaviours (Platt and Retallack, 2009).

11.3.4 Innovation in Technology and Materials

New technologies and innovative materials pave the way for new designs and methods of construction. As already noted, it is not desirable to continue to build as it has been done in recent times, because of the need to address the issues of finite resources and because of the construction industry's effect on climate change. Consequently, there is a need for new, low embodied energy products to decrease domestic energy use and reduce the dependence of buildings on fossil fuel energy (Kibert, 2007). That said, building design and materials used still need to provide an acceptable indoor climate against more hostile external weather, without excessive heating or mechanical cooling. For instance, allowances needs to be made for periods of summertime overheating. We should not use air conditioning and must consider a compromise between modern lightweight and traditional heavyweight construction, both of which could be used within the same dwelling. Exposed heavyweight thermal mass walls subject to overnight cooling can be used in north facing rooms to mitigate overheating. Together with passive ventilation and the use of different rooms at different times of the year, it returns us to the values of vernacular architecture.

Therefore, innovation will require examination of vernacular architecture for a rethink on design and materials. Similarly, there is also the need to develop solutions that are not over-designed, such as complex mechanical, electrical systems that the consumer cannot understand or operate correctly. Green design must be simple and inexpensive, with practical features that the consumer can understand and operate. Society should not aim to increase dependence on technologies, unless they are cost effective, practical, simple and reliable in use. Such a philosophy will ensure that benefits are lasting and sustainable.

11.4 Rethinking Construction

Sustainable construction involves a balance of three primary aspects: economic, environmental and social-cultural. Other chapters attest to the developments in two out of three aspects. However, it seems there is minimal current action being taken that addresses the social-cultural aspects of sustainable construction, so one third of the requirements are not being met. This requires a complete rethink about old accepted systems. A new efficient approach to construction that also introduces the social aspect will start with a charrette (Poremba, 1998). The charrette is a pre-planning consultation that brings together all interested parties, ranging from individual public members and upwards. The charrette is used to hammer out an acceptable solution before applying for planning permission. Everyone in the community gets an input to satisfy the hitherto ignored social aspect of sustainable construction. The charrette is not about whether it will be done; it is about how it can be done in an acceptable way (e.g. a wind farm could be the subject of a charrette with the community acceptance based on agreed preferential domestic electricity prices or rebates) (Gibson and Gebken, 2003).

Another way of increasing the sustainability of design could be the introduction of Extended Post Occupancy Evaluation (EPOE)-based performance fees. Under such schemes, clients would make new demands on building designers for lower initial fees and the introduction of a performance-based income stream so the designers are able to profit from the energy savings made by their buildings. They should accept a lower initial fee and then receive a percentage of money saved per annum to provide an incentive. The lower initial fee benefits the client and stimulates industry. Monitoring of energy saved relative to benchmarks for buildings would provide the income stream. Valuable data on the actual in use performance of the features incorporated into energy saving buildings could be collected and used for future designs. This (EPOE) would also highlight latent defects that could be designed out of future projects. The EPOE data would be valuable to facilities management, thereby ensuring that the best possible results were being derived from energy saving measures, the designers will be very keen to monitor and advise when they share in the savings. EPOE stops at deconstruction when the last profit shares are paid to the designers based on the value of materials that can be reused and recycled (Innes, 2005).

The Building Research Establishment (BRE; see http://www.bre.co.uk), a former UK government establishment but now a private organisation that carries out research, consultancy and testing, owns and operates the Building Research Establishment Environmental Assessment Model (BREEAM), SAP and CfSH assessment tools that the government has adopted in order to guide the construction industry (http://www.breeam.org). These complex assessment tools have been shown to be inaccurate when applied to some commercial buildings and even when such buildings achieve poor SAP ratings the 'in use' energy consumption is allegedly much higher than estimated (Spring, 2010). The National House Building Council (http://www.nhbc.co.uk), the UK standard setting body, now has post-occupancy research data for a house built to CfSH level 5 at the BRE Innovation Park, Watford (Sigma Homes). This exemplar, high-profile dwelling gave an as-built heat loss 40% worse than predicted at design stage. The solar panels produced hot water at the wrong times, and the roof-mounted turbine was useless. The photovoltaics worked well until a house that was built next door shaded them. Complex operating systems meant that the occupiers needed almost constant support from helplines. Furthermore, out of 25 windows only four were found necessary (NHBC, 2009). It seems that we may need an assessment system that is based on actual energy consumed by the buildings.

The current material selection process outlined by the Green Guide (http://www.thegreenguide.org.uk) and used for BREEAM and the CfSH, needs further refinement. If all items in one particular group are poor performers then the best of the poor performers gets the 'A' rating as groups are assessed in isolation. As operational energy decreases, there also needs to be an assessment tool that focuses more on the embodied energy within materials and construction elements. The current assessment tools appear to be based on the process of designing the building first and then retrofitting elements to satisfy requirements. The CfSH awards two points for efficient external lighting and two for cycle racks. Then, with regard to low-energy building design, the fabric heat loss parameter also rates at only two. Surely this is out of proportion in the light of the importance the UK government is placing on reduction of fabric heat losses to decrease energy consumption.

The European Energy Performance of Buildings Directive (EPBD) has been the legislation that demonstrated how poorly commercial buildings were performing. For larger public buildings the EPBD requires a displayed energy certificate (DEC) to publicly display the actual energy use, carbon emissions and a rating on a scale from A to G for the building in use. Actual energy use in many instances was far greater than predicted by assessment software. Unfortunately, the DEC is not used in private commercial buildings or homes; instead an estimated energy performance certificate (EPC) is used (Dixon et al., 2008).

11.5 Thoughts for Change

Time before irreversible climate change is upon us is limited, and society needs to act now against carbon pollution. We must use our expertise and innovative designs to reduce domestic energy consumption, whilst making the most efficient use of materials. Designers need training to enable the superior properties of modern materials and technology to be integrated with proven vernacular architecture. As well as energy efficiency in homes, we must also bear in mind that due to prior carbon dioxide emissions the planet will continue to warm and this may produce extreme weather events.

Sustainable construction must be addressed at the very start of a project. A new streamlined process is needed that starts with the charrette to reduce the bureaucracy (Gibson and Gebken, 2003). Lower initial fees for clients and design fees paid on actual performance are more efficient. The Extended Post Occupancy Evaluation will both close the materials loop and provide some much needed data on the efficiency of energy saving designs as built and used (Innes, 2005). Complex assessment tools that produce assessments 40% adrift are a waste of our resources (NHBC, 2009). Industry needs a step change in design allied to a simple tool to balance energy, lifespan, cost and recyclability. This needs to be backed up by tests on houses 'in use' and a domestic DEC to show actual energy use, not fudged EPCs based on design estimates. Equally important is the need to educate and inform professionals and the public so that they will understand the importance of sustainable construction and become aware of how to use and operate buildings to safeguard our common future (Pucket, 2010). This is the realistic way to progress towards ambitious government targets.

Construction now has to reflect the government aim to reduce demand for energy used in domestic space heating. How long will it be before building control departments are profiting at the expense of individual assessors, by using free training and software to provide CfSH assessments that can be sold to private applicants?

Without doubt, the recent government action to introduce 'feed in tariffs' (FITs) is a major step towards the introduction of small-scale energy generation using renewables. It will give UK companies the confidence to invest in renewable technology. The building owner will be able to take advantage of the government incentives and invest in energy saving technologies that can offer, in some instances, a tax-free return of up to 8%, which is a very good offer in the light of current interest rates. Unfortunately, some problems may arise with the many companies already eagerly cashing in on the government scheme. These companies offer the property owner free photovoltaic installations in return for the handover to them of the generation tariff for the full 25 years. The agreements made with these companies have no escape or opt out clause for the property owner. That problem aside, this positive government action is sending out a clear message to the public. FITs may be only the first step of many to be made by UK government towards reversing the public cynicism surrounding its motives.

11.6 Concluding Remarks

Some people may view the arguments presented in this chapter to be somewhat negative insights and perceptions of sustainable design and construction, particularly towards government leadership, but it should be noted that there are an array of new innovations and developments (http://www.greenbuildingbible.co.uk) readily available to the public, the construction industry and government if they demand them. Therefore, with an embracement for change, emerging sustainable design and construction technologies (see Chapters 12 and 13) should be viewed as an opportunity for society to be receptive to the challenges that climate change are predicted to bring ... without the need to hug a tree!

References

Addis, W. & Schouten, J. (2004) *Design for Deconstruction: Principles of Design to Facilitate Reuse and Recycling*. CIRIA Publication, London.

Department of Business, Enterprise and Regulator Reform (BERR) (2008) *Strategy for sustainable construction*. Department of Business, Enterprise and Regulator Reform, London.

Department for Communities and Local Government (DCLG) (2009) *Code for Sustainable Homes Technical Guide Version 2*. RIBA Publishing, London.

Department for Communities and Local Government (DCLG) (2010) *Code for Sustainable Homes: A Cost Review*. Communities and Local Government Publications, Weatherby, UK.

Department of the Environment, Food and Rural Area (DEFRA) (2008) *Climate Change Act*. HMSO, London.

Department of the Environment, Transport and the Regions (DETR) (2000) *Building a better quality of life: a strategy for sustainable construction*. Department of the Environment, Transport and the Regions, London.

Dixon, T., Keeping, M. & Roberts, C. (2008) Facing the future: energy performance certificates and commercial property. *Journal of Property Investment and Finance*, **26**, 96–100.

Dye, A. & McEvoy, M. (2008) *Environmental Construction Handbook*. RIBA Publishing, London.

Edwards, B. (2010) *Rough Guide to Sustainability*, 3rd ed. RIBA Publishing, London.

Eves, C. & Kippes, S. (2010) Public awareness of 'green' and 'energy efficient' residential property: An empirical survey based on data from New Zealand. *Property Management*, **28**, 193–208.

Foster, J. & Harrington, R. (2000) *Mitchell's Structure and Fabric Part 2*, 6th ed. Pearson, Harlow.

Gibson, E. & Gebken, R. (2003) Planning charrettes using the project Definition Rating Index: architectural engineering, building integration solutions 2003. In: *Proceedings of the Architectural Engineering 2003 Conference*, 177–183.

Halliday, S. (2007) *Sustainable Construction*. Butterworth-Heinemann, Oxford.

Harrison, J. (2006) The role of materials in sustainable construction. *Materials Forum*, 30, 110–117.

Innes, S. (2005) Action is key to building a sustainable future. *Building Engineer*, 80, 16–17.

Intergovernmental Panel on Climate Change (IPCC) (2007) *Climate Change 2007: The Physical Science Basis. Contribution of Working Group I to the Fourth Assessment Report of the Intergovernmental Panel on Climate Change*, ed. Solomon, S., Qin, D., Manning, M., Chen, Z., Marquis, M., Avery, K.B., M. and Miller, H.L. Cambridge University Press, Cambridge.

International Standards Organisation (2008) *BS ISO 15392-Sustainability in Building Construction*. IIMSO, London.

Kibert, C. (2007) The next generation of sustainable construction. *Building Research and Information*, 35(6), 595–601.

King, C. & Weeks, C. (2010) *Information Paper, Sustainable Refurbishment of Non-traditional Housing and Pre-1920s Solid Wall Housing*. BRE Press, Watford.

Lomas, K. (2010) Carbon reduction in existing buildings: a transdisciplinary approach. *Building Research and Information*, 38(1), 1–11.

McDonough, W. & Braungart, M. (2002) *Cradle to cradle: remaking the way we make things*. North Point Press, London.

National House-building Council (NHBC) (2009) The sigma home. *National House-building Council Technical Newsletter*, Standards Extra **46**, February; Sustainability Extra, December, 1–4.

Pitt, M., Tucker, M., Riley, M. & Longden, J. (2009) Towards sustainable construction: promotion and best practices. *Construction Innovation: Information, Process, Management*, 9, 201–224.

Planning Advisory Service (2009) Setting local requirements for sustainable buildings. Available from http://www.pas.gov.uk/pas/core/page.do?pageId=119232.

Planning Portal (2010) Part L (conservation of fuel and power). Available from http://www.planningportal.gov.uk/buildingregulations/approveddocuments/partl.

Platt, R. & Retallack, S. (2009) How the;Public thinks lower-carbon behaviour could be made mainstream. Institute for Public Policy Research, London.

Poremba, A. (1998) Early cost intelligence for renovation projects. *Proceedings of the 1998 42nd Annual Meeting of AACE International, June 28-July 1, 1998.*

Pout, C.H., MacKenzie, F. & Bettle, R. (2002) *Carbon Dioxide Emission from Non-domestic Buildings: 2000 and Beyond*. Building Research Establishment, Watford.

Pucket, K. (2010) Learning the lessons on green education. *Construction Manager*. Available from http://construction-manager.co.uk/construction-proffessional/learning-lessons-green-ed.

Rowlands, I.H., Parker, P. & Scott, D. (2002) Consumer perceptions of 'green power'. *Journal of Consumer Marketing*, 19, 112–129.

Sassi, P. (2008) Defining closed-loop material cycle construction. *Building Research and Information*, 36, 509–519.

Spring, M. (2010) Scottish Parliament: Miralles' magnificent mess revisited. *Building* (online), 29 January. Available from http://www.building.co.uk/scottish-parliament-miralles%E2%80%99-magnificent-mess-revisited/3156995.article.

Stern, N. (2006) *Stern Review: The Economics of Climate Change*. TSO, London.

Treloar, G.J., Love, P.E.D. & Faniran, O.O. (2001) Improving the reliability of embodied energy methods for project life-cycle decision making. *Logistics Information Management*, 14, 303–318.

United Nations Environment Programme (2003) Drivers for sustainable construction. *Industry and Environment*, 26, 22–25.

Progress in Eco and Resilient Construction Materials Development

Jamal M. Khatib

12.1 Introduction

Over the last 50 years, there is increasing evidence that the world's is climate has been changing and the Earth's temperature has been rising. The increase in earth surface temperature is attributed to the increased levels of CO_2 (carbon dioxide) emissions into the atmosphere. The small rise in temperature is likely to have caused the loss of ice cap and the melting of Himalayan glacier (Shiva, 2009). This is manifested through flash flooding, rising water level and hurricanes bringing misery to many people around the globe. Countries and places like Holland, Bangladesh and New York City are particularly at risk due to the rise in sea level (Bremner, 2010). Flash flooding due to rising river water level is becoming frequently in many parts of the world including the United Kingdom (*Guardian*, 2009). We, as a society, can not continue to consume more that we can generate. This is becoming increasingly problematic with the increase in world population and the reliance more and more on fossil fuels.

Human activities including construction may be partly to blame for this change, through deforestation, quarrying and the associated increase in CO_2 emissions. Construction and building activities consume large amount of raw materials. The world resources and reserves are limited and can not supply the construction market at the same rate of consumption indefinitely. Raw materials for construction applications have to be obtained, processed and transported before installation. In each of these stages, energy input is required leading to CO_2 emissions. Governments across the globes have started to take measures for reducing or limiting CO_2 emission and other harmful substances into the atmosphere. Therefore, and in order to play a

---

Solutions to Climate Change Challenges in the Built Environment, First Edition.
Edited by Colin A. Booth, Felix N. Hammond, Jessica E. Lamond and David G. Proverbs.
© 2012 Blackwell Publishing Ltd. Published 2012 by Blackwell Publishing Ltd.

part in alleviating disasters, the future of construction should embrace the principles of sustainable development by consuming less, making efficient utilisation of available resources, reusing and recycling. The production of sustainable construction materials can play an important role in reversing the devastating effect of climate change. Also, in order to deal with the immediate consequences of flooding, existing housing stock should be maintained, new houses should be designed and built to be more climate resilient and the choice of appropriate construction materials may play a role in achieving this aim.

Materials commonly used in construction applications include timber, concrete, glass, masonry, plastic, metals and alloys. Energy at various levels is required to produce each of these materials. Some are a finite resource while others are replaceable, yet are of lower strength or durability. This chapter will describe briefly each of these materials and appropriate measures that can be taken in order to make these materials more eco-friendly and resilient. Also, the benefits of nanotechnology in producing sustainable materials will be explained.

12.2 Concrete

Concrete is the second most consumed material in the world with water being the first (Concrete Centre, 2003). The average consumption of concrete is 1 m^3 per person per year (Aitcin 2000). It is a versatile material, exhibiting good properties including compressive strength, can be made in various forms and shapes and is relatively cheap. Concrete can be used as plain concrete or reinforced concrete in numerous construction applications. Steel, due to its high tensile strength, is used with the concrete, so that the combined materials (i.e. reinforced concrete) will possess good tensile and compressive properties, which will make it useful as an efficient structural material allowing us to build large structures. These include high-rise building and large-span structures (e.g. large halls and bridges).

Conventional concrete consists of cement, water, coarse and fine aggregates. Water is generally available and does not require excessive energy when used in concrete. Coarse aggregate generally requires quarrying, crushing to appropriate sizes so that it can be used in concrete. As a result, large amount of fines are generated and often discarded (Langer, 2009). Fine aggregate are normally extracted from appropriate source and crushing is generally not required. Cement is the most expensive component in concrete and is also energy intensive requiring a calcining temperature of 1450°C during manufacture. The cement industry contributes more than 5% to global CO_2 emissions, due to the burning of fossil fuels and due to decomposition of limestone used in cement production (Gigaton Throwdown Initiative, 2009; Huntzinger & Eatmon, 2009).

From the above description, there follows a challenging question: how to make concrete an eco material without compromising the design life of a structure or without having an impact on resilience. In other words, the properties of concrete should not be compromised with the various eco measures that can be undertaken.

Over the last few decades, attempts have been made to use waste materials or industrial by-products in construction. Fly ash, for example, a by-product of the coal power industry, has been used in concrete for partial replacement of cement (Sear, 2001). Large amounts of the fly ash produced in the world are sent to landfill and thus cause environmental problems. China generates as much as 300 million tonnes of fly ash every year, followed by India with 120 millions. In the United Kingdom (UK), for example, half of the 9 million tonnes of fly ash generated ends in landfill despite attempts made to allow greater utilisation of fly ash in construction applications. Typically, 30% of cement can be replaced with fly ash. Recent research shows that it is possible to use higher volumes of fly ash far in excess of 30% (Siddique and Khatib, 2010). Provided that proper curing regime is adopted and an effective mix design, the long-term mechanical and durability performance of such concrete can match and in some cases exceed those of traditional concrete. However, the early strength gain of concrete containing fly ash may be less than traditional concrete and contractors view the early strength as crucial, as they aim to strike the formwork early and speed up construction. Recently, attempts have been made to allow greater utilisation of fly ash in construction materials by using activators but there are some practical problems issues associated with this type of materials, thus emphasising the importance of research in this area.

Another cement replacement material, which has successfully been used in concrete as partial substitution of cement, is ground granulated blastfurnace slag, a by-product of the steel making industry, and replacement up to 70% of the cement can be used. Other waste materials that can be used as cement materials are desulphurised waste (Khatib *et al.*, 2008), rice and wheat husk ashes (Zhang and Khatib, 2008) and silica fume (Wild *et al.*, 1995). Also the waste fine from aggregate processing can be used as cement substitute. Recently, there has been increasing interests to produce novel materials such as metakaolin to partially replace the cement (Khatib, 2009). Although metakaolin is currently more expensive than cement, the energy input to produce it is roughly half of that used to produce cement. Additionally, there is the potential that metakaolin can be produced more cheaply if demand for it increases.

Reducing the consumption of virgin raw materials can be achieved by using recycled aggregate as a substitute to the conventional coarse and fine natural aggregate. With careful concrete mix design, adequate performance can be achieved using recycled aggregates. Examples of recycled aggregates include recycled concrete, recycled brick, recycled glass, incinerator bottom ash, foundry sand (sand used in metal casting) and recycled waste rubber (Khatib *et al.*, 2010; Morales-Fernandez *et al.*, 2005; Khatib, 2005; Oikonomou and Mavridou, 2009). In addition, concrete while in service and during demolition can contribute to carbon sequestration which is another added advantage (Glavind, 2009).

Natural fibres can be used in cement-based materials to produce building materials. These would be particularly useful for the economic development of farming and rural areas in the developing world where building components for use as non-load bearing and roofing tiles can be produced at lower cost than traditional building materials (Savastano et al., 2009). Examples of natural fibres include fibres extracted from sisal, bamboo, jute, sugar cane and banana.

It is envisaged that future houses in floodplains will consists of subfloors that can be used as garages. Future trend may include the use of economical lightweight concrete with low permeability which would be suitable for certain parts such as those subjected to rising water level and hence providing resilience to houses subjected to flood. Also the light-weight concrete can be used as insulating materials in buildings.

Other interesting future trend in concrete production is the use of self-compacting concrete. If designed properly, self-compacting concrete can be a sustainable construction material as it can provide adequate engineering performance and reduce noise on construction site. However, self-compacting concrete is still expensive and attempts in various places in the world are being made in order to reduce cost of production. This would incorporate industrial by-products, waste fines that are normally discarded and cost-effective admixtures so the water content in the concrete mix is kept to a minimum thus achieving the required mechanical and durability performance (Khatib, 2008).

12.3 Brick and Masonry

The manufacture of brick is an energy intensive process and requires a temperature between 900°C and 1050°C. Brick is mainly made out of clay. Industrial wastes and by-products such as PFA, ground glass and sewage sludge have been used in brick manufacturing which would otherwise go to landfill. Using these materials can improve the properties of the brick, reduce the firing temperature and act as a source of fuel (Bingel and Bown, 2009). Brick is regarded as durable construction material that can last for many decades. About 6 million residential properties, which are built out of either brick or stone, in the UK are more than 160 years old (Department of Communities and Local Government, 2006). Stone structures can last much longer; typical examples include the pyramids and the tower of London. Generally, clay bricks have high thermal mass resulting in a relatively steady internal home temperature and less energy required for heating or cooling.

Brick and masonry can be reclaimed and reused although the cement-based mortars in brick or masonry can not easily be removed and so this may not be cost effective. Alternatively, bricks can be crushed and used in construction applications such as hardcore, capping layers in road construction and also aggregates in concrete or in asphalt mixtures. Ground brick can be used as partial substitute to cement in concrete due to its pozzolanic action (Wild et al., 1996, Khatib, 2005). It must be borne in mind, however, the energy required for crushing and grinding and also the long-term performance of construction materials containing recycled or ground brick. As will be mentioned later that there is also potential of using crushed and ground brick in polymer-based composites.

Recently, there has been an interest in using unfired-clay bricks. These brick have 14% embodied energy compared with fired brick, thereby drastically reducing the amount of CO_2 required by about 85%. However, the mechanical and durability performance of unfired-clay bricks are lower than those of fired brick. These include deterioration due to wetting and drying which make them unsuitable for buildings subjected to flood. Rather they are suitable to be used in internal non-load bearing walls.

Rammed earth is a very similar material to non-fired brick that has been used extensively in the past for construction. Now there is renewed interest in using it in construction as a sustainable construction material (Hall and Djerbib 2006). In order to improve the mechanical performance of the material, cement or lime can be used to improve compressive strength while the addition of fibres can improve tensile strength and resistance to cracking. As with unfired clay brick, rammed earth materials may not be suitable in areas subjected to rising water level and adequate measures need to be taken in such environment.

12.4 Glass

Glass is manufactured from mainly sand, sodium carbonate and limestone. It requires heating to a temperature of 1600°C, which is an energy intensive process (Atkins, 2009). However, glass may be seen as sustainable material due to the wider availability of raw materials. Glass is a durable construction material and is used in many construction applications such as cladding and windows. Glass possesses good engineering properties including good tensile strength. However, the use of glass as a structural material is limited mainly due to the brittle nature of the glass and surface defects. Steel mesh and fibres can be included in the glass to reduce its brittleness and improve its properties. Recent attempts have been made to use glass in more construction applications such as staircases and even bridges (Civil Engineering Portal, 2010).

Glass can be regarded as an eco-resilient material as it is impermeable and resistant to weathering conditions. This may make it suitable for use in area subjected to rising water level. In addition, its transparent properties may allow it to be used as a barrier material where concrete or brick may be inappropriate to conserve the aesthetic qualities of historic or natural environments. Glass can be reused and recycled for many uses such as aggregate replacement in concrete, bituminous mixtures and even in water treatment plants as a filtering medium.

12.5 Timber and Bamboo

Timber is regarded as one of the oldest construction materials used by humans and is still used extensively today. It is a versatile material and widely available. If managed properly, timber can be considered as an eco-friendly construction material in that timber which is extracted from forests can be replaced by planting more trees. According to Asif *et al.* (2005), the embodied energy for timber is substantially lower than other construction materials. The added benefits of forestation are the consumption of CO_2 and the release of oxygen through photosynthesis, which plays a vital part in the sustainability of our environment. Planting trees also reduces soil erosion, landslides and potentially flooding in mountainous areas. Timber can broadly be classified into softwood and hardwood based on its engineering characteristics. Hardwood is stronger, cost more and takes longer period to grow and mature than softwood. Timber is used in many construction applications including joists, framing and stud walls.

Sustainability is going to shape our future and reducing waste and recycling are keys toward achieving this aim. With this in mind timber producers have opted for strategies for reducing, reusing and recycling. There has been an increase in the production of engineered wood products, which would allow greater utilisation of timber and minimise wastage. These wood products would normally involve adhesive bonding and include parallel strand lumber (PSL), structural I-beams, chipboard, plywood, oriented strand board (OSB), laminated veneer lumber (LVL) and structural glulam. Engineered wood products allow utilisation of less mature trees and of physically shorter and smaller trees, which require a shorter growing time before harvesting (Milner, 2009). The design life for building elements made with engineered wood products depends not only on the treatment and type of timber but also on the quality of adhesives used. With the advancement of polymer science, 50-year life of adhesive appears to be achievable. This coincides with the rotation period for timber harvesting (Milner, 2009). Examples of construction projects made with engineered wood products include, the superior dome in Michigan, United States, the winter garden in Sheffield, United Kingdom and the Australian Maritime Museum.

Bamboo is another material that can be used construction and is similar to timber. It is fast growing and due to its extensive root system, new bamboo shoots grow spontaneously and therefore no replanting is required. Harvesting takes place every 3–5 years compared with 25 years for softwood and 50 years for hardwood (Bamboo Inspiration, 2010). Bamboo has high strength-to-weight ratio and is used in various construction applications such as, scaffolding, roofing, decking and sheeting. Like timber, bamboo is regarded as a sustainable building material as it can be harvested and replenished more frequently than timber. Also the bamboo plant because of photosynthesis can consume CO_2 and release oxygen.

12.6 Steel

Manufacturing steel is an energy intensive process requiring high temperatures. Despite its high embodied energy, steel is very useful material for construction applications. It can be used as the main construction material by itself or in conjunction with other materials such as concrete.

Steel can be used to construct skyscrapers such as the tallest building in the world with a height of 880 m and suspension bridges spanning more than 1000 m. There is great demand on steel around the world and this is expected to increase. Efficient utilisation of steel is important if the society is to continue to benefit from it. Steel can be recycled but as far as sustainability is concerned, protecting steel while in service should be given a priority as this can lead to prolonging the life of structure and contribute towards a sustainable environment.

Corrosion of steel represents the biggest threat to its service life. Water and oxygen are necessary ingredients for the corrosion to occur and the presence of salt would speed up corrosion. If proper measures are taken during the design stage of the structure steel can be protected and last for a long time. These measures can include coating of steel and cathodic protection. Cathodic protection has been used for many years to protect reinforced concrete bridges, and recently it has been used to preserve historic steel-framed buildings (Lambert, 2009). Also in reinforced concrete structures, proper cover to reinforcement and good quality concrete can protect the reinforcement to last for a long time depending upon the severity of the exposure conditions. Therefore, more attention is required if the building or structure is subjected to chloride ingress such as those exposed to the sea wind and de-icing salt. Another measure, which can be employed, is the use of stainless steel where higher corrosion resistance can be obtained. Although stainless steel is more costly than conventional steel, if service life is considered and life cycle analysis is conducted, the use of stainless steel in construction may be economically viable.

12.7 Polymer-based Materials

Recently, the use of polymer and polymer-based materials in construction has been receiving attention. Polymer can be used in conjunction with other materials such as waste products and fine aggregates to produce a new composite material that can have an adequate strength and durability (Czarnecki et al., 2010). Although the use of resin leads to less workable materials, it offers a material with much reduced permeability. This type of materials can be suitable as a construction material in building houses subjected to floods.

Emerging technologies will see the use of new dwelling made entirely out of polymer-based materials such as fibre-reinforced plastics. Singleton and Hutchinson (2010) attempted to use a new process to produce fibre-reinforced polymer composites as the predominant materials in building houses. Building elements from this new material even included walls and floors. The new method of fibre-reinforced plastic production allows the manufacture of hollow and solid sections such as hollow tubes and I-sections. Provided that the material is durable, this technology would be ideal for building houses located in floodplains as it has the potential to make the house waterproof. The weight of the fibre-reinforced plastic is much less than traditional materials such as brick, concrete or steel so it would be suitable for floating houses. As most of the building elements are prefabricated, the construction time is faster and there is no need for skilled labour. It is indicated that the fibre-reinforced plastic can lead to a reduction in CO_2 emission (Singleton and Hutchinson 2010).

The use of polystyrene materials in conjunction with reinforced concrete in the production of walls and floors is becoming more common. This allows efficient utilisation of material by reducing self-weight thereby reducing the amounts of building materials required. The thermal performance of the building will also be enhanced with the inclusion of polystyrene. This composite material is lighter than traditional material and can be used for walls as well as floors and staircases (EMMEDUE, 2010).

12.8 Nanotechnology

Nanotechnology is being explored in many areas of science and construction materials is no exception. Nanoscience is concerned with controlling matters at nano level (10^{-9}m). Nanotechnology is expected to restructure our future by designing and fabricating materials to suit our own needs. This may lead to the production of eco and resilient construction materials.

Nanotechnology can be used to produce materials that are load bearing and non-load bearing. The materials can be cement and wood based products, polymer and other composites. Other materials may include high-performance insulating materials that can substantially contribute to the energy efficiency of buildings.

Cementitious mixtures (e.g. concrete) can incorporate nanosized particles, which can act as nucleation sites for the hydration products, resulting in materials that have superior mechanical and durability properties (Lindgreen *et al.*, 2008). Clay particles and limestone fines are examples of these materials and as explained previously, this will lead to a reduction in cement used and thus contribute to the reduction in CO2 emission. Nanoparticles can be used in cement-based mixtures containing steel reinforcement. The purpose of these materials is to prevent the corrosion of reinforcement and increase the lifespan of building materials, thus producing eco and resilient materials.

Materials surfaces incorporating nanomaterials can be used to renovate existing building and infrastructure. Surfaces can be produced so that they are self-cleaning resulting in the reduction of maintenance cost of a building. These materials incorporate photocatalytic substances, which are active under the exposure to UV radiation (Geiker and Anderson, 2009). They can be used to clean concrete and glass surfaces.

Nanotechnology also presents the opportunity to produce wireless sensors to monitor and control the interior environment of buildings and can also monitor the performance of building materials while in service. A sensor, for example, can be placed inside building elements to monitor dimensional changes or monitor defects.

12.9 Future Trends

Sustainable development is going to drive our future. We must plan and act in a sensible and sustainable manner. Regarding construction materials, it is anticipated that there will be greater utilisation of waste and recycled materials as well as the production of novel and innovative materials for use in construction. Also there will be an increase in the use of high performance materials, which would allow the use of thinner members in construction and thus reduce the quantity of the materials that would normally be required. New ways of designing our buildings and structures will allow efficient utilisation of resources. For example, reducing the self-weight of building element by using lighter material such as polystyrene in some parts of the building element can be advantageous. Also more utilisation of fibre-reinforced plastic will be used, either as building components or in conjunction with other materials such as reinforced plastic rods inside concrete or timber. More high-grade products will be constructed out of waste timber. Steel has high embodied energy and has dwindling supply, so serious attempts will be made in the future to reduce the quantities of steel that is normally used in construction while ensuring the maximum lifespan for the steel uses which are not replaceable. More utilisation of glass in construction applications is a possibility by the manufacturing of new building elements made of glass such as bridges, thus reducing the need for steel. We may see in the future more utilisation of floating houses, and this can be beneficial in areas where water level is expected to rise. The use of fibre-reinforced plastic can be beneficial in these situations. In areas where there is pressure on land space because of population growth, constructing building offshore may become a necessity. There are plans in Singapore to construct buildings offshore, and on these occasions adequate design is required as well as using high-performance materials, which should last the lifespan of the building.

References

Aitcin, P.C. (2000) Cements of yesterday and today: concrete of tomorrow. *Cement and Concrete Research*, 30(9), 1349–1359.

Asif, M., Muneer, T. & Kubie, J. (2005) Sustainability analysis of window frames, *Building Services Engineering Research and Technology*, 26(1), 71–87.

Atkins, C. (2009) Sustainability of glass in construction. In: *Sustainability of Construction Materials*, ed. Khatib, J.M., Woodhead Publishing, Cambridge.

Bamboo Inspiration (2009) *Bamboo construction*. Available from http://www.bamboo-inspiration.com/bamboo-construction.html.

Bingel, P. & Bown, A. (2009) Sustainability of masonry in construction. In: *Sustainability of Construction Materials*, ed. Khatib, J.M., Woodhead Publishing, Cambridge.

Bremner, T.W. (2010) The future of construction materials in a sustainable world. In: *2nd International Conference on Sustainable Construction Materials and*

Technologies, Ancona, Italy, 28–30 June, ed. Naik, T.R., Canpolat, F., Claisse P. & Ganjian, E. Honouree vol., 59–68.

Civil Engineering Portal (2010) Skywalk-Glass bridge. Available from http://www.engineeringcivil.com/skywalk-glass-bridge.html.

Concrete Centre (2003) Sustainable development in the cement and concrete sector, Concrete Centre, Project Summary 2003. Available from http://www.concretecentre.com.

Czarnecki, L., Garbacz, A. & Sokołowska, J. J. (2010) Fly ash polymer concretes. In: *2nd International Conference on Sustainable Construction Materials and Technologies*, Ancona, Italy, 28–30 June, ed. Naik, T.R., Canpolat, F., Claisse, P. & Ganjian, E. Honouree vol., 127–138.

Department of Communities and Local Government (2006) Housing statistics. Available from http://www.communities.giv.uk/housing/.

EMMEDUE (2010) Advanced building system. Available from http://www.mdue.it.

Geiker, M.R. & Anderson, M.M. (2009) Nanotechnology for sustainable construction. In: *Sustainability of Construction Materials*, ed. Khatib, J.M. Woodhead Publishing, Cambridge.

Gigaton Throwdown Initiative (2009) Gigaton pathways in the construction materials sector. Available from http://www.gigatonthrowdown.org.

Glavind, M. (2009) Sustainability of cement, concrete and cement replacement materials in construction. In: *Sustainability of Construction Materials*, ed. Khatib, J.M. Woodhead Publishing, Cambridge.

Guardian (2009) UK flooding – as it happened. Available from http://www.guardian.co.uk/news/blog/2009/nov/20/flooding-live.

Hall, M. & Djerbib, Y. (2006) Moisture ingress in rammed earth: part 3 – sorptivity, surface receptiveness and surface inflow velocity. *Construction and Building Materials*, 20(6) 384–395.

Huntzinger, D.N. & Eatmon, T.D. (2009) A life-cycle assessment of Portland cement manufacturing: comparing the traditional process with alternative technologies. *Journal of Cleaner Production*, 17(7), 668–675.

Khatib, J.M. (2005) Properties of concrete containing fine recycled aggregates. *Cement and Concrete Research Journal*, 35(4), 763–769.

Khatib, J.M. (2008) Performance of self-compacting concrete containing fly ash. *Construction and Building Materials Journal*, 22(9), 1963–1971.

Khatib, J.M. (2009) Low curing temperature of metakaolin concrete. *American Society of Civil Engineers (ASCE) – Materials in Civil Engineering Journal*, 21(8), 362–367.

Khatib, J.M., Baig, S., Bougara, A. & Booth, C. (2010) Foundry sand utilisation in concrete production. In: *2nd International Conference on Sustainable Construction Materials and Technologies*, Ancona, Italy, 28–30 June, ed. Naik, T.R., Canpolat, F., Claisse, P. & Ganjian, E. vol. 1, 931–938.

Khatib, J.M., Mangat, P.S. & Wright, L. (2008) Sulphate resistance of blended binders containing FGD waste. *Construction Materials Journal – Proceedings of the Institution of Civil Engineers (ICE)*, 161(CM3), 119–128.

Lambert, P. (2009) Sustainability of metals and alloys in construction. In: *Sustainability of Construction Materials*, ed. Khatib, J.M. Woodhead Publishing, Cambridge.

Langer, W.H. (2009) Sustainability of aggregates in construction. In: *Sustainability of Construction Materials*, ed. Khatib, J.M. Woodhead Publishing, Cambridge.

Lindgreen, H., Geiker, M., Kroyer, H., Springer, S. & Skibsted, J. (2008) Microstructure engineering of Portland cement pastes and mortars through addition of ultrafine layer silicates. *Cement and Concrete Composites*, 30, 686–699.

Morales-Hernandez, B., Khatib, J.M. & Gardiner, P. (2005) Use of municipal solid waste incineration bottom ash (MSWI-BA) in cement mortar. In: *Proceedings of the 1st Global Slag Conference – From Problem to Opportunity*, Dusseldorf, Germany, 14–15 November.

Milner, H.R. (2009) Sustainability of engineered wood products in construction. In: *Sustainability of Construction Materials*, ed. Khatib, J.M. Woodhead Publishing, Cambridge.

Oikonomou, N. & Mavridou, S. (2009) The use of waste tyre rubber in civil engineering works. In: *Sustainability of Construction Materials*, ed. Khatib, J.M. Woodhead Publishing, Cambridge.

Savastano, H., Santos, S.F. & Agopyan, V. (2009) Sustainability of vegetable fibres in construction. In: *Sustainability of Construction Materials*, ed. Khatib, J.M. Woodhead Publishing, Cambridge.

Sear, L.K.A. (2001) *The properties and use of coal fly ash*. Thomas Telford, London.

Shiva, V. (2009) Vandana Shiva on climate change, at the third pole: the Himalayas. Available from http://isiria.wordpress.com/2009/05/19/vandana-shiva-on-climate-change-at-the-third-pole-the-himalayas/

Siddique, R. & Khatib, J.M. (2010) Mechanical properties and abrasion resistance of HVFA concrete. *Materials and Structures – RILEM Journal*, 43(5), 709–718.

Singleton, M. & Hutchinson, J. (2010) Development of fibre-reinforced polymer composites in building construction. In: *2nd International Conference on Sustainable Construction Materials and Technologies*, Ancona, Italy, 28–30 June, ed. Naik, T.R., Canpolat, F., Claisse, P. & Ganjian, E. vol. 1, 151–161.

Wild, S., Sabir, B.B. & Khatib, J.M. (1995) Factors influencing strength development of concrete containing silica fume. *Cement and Concrete Research Journal*, 25(7), 1567–1580.

Wild, S., Khatib, J.M., Sabir, B.B. & Addis, S.D. (1996) The potential of fired brick clay as a partial cement replacement material. In: *International Conference – Concrete in the Service of Mankind: Concrete for Environment Enhancement and Protection*, Dundee, 24–28 June 1996, ed. Dhir, R.K. & Dyer, T.D. 685–696.

Zhang, J.S. & Khatib, J.M. (2008) Using wheat husk ash in the production of a sustainable construction material. *Real Estate and Development Economics Research (READER) Journal*, 1(1), 18–29.

Energy Efficiency: Alternative Routes to Mitigation
David Coley

13.1 Introduction

Later, in Chapter 23, we can see the scale of the reductions we need to make and introduce the plethora of renewable energy technologies that could be deployed. In this chapter we will look at alternatives to switching fuels by considering energy efficiency and the direct engineering of the climate. We will then present one of many possible combinations of changes to our energy supply that has the potential to make major carbon savings, together with an abatement curve that will allow you to build a costed mitigation strategy of your own.

13.2 Energy Efficiency

Often termed 'the silent renewable', energy efficiency, or conservation, is the most promising sustainable energy strategy in the short to medium term, offering the promise of large carbon savings for relatively little expenditure and little environmental impact. In general energy efficiency is cheap, the technology is fully developed, there are unlikely to be public concerns over its use, and construction and implementation times are very short. The power of the approach when taken seriously can be gleaned from the observation that although Japan's wealth (in the form of its per capita gross domestic product) is similar to the United States', it is achieved by emitting less than half the mass of carbon dioxide per capita.

Solutions to Climate Change Challenges in the Built Environment, First Edition.
Edited by Colin A. Booth, Felix N. Hammond, Jessica E. Lamond and David G. Proverbs.
© 2012 Blackwell Publishing Ltd. Published 2012 by Blackwell Publishing Ltd.

13.2.1 Cogeneration

Cogeneration, or combined heat and power (CHP) schemes, try to circumvent the poor efficiency seen in traditional electrical generation technologies. This is typically only 30% with the remaining 70% of the energy being lost to the surroundings as heat. This efficiency is so low that the heat loss from UK power stations exceeds the total UK domestic heating requirement (Atkins, 1986). There is nothing to stop us using this waste heat to warm our homes or offices. When the electrical generator is at some distance from the building to be heated, such a scheme is referred to as *district heating* and has proved successful at various locations around the world, for example in Sweden. More popular have been CHP schemes based around a single facility, such as a hospital, with a large and reasonably constant need for both electrical power and heat.

A small cogeneration plant typically uses a diesel engine (or gas turbine) to generate the electricity and then utilises the heat from the exhaust gases to heat water for distribution to the heating system. Because there is no need to replace the heating system itself, such plants can be fitted without the need for major refurbishment and often in the existing boiler room. In future the use of fuel cells for cogeneration might become a reality.

Whether cogeneration makes economic or environmental sense has to be considered on a case by case basis. This is because although excess electricity can be exported from the site via the national grid, heat cannot be; so a site that uses much more electricity than heat will see little benefit from introducing cogeneration. A heat:power ratio of approximately 1 is typically considered an ideal situation.

13.2.2 Energy Efficiency in Buildings

Over one third of carbon emissions in the developed world are associated with energy use in buildings. Heat loss through the fabric of a building is proportional to the temperature difference between the internal and external environment. Typically the amount of energy required to heat a building will reduce by around 7% for each 1°C drop in internal temperature. Hence setting thermostats as low as possible is an obvious first step. It is worth noting that we typically heat buildings to around 20°C, whereas body temperature is around 37°C. Hence buildings never warm us, we always warm buildings. The heating system is only used to reduce our rate of heat loss. Surely a better way to do this is to use localised insulation (i.e. more clothes). To overly heat the many thousands of tonnes of concrete in the shell of a building rather than put a jumper on is nonsensical.

The Passivhaus movement is attempting to make energy efficiency the key to low-carbon design. Passivhaus buildings (both houses and commercial building) have low fabric losses and have much lower infiltration losses than normal buildings. They also make use of a simple mechanical ventilation system that recovers heat from stale air and uses it to warm (with an efficiency of over 80%) incoming fresh air. Such buildings have been built in many parts of the world. To be credited as a Passivhaus, the building must be designed to use no more than 15 kWh/m² of floor area for heating. This is typically one tenth of that used by homes in the United Kingdom and reduces a domestic heating fuel bill to around £70 per annum at an increased build cost of only 6%. Given that many thousands of such buildings have been built and shown to be successful, it is difficult to see why many countries are focusing on building mounted renewables rather than energy efficiency.

Energy efficiency in use is not the only consideration. The energy used to manufacture the materials used in the building and transport them to site also needs to be minimised in order to create a sustainable design. Traditionally only local materials were available and these were transported no more than a few hundred metres. Most of today's designs no longer keep to this tradition with modern, high embodied energy, materials such as aluminium, glass and plastic being extensively used. This need not be the case, with many architectural practices and enthusiasts constructing buildings from rammed earth, straw, cob, local stone and wood from sustainable forests.

13.3 Carbon Sequestration and Climate Engineering

Although the production of carbon dioxide is a necessary result of burning fossil fuels, there are alternatives to simply venting any carbon dioxide to the atmosphere. And even if we do release it into the atmosphere there exists the possibility of capturing carbon dioxide and storing it. To do this, two problems need to be solved: firstly, how to capture the carbon dioxide and, secondly, how to store it for a considerable length of time (possibly thousands of years). Several ways have been proposed to solve both problems. Two methods, the direct injection of carbon dioxide into geologic features and the planting of trees, have already been deployed. Others are at various stages of development and at present it is unclear which ones are likely to be realised on a substantial basis. However, what is clear is that, at least in theory, the storage capacity of the biosphere and the deep Earth is vast.

In addition to sequestration, another proposed approach to engineering the climate is to reflect incoming sunlight back into space before it reaches the planet, thereby reducing the global temperature.

13.3.1 Capture Technologies

One possibility is to combine carbon dioxide capture with a coal gasification combined cycle plant. Such plants gasify the coal to produce a synthesis gas of carbon monoxide and hydrogen. The carbon monoxide could then be reacted with water to produce carbon dioxide for storage and the hydrogen burnt within a gas turbine. This approach has the advantage that the hydrogen would also make a suitable transportation fuel. An example

(Herzol and Golomb, 2004) is the Great Plains Synfuel Plant in North Dakota, United States which gasifies 16 000 tonnes of lignite coal per day, captures 2.7 million m³ of carbon dioxide and pipes this 325 km to Weyburn, Saskatchewan where it is injected into the ground as part of an enhanced oil recovery process.

13.3.2 Storage Technologies

Three storage technologies have been suggested: geological storage deep underground; storage within the oceans at depths of over 1 km and biological storage within forests or marine organisms.

Table 13.1 lists the estimated storage potential of these various options. There are some concerns over the safety of geologic storage, particularly the possibility of large quantities of carbon dioxide returning to the surface rapidly, and the potential for induced seismicity. There has been at least one large-scale natural release of carbon dioxide in recent history – that at Lake Nyos in Cameroon. This caused 1800 human fatalities in 1986 when an upwelling of carbon dioxide displaced the surrounding air (BBC, 2002).

Oceanic storage at depths in excess of 1 km offers the greatest storage facility for anthropogenic carbon dioxide. Two storage mechanisms have been proposed, firstly dissolving the carbon dioxide at mid-depths (1.5–3.0 km) and secondly injecting the carbon dioxide at depths in excess of 3 km where it would form lakes of liquid carbon dioxide.

Using the biosphere for carbon storage offers several advantages: firstly there would be no need to capture the carbon dioxide at the point of production, and secondly, if reforestation were used, many would see the approach of net environmental benefit. This is unlikely to be the case for geologic or oceanic storage. Possibilities include growing tress and fertilising the oceans to increase algae growth. Many are unhappy with the latter largely because its impacts are little understood. Others question the mass of minerals needed. The effect of mass fertilisation on the ecology of the oceans is also unknown; as is the percentage of carbon that would finally be naturally exported to the deep ocean by such fertilisation. Yet it has been

Table 13.1 Total planetary carbon dioxide storage potential.

Option	Approximate worldwide capacity
Ocean	1000–10 000+ GtC
Deep saline aquifers	100–10 000 GtC
Depleted oil and gas reservoirs	100–1000 GtC
Coal seams	10–1000 GtC
Terrestrial (reforestation)	10–100 GtC
Utilisation	currently < 0.1 GtC per annum

Note: Total planetary carbon dioxide storage potential. For comparison, worldwide carbon emissions are around 8 GtC per annum (from Herzol & Golomb, 2004).

suggested the cost of implementing a storage programme based on oceanic fertilisation could be extremely modest compared to other capture and sequestration technologies, at €0.77–7.77 per tC.

13.3.3 The Reflection of Solar Radiation

Rather than trying to absorb excess carbon dioxide, we might attempt to reflect some of the sunlight striking the planet back into space. This would have an obvious cooling effect. It was first suggested (Budyko, 1974) that rockets could be fired into the stratosphere to carry the necessary aerosols aloft. Later it was postulated this might be better achieved by a modification to commercial jet fuel. At an even higher altitude, giant reflectors could be assembled in space. They would cast a shadow on the planet that would transverse the surface, reflecting around 1% of the overall incident light. The cost, or even feasibility, of doing this is unknown.

13.4 A Sustainable, Low-Carbon Future?

In this chapter (and Chapter 23), we have looked at a variety of ways in which some of our future energy demands might be met by sustainable means. For many of the technologies studied, the size of the resource is such that it would, in theory, be possible to meet all our requirements using a single resource. Large carbon reductions could be surprisingly easy, and Pacala and Socolow (2004) have shown that it could in part be achieved without the need for new, unproven technologies. Pacala and Socolow identify 15 technologies or adaptations each capable of saving 1 GtC per annum by 2054, as shown below.

13.4.1 Improved Car Fuel Efficiency

Assume car ownership grows to 2 billion (roughly four times today's number) and average annual distance travelled stays at 16 000 km (10 000 miles), but fuel efficiency doubles from 13 km/l to 26 km/l (30 mpg to 60 mpg). Although there are many cars that are already capable of such fuel economy, there is a history of efficiency improvements being lost to changes in driving style, with drivers opting for higher speeds and greater acceleration rather than lower operating costs.

13.4.2 Reduced Reliance on Cars

Assume car ownership grows as in Option 1, but they were only used half as much, that is, annual average travelled distance falls to 8000 km (5000 miles). Unfortunately in much of the developed world the trend is in the opposite direction with annual distance travelled by car increasing.

13.4.3 Increased Energy Efficiency in Buildings

Assume energy use in buildings drops by 25%. This is easily achievable if governments invest in such improvements to the degree they do with wind or nuclear power. Unfortunately the temperature that buildings are heated to is increasing not reducing and much of the gains that should be being realised through improved insulation are being lost to such comfort taking.

13.4.4 Improved Power Plant Efficiency

Assume coal-based generation efficiency were to grow from 40% to 60%, whilst the amount of electricity generated by coal doubles. This is a straightforward technological fix only requiring the wealthier nations paying for the replacement of aging power stations in the less developed world.

13.4.5 Replace Coal-Based Generation by Gas-Based Generation

Replace 1400 GW of coal-based power stations by natural gas-based ones (even though coal-based generation efficiency is assumed to increase to 50%). The savings arise from the lower emissions of carbon per joule of energy from gas-based generation. Again, this simply requires the transfer of money and technology.

13.4.6 Carbon Storage from Power Plants

Assume storage and capture at 800 GW of base-load coal power stations (or 1600 GW of gas power stations) and that this prevents 90% of carbon dioxide being released from these plants. Pacala and Socolow consider that the most likely technology would be pre-combustion capture of carbon dioxide with hydrogen being burnt to generate electricity. Unlike the previous five approaches this is an untried technology, at least at the scale required. Pilot projects are well underway, and these should lead to a much better understanding of the costs.

13.4.7 Carbon Capture from Hydrogen Generation

Assume 250 Mt of hydrogen per annum is generated from coal and storage is used for the resultant carbon dioxide, however the hydrogen is not burnt to generate electricity, but to replace other fossil fuel use – for example, that within cars. The production of the hydrogen is not the issue here, but the use in vehicles where chilling and pressurisation require additional energy expenditure and hydrogen leakage remains an issue. (The latter is not a question of safety, but of finding one's fuel tank empty after a week due simply due to evaporation.)

13.4.8 Storage of Carbon Captured in Synfuels Plants

Assume oil-like fuels are produced from coal on a large scale (30 million barrels per day) with half the carbon entering the plant being sequested. The production of synthetic oil is well developed (in part because of developments in South Africa during apartheid), and the sequestration would be no more difficult than for electricity generation.

13.4.9 Fission

Assume nuclear power capacity is tripled to 700 GW. Again, this is a straightforward technological solution, although unless fast breeder technology is introduced more expensive uranium resources will have to be accessed toward the end of the twenty-first century.

13.4.10 Wind Electricity

Assume 2000 GW of wind power capacity replaces coal generation (i.e. the installation of 1 million 2 MW turbines). This allows for each GW of wind power still requiring 0.33 GW of base-load generation from other sources. This is 50 times today's level of deployment. This would be very easy to achieve with no technological developments but would require a large increase in the world's rate of turbine, tower and blade production and a decrease in on-site construction times.

13.4.11 PV

Assume PV deployment increases 700-fold to 2,000 GW. This would require 2 million hectares of land (or 2–3 metres squared per person). Roofs would be ideal for this level of adoption. As for wind, this can be achieved with today's technology.

13.4.12 Green Hydrogen

Assume, for example, 4000 GW of wind power were used to produce hydrogen for vehicle use (with the hydrogen being used in fuel cells). This is twice the number of turbines required by Option 10 and suggests that we are better off using hydrogen for electricity generation than as a vehicle fuel. It also suffers from the basic hydrogen storage problem mentioned above.

13.4.13 Biofuels

Assume the production of 34 million barrels a day of biofuels replacing an equivalent amount of gasoline use. This is 50 times larger than current world

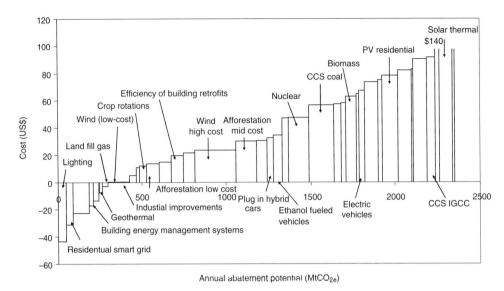

Figure 13.1 Abatement cost curve for the United States (data from Bloomberg energy finance). The total 2400 MtCO$_{2e}$ reduction is equivalent to a 40% reduction on 2005 emissions. The height of any rectangle gives the cost in US$ per tonne of carbon dioxide equivalent saved. For a particular technology, the width represents the mass that could be saved by the wide-scale adoption of the technology. (An equivalent for the world as a whole can be obtained from http://www.mckinseyquarterly.com.)

production. This would require 250 million hectares of land – an area equal to one sixth of the world's cropped land – and assumes no fossil fuel inputs are required.

13.4.14 Forest Management

If tropical deforestation is reduced to zero (rather than halved) by 2054, then 0.5 GtC per annum would be saved. An equal amount of carbon could be saved by reforesting or afforesting 250 million hectares in the tropics, or 400 million hectares in the temperate zone. Current areas of tropical and temperate forest cover are 1500 and 700 million hectares respectively.

13.4.15 Agricultural Soils Management

The tilling of previously natural land releases up to half the stored carbon in the soil, primarily because such tilling increases the rate of decomposition of organic matter in the soil. *Conservation tillage* (i.e. drilling seeds into the soil rather than ploughing) and erosion control can reverse this loss of carbon. By 1995, 110 million hectares of the world's 1600 million hectares of cropland were being planted in this way. 1 GtC per annum would be saved if these practices were expanded to all cropland.

By deploying 7 of the 15, 7 GtC per annum would be saved. (Current carbon emissions are 8 GtC per annum, but rising steeply.) Further reductions could then be made using some of the other technologies outlined above. Although several have questions surrounding their deployment, most of these hinge as much on willingness to adopt or adapt as on true technological issues and the implementation of several could be very rapid.

13.5 Solutions: Abatement Costs

Not all the technologies discussed in this chapter have equal cost. Figure 13.1 shows the cost of various technologies and the saving in carbon they might deliver. The graph creates a natural time line for action: the technologies on the left are the most cost-effective ones and should therefore be invested in immediately. Those below the horizontal have negative costs (i.e. they save money and are connected with efficiency), reinforcing the thoughts at the beginning of the chapter. By picking a range of technologies from the figure, one can create a tailored mitigation strategy and estimate its cost.

From Figure 13.1 it is obvious that the most cost effective measures are based on energy efficiency rather than alternative generation. This has been known for some time; however, society has found it very difficult to translate this into action. Nowhere is the difference in cost, and its implications, more blatant than in the comparison of the costs of wind power (on the right of Figure 13.1) to the cost of insulating buildings (on the left of Figure 13.1). Yet many governments subsidise wind power (or nuclear power) to a far greater extent than insulation. Clearly curves such as Figure 13.1 are not influencing policy, and until they do policy is unlikely to be based on finding a low cost solution to climate change.

References

Atkins, G. (1986) The advantages of CHP systems. In: *Proc. I Mech E Symposium on CHP*, Sheffield, UK, 1986.

BBC (2002) *Killer Lakes*. BBC 2, 4 April. Available from http://www.bbc.co.uk/science/horizon/2001/killerlakes.shtml.

Budyko, M.I. (1974) *Climate Changes*. American Geophysical Union, Washington, DC.

Herzol, H. & Golomb, D. (2004) Carbon capture and storage from fossil fuel use. In: *Encyclopaedia of Energy*.

Pacala, S. & Socolow, R. 2004. Stabilization wedges: solving the climate problem for the next 50 years with current technologies. *Science*, **305**, 968–972.

The Benefits of Green Infrastructure in Towns and Cities

Susanne M. Charlesworth and Colin A. Booth

14.1 Introduction

The built environment is a swathe of mostly grey concrete that has deleterious impacts, both aesthetically and climatically. However, by greening the outside of buildings and surrounding pavements, a wealth of benefits can accrue to the environment, society and urban living. Vegetation (green infrastructure) can intercept rainfall and reduce flooding risk, enhance urban biodiversity, limit the overheating of buildings and improve human health and wellbeing and, concomitantly, will sequester carbon to offset CO_2 emissions.

External greening of buildings is not a new phenomenon because green roofs were a common element of vernacular styles of buildings until the early 1900s (Early *et al.*, 2007). However, in recent times, particularly with the development of prevention and site control strategies for stormwater using sustainable drainage systems (SUDS) (see Chapter 19), planting vegetation on the roofs and walls of buildings or as part of street-level pavement designs is recognised to offer many environmental and social benefits. Moreover, it is anticipated these rewards, or Ecosystem Services, will be most notable with the onset of predicted climate changes, and this chapter introduces insights into the benefits of greening towns and cities, which can address some of the predicted changes to both rainfall and temperature regimes.

14.2 Integrating Vegetation into the Built Environment

Studies of urban form and the spatial distribution of green space in the urban environment (e.g. Rose *et al.*, 1999; Akbari *et al.*, 1999, 2003) have shown that the densest areas in cities have the least vegetation, with Whitford

---

Solutions to Climate Change Challenges in the Built Environment, First Edition.
Edited by Colin A. Booth, Felix N. Hammond, Jessica E. Lamond and David G. Proverbs.
© 2012 Blackwell Publishing Ltd. Published 2012 by Blackwell Publishing Ltd.

et al. (2001) finding that less affluent residential areas also lack green space. Unfortunately, once a development is built, it is not normally possible to allocate space for greening (Wilby, 2007). However, where it is possible to make minor changes to vegetation cover, there are vast benefits to be gained from integrating green roofs and walls and pavement planting into the cityscape. The following sections, therefore, explore the benefits accrued from this approach.

14.2.1 Green Roofs and Walls

Green roofs are normally constructed on a flat rooftop (slope <10°C), but the term *green roofs* also covers balconies and terraces, which are designed as either purposely inaccessible or as spaces to be used as public open space. There are various names and descriptions used, but in essence there are two types of green roofs, (1) extensive and (2) intensive, each with different characteristics (Table 14.1). Figure 14.1 shows a cross-section of a typical green

Table 14.1 A comparison of the key characteristics of both extensive and intensive green roofs (after Oberndorfer *et al.*, 2007).

Characteristics	Extensive roof	Intensive roof
Purpose	Functional; storm water management, thermal insulation, fireproofing.	Functional and aesthetic; increased living space.
Structural requirements	Typically, these are within standard roof weight bearing parameters; additional 70–170 kg per m^2.	Planning is required in the design phase or structural improvements are necessary; additional 290–970 kg per m^2.
Substrate type	Lightweight; high porosity, low organic matter.	Light to heavyweight; high porosity, low organic matter.
Average substrate depth	2–20 cm	>20 cm
Plant communities	Low growing communities of plants and mosses selected for stress-tolerance qualities.	There are no restrictions other than those imposed by substrate depth, climate, building height and exposure, and irrigation facilities.
Irrigation	Most roofs require little or no irrigation.	Often requires irrigation.
Maintenance	Little or no maintenance is required; some weeding or mowing as necessary.	These have the same maintenance requirements as a similar garden at ground level.
Cost (above waterproofing membrane)	~$100–300 per m^2.	~$200 per m^2.
Accessibility	Generally functional rather than accessible; but they do need basic accessibility for maintenance.	Typically accessible; bylaw considerations.

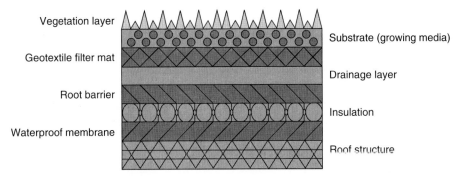

Figure 14.1 Typical structural profile of a green roof.

roof system. Beginning from the base upwards, the various components used are suitable for plant growth without damaging the building fabric: (1) a root barrier layer (waterproof geomembrane to restrict plant penetration), (2) a drainage layer (controlling water retention), (3) a substrate layer (the growing medium, e.g. soil or compost) and (4) a vegetation layer (the surface plants).

The establishment of all green roofs has structural loading implications, which need to be considered during the design stage, whether as a new build or retrofit. The structural capacity can often limit the available options involved in the design of any green roof system. That said, a further consideration, normally at the design stage, is the maintenance of the green roof once installed. This is because, typically, intensive green roofs require greater maintenance than extensive green roofs, but environmental factors (e.g. sunlight, temperature and rainfall) will also influence the necessity for maintenance.

Unused roof space can represent up to 50% of the impermeable surfaces of cities (Mentens et al., 2006). If only half of this currently unused space was greened, Rosenzweig et al. (2006) have computed that the 'Urban Heat Island Effect' (see Section 14.5) in a city such as New York could be reduced by as much as 0.8°C.

Whilst green walls have been used less than green roofs, they have the same benefits of heat reduction, storm peak attenuation and insulation, for both maintaining heat inside buildings in the winter and cooling the building during the summer. Ip et al. (2010) studied a 'vertical deciduous climbing plant canopy' in the United Kingdom and found seasonal benefits due to shading in the summer reducing the internal building temperature by 4–6°C and in the autumn, after leaf fall, incident solar radiation was allowed through the windows, heating the room inside. There are many ways of constructing a green wall, but most are similar to the one shown in Figure 14.2a, demonstrated at the Building Research Establishment (BRE) site at Watford, United Kingdom, in which a frame is attached to the wall leaving a small gap as shown in Figure 14.2b. The gap enables recycled water to be piped up through the structure and is also a means of insulating the building behind the wall. The plants making up the wall at BRE included

Figure 14.2 (a) Green wall at the Building Research Establishment, Watford, UK. (b) Gap between wall and frame. (c) Lettuce and clover planted in the wall. (d) Recycled water used to water the green wall.

grass and clover over the upper two thirds and then cabbages and lettuces at pedestrian level (Figure 14.2c). Roof runoff water and other sources of greywater can be used to water green walls, as is shown in Figure 14.2d.

14.2.2 Vegetated Porous Paving

Porous paving incorporating vegetation is being used increasingly for car parks that are fairly lightly trafficked. Figure 14.3 shows one of the largest installations of this type in Western Europe. Situated in Gijon, Northern Spain, comprises >70 spaces to service a busy sports and leisure centre.

Figure 14.3 Vegetated porous paving in Gijon, Northern Spain.

In order to provide protection for the vegetation used, and load support, the paving is structured using various types of interlocking blocks. These can be made of plastic or concrete and examples are shown in Figure 14.4. Depending on the supporting structure used, the vegetated surface can either be underlain by aggregate or the interlocking blocks can be placed directly onto the soil. The type used is obviously dependant on local conditions as well as the type of running surface required.

The types of vegetation used in this sort of porous paving is dependant on local conditions, although seed mixes are recommended (Woods-Ballard *et al.*, 2007) and available through suppliers. Generally, these mixes will germinate, but will eventually be out-competed by indigenous species. Figure 14.4c illustrates this whereby virtually each small square is occupied by a different species; this shows the biodiversity benefits of using vegetation even at this small scale (see Section 14.4).

14.3 Intercepting Rainfall and Reducing Flood Risk

Whilst Wheater and Evans (2009) state, 'the effects of urbanisation are well understood', the characteristics that are unique to urban areas lead to unique impacts during flooding events. These impacts are best illustrated using the storm hydrograph (see Charlesworth *et al.*, 2003) in which the storm peak is increased and lag time is decreased due to the smooth, straight channels associated with urban areas and the increased drainage density brought about by the storm sewer network collecting surface water flows from housing and industrial estates (World Meteorological Organization, 2008;

Figure 14.4 Vegetated porous paving: (a) and (b) Using concrete blocks (Santander, Northern Spain). (c) Using plastic blocks (Gijon, Northern Spain).

Charlesworth *et al.*, 2003; Campana and Tucci, 2001). Increased impermeable areas and removal of vegetation increase surface runoff and decrease evapotranspiration (Butler and Davies, 2004; Jones and Macdonald, 2007), all of which exacerbate the 'flashy' hydrology of urban environments. According to Mentens *et al.* (2006), simply extensively greening 10% of Brussels (Belgium) roofs would lead to a reduction in runoff of 54% per individual building, and of 2.7% regionally. Increasing green cover in general would, according to Mentens *et al.* (2006), increase the amount of water evaporated in an urban area, a subject covered in Section 14.5 in which the evaporative cooling benefits of areas of vegetation are covered, and also increase the infiltration of excess water and, hence, reduce flooding.

14.4 Enhancing Urban Biodiversity

Removal of natural and seminatural landscapes to make way for urbanisation has an obvious detrimental impact on habitats and survival of ecosystems and biodiversity (see Chapters 9 and 10). Whilst natural environments must be protected, wherever necessary, limiting urban progress by outright opposition to new constructions and/or redevelopments does not follow the balanced view of sustainability. Therefore, greening the built environment at the architectural design stage and/or retrofitting existing buildings can offer considerable biodiversity benefits and opportunities to conserve many species and, furthermore, a suitable habitat to re-attract those species already lost from towns and cities. For instance, the roof area of the 32-storey building of the headquarters for Barclays Bank in London was converted into a green roof, and, soon after completion, it was found that ~10% of the invertebrate species identified on the roof were considered nationally rare and, moreover, 2 of the 20 beetle species found were very rare and had previously only been recorded six times before in the United Kingdom (Warwick, 2007).

It is possible, in theory, that almost any habitat or planting design could be recreated on the roof of a building, but all too often, in practice, technical and financial constraints are the overriding determinants influencing the flora and fauna living on a green roof or a green wall. That said, a wide variety of plants can be grown (e.g. mosses and ferns, herbaceous perennials, annuals, grasses, bulbs, trees and shrubs), which can provide a habitat for numerous invertebrate species (e.g. spiders, beetles, bees and wasps) and nesting for a range of bird species (e.g. swallows, house martins, pied wagtails, black redstart and greenfinch) (Early et al., 2007). For instance, the city hall in Chicago, United States, is an 11-storey building that has had its roof area planted (in 2000) with 20,000 herbaceous plants from 158 different species (including shrubs, vines and trees). Similarly, the stepped terraces and walls of the 14-storey Fukuouka Prefectural International Hall (ACROS building) in Japan were planted with 38,000 plants from 76 different species (Warwick, 2007).

Insights into the value of urban ecology are covered elsewhere in this book (see Chapter 9), which highlights that the biodiversity potential of Brownfield sites is often under-estimated and that, in fact, they can sometimes host rare wildlife assemblages. Therefore, mitigation of wildlife losses from brownfield sites can also be achieved, where greenery, such as green roofs or walls, is incorporated into the development and regeneration of former brownfield landscapes. This, itself, is not without challenges and will not provide a direct replacement but, nonetheless, the greening of buildings does provide an opportunity to restore biodiversity into the built environment. Furthermore, this accords with the UK government's 'Planning and Policy Statement 1' (PPS 1: Delivering Sustainable Development), which details that when preparing development plans, 'planning authorities should seek to enhance, as well as protect biodiversity, natural habitats, the historic environment and landscape and townscape character' (see http://www.communities.gov.uk).

14.5 Limiting the Overheating of Buildings

All urban areas will obviously be affected by global climate change to some extent, but nearly two centuries ago society was already having a profound impact on climate at the city scale. The 'Urban Heat Island Effect' (UHIE) was first found in London in 1819 (Greater London Authority [GLA], 2006) and is peculiar to cities where, even in winter, urban areas can be several degrees warmer than the surrounding countryside. The UHIE has the potential to have adverse impacts on human comfort, and even has health issues (Coutts *et al.*, 2007), especially during extreme events, for instance night-time temperatures in London can be some 6–9°C higher than those in rural areas (GLA, 2006).

Distribution of this heat is related to the reasons for the existence of so-called heat islands. This is due to a combination of factors including: lack of vegetation, anthropogenic activities due to transport, heating, cooling and the thermal properties of the fabric of urban structures, which store and then release heat (Memon *et al.*, 2008). Hence, remote sensing of urban areas (Wilson *et al.*, 2003) has revealed a patchwork of discrete heat islands related to the distribution and structure of buildings and streets, as well as areas within the city that can be the same temperature as surrounding rural areas and which are usually associated with parks and green space (Yu and Hien, 2006). Warm air and associated pollutants (such as ozone) can become trapped due to the lack of convective overturn. Energy is, therefore, used to cool building interiors (air conditioning) so that people can live and work in them in comfort, but the very act of trying to reduce temperatures increases it outside, as excess energy is released from the building and into the urban environment.

According to Yu and Hien (2006), the incorporation of vegetation in urban areas is one of the main ways of mitigating the UHIE, creating what they called the 'oasis effect', whereby temperatures are reduced at the local level near the planted areas, whether for buildings surrounding a park, or with vegetation planted around the individual construction. That said, green roofs have been used extensively for this purpose and they have the added advantage that they can be retrofitted to suitable buildings without the need for extra space. Augmentation of vegetative cover in cities at the *local* level (i.e. mitigation of the UHIE) can also have *regional* benefits (i.e. climate change mitigation) (Sailor, 1998). In fact, a study modelling hypothetical cities computed that increasing vegetative cover by just 6.5% could reduce summertime temperatures by 3–5% (Sailor, 1998).

'Arguably one of the most efficient ways of passive cooling for buildings and urban spaces in hot regions' is evaporative cooling (Robitu *et al.*, 2006, p436), which can be carried out biologically in plants or porous paving systems, the latter declared by Asaeda and Ca (2000, 363) as 'the most effective method to moderate the thermal conditions of the pavement surface'. Evaporative cooling occurs from a wet surface when moisture is evaporated into the overlying air, releasing latent heat and cooling the atmosphere and the following sections briefly consider the efficacy of this process utilising vegetation to promote this effect.

The evaporative cooling benefits from vegetated devices are due to the process of evapotranspiration from the leaf surface into the overlying air, thus cooling it. Wanphen and Nogano (2009) suggest that green roofs, therefore, can reduce building surface temperatures as well as those in the surrounding atmosphere and, hence, reduce the need for air conditioning.

This has been proven not only by field trials, but also by computer modelling (e.g. Onmura *et al.*, 2001) and is a technique that has been used globally with particular reference to mitigating the UHIE. However, one of the main reasons in the United Kingdom for the lack of uptake is anxiety over maintenance, in particular that of vegetated devices, such as green roofs. Whilst not arduous, nonetheless, a certain amount of care is needed for the green roof to be kept in optimal condition.

Wanphen and Nagano (2009) tested a variety of porous and nonporous materials for use on roofs without plants, mainly for their evaporative properties and found that siliceous shale had the ability to reduce daily average surface temperature by up to 8.6°C. Whilst these materials are lightweight and of simple construction, the plants making up a green roof anchor the substrate into position and whilst Wanphen and Nagano (2009) suggest using a net to hold the unconsolidated shale in place, as water passes through during a storm, it still may not be robust enough to avoid particles being dislodged to fall into the street below. However, the simple design and structure, allied with the lack of plant maintenance, make this a positive addition to the SUDS 'menu' of available techniques (Environmental Protection Agency [EPA], 2000).

14.6   Improving Human Health and Wellbeing

It has been proven by various authors (de Vries *et al.*, 2003; Groenewegen *et al.*, 2006; Maas *et al.*, 2006) that proximity to green space in an otherwise dense urban area has a positive impact on perceptions of health and wellbeing. Specifically in times of intense heat stress, a study by Lafortezza *et al.* (2007) demonstrated a relationship between green space and wellbeing. Furthermore, they recommended it should be enshrined in UK policy that green space should be 'adapted for climate change by providing access to water and shade' (p106).

The 2003 heat wave across Europe cost 17 billion euros in damage and is thought to have caused up to 50 000 additional deaths (AMICA, 2007). Nicholls and Alexander (2007) cite intensification of heat stress due to both global climate change and amendments to the local climate (such as UHIE) in cities as being one of the prime factors negatively impacting residents' quality of life. As a result, human health is affected (EEA, 2005) and perceptions of wellbeing are similarly reduced. The Meteorological Office (2009) reports that by 2040 more than half of the summers in the United Kingdom will be warmer than 2003, and that the temperatures in 2003 would be classified as cool by 2040; health and wellbeing in such a scenario can, therefore, only decline.

In a comprehensive review of the literature by Tzoulas *et al.* (2007), a complex relationship between 'Green Infrastructure', ecosystem and human health is found. It is further suggested that additional integration of 'Green Infrastructure' in urban areas would have positive economic impacts since provision of health care could be reduced. Tzoulas *et al.* (2007) quote a study by Bird (2004) in which savings to the UK National Health Service

(NHS) could be as much as £1.8 million (2.7 million euros) per year should 20% of the population live within 2 Km of between 8 and 20 hectares of greenspace and take part in 30 minutes of activity there on five days a week. However, the concepts of both ecosystem and human health are difficult to define precisely, although their close relationship is clear; there are, therefore, opportunities for interdisciplinary study which models these relationships so the benefits can be enshrined in national policies.

14.7 Sequestering Carbon to Offset CO_2 Emissions

Since this chapter considers the use of vegetated devices to address global climate change, no account will be taken of the capabilities of urban soil to sequester carbon (see Chapter 10), but obviously the replacement of impervious surfaces with a growing medium will impact on the carbon sequestration capacity of the subsequent surface and dead vegetation will be incorporated into the growing medium. Schlesinger (1999) states that the soil carbon cycle is the least well known of all the carbon sinks. However, Pouyat et al. (2006) quote values of 1.9×10^3 million tonnes (Mt) of carbon stored in urban soils in the United States, but Pataki et al. (2006) express the difficulties of using such calculations. This is because it depends on soil type and soil forming factors such as climate and underlying lithology (Booth et al., 2008).

The study of carbon sequestration by 'Green Infrastructure' in urban environments is in its infancy, although Pataki et al. (2006) state that urban trees are probably the most studied. There is a considerable literature on the carbon sequestration abilities of constructed wetlands, but since these are relatively unlikely to figure within city boundaries, they will not be considered here. Nowak and Crane (2002) estimate carbon storage in urban trees in parks and on streets of ~700 Mt in the coterminous United States, with sequestration averaging 22.8 Mt C yr^{-1}. Pataki et al. (2006) listed seven US cities with numbers of urban trees varying from ~17 000 up to ~200 000 and sequestration rates per tree of between 33 and 126 kg yr^{-1}, leading to reductions in overall CO_2 of 80–250 kg per tree. It is difficult to assign a definite amount of carbon released from a building, since these vary a great deal due to use, construction and other factors. However, using a standard family car can release 1 tonne carbon yr^{-1}, therefore using the calculations by Whitford et al. (2001), of the carbon stored in residential areas in Merseyside by trees using Formula 1, and carbon sequestered using Formula 2:

Formula 1: carbon storage (tonnes ha^{-1}) = 1.063 × % tree cover
Formula 2: carbon sequestered (tonnes ha^{-1} yr^{-1}) = 8.275 × % tree cover

Whitford et al. (2001) found that areas of dense tree cover were capable of storing >16 tonnes C ha^{-1}, and sequestering up to 0.13 tonnes C ha^{-1} yr^{-1} and, thus, one hectare of urban tree cover or ~160 trees at 100 kg C stored per tree, could account for the emissions of 16 family cars per year. Crucially, areas storing most carbon were the most affluent due to their greater coverage of green space leading to these areas performing better ecologically in comparison to less affluent areas. Some cities have implemented tree planting schemes amounting to urban forests (Chicago, United States), which have subsequently been found useful in terms of carbon sequestration. Brack (2002) outlines the pollution mitigation and carbon sequestration value of an urban forest (Canberra, Australia) and also lists many other benefits of such a scheme, including shading, visual amenity and control of urban glare and reflection. The study predicted the amount of carbon that would be sequestered between 2008 and 2012 by the 452.2×10^3 trees would be 30.2×10^3 tonnes. This could translate into a financial value of the whole forest of over \$US 20 million due to reduction in energy consumption and atmospheric pollution amelioration.

There are very few studies of green roofs estimating their carbon sequestration capabilities, but the City of Los Angeles Environmental Affairs Department (2006) quotes an area of prairie grass that sequestered 700 tonnes of carbon in 2000; however, the numerical area was not given. Getter and Rowe (2009) report a study in which they assess the carbon sequestration ability of extensive green roofs over 2 years of monitoring. They admit that green roofs have often been studied from their energy saving and heat island mitigation abilities, but rarely in their climate change mitigation role. They detail the 'terrestrial carbon sequestration' pathway via vegetation from photosynthesis taking up CO_2 to transfer of the carbon eventually into the substrate due to incorporation of plant litter. They, therefore, sampled aboveground biomass, belowground biomass and substrate carbon, finding $167.9\,g\ C\ m^{-2}$, $106.7\,g\ C\ m^{-2}$ and $912.8\,g\ C\ m^{-2}$, respectively, and the entire roof system sequestered $375\,g\ C\ m^{-2}$. Taking the latter figure as an average, they calculate that if the city of Detroit, United States, greened its ~15 000 hectares of rooftop, then potentially 55 252 tonnes of carbon could be sequestered.

14.8 'Green Infrastructure' Solutions for Climate Change Challenges

City structure is complex; it is therefore difficult to prescribe greening approaches that are suitable across the city as a whole. Treating the built environment as a bull's eye (e.g. city centre, suburbs and urban periphery; see Charlesworth, 2010), the following solutions are identified.

14.8.1 City Centre

The terminology used here is less than satisfactory, but is used in the absence of more suitable nomenclature. Hence, 'city centre' describes an area with the highest density of buildings, the least amount of vegetative cover and is, therefore, the most impermeable. It can include the 'central business district' (CBD), retail areas and even residential areas where gardens may have been sealed and the houses built in close proximity to one another, perhaps as

Figure 14.5 A street planter incorporated into the urban landscape, Berlin, Germany.

terraces or high-rise blocks. Investigations have shown (e.g. Ca *et al.*, 1998; Yu and Hien, 2006) that temperatures in urban parks can dip almost to those of the surrounding countryside. It is vital, therefore, existing green areas are preserved and that maximum use is made of any city's green space including gardens, roadside grass verges and so-called pocket parks.

Green roofs have the advantage that they require no extra space, in fact intensive green roofs have aesthetic and amenity appeal, can release extra space, can be easily retrofitted to existing buildings if they are structurally sound and are, therefore, eminently suitable for use in a city centre. Most cities that have used green roofs have done so in order to successfully mitigate the UHIE (see Charlesworth, 2010).

Street trees can also be retrofitted (Antonelli, 2008) and provide shade for buildings, both from the sun and also from the prevailing winds, their benefits in terms of carbon sequestration are outlined in section 14.7. Rain gardens or street planters (Figure 14.5) can also be retrofitted at street level; made of stone, they integrate well into the built environment (DTI, 2006). Whilst providing storm peak attenuation, they also provide visual amenity and, with street trees, will roughen the profile of the street surface, cutting down on wind canyoning and, hence, increasing physical and thermal comfort

Figure 14.6 Detention pond incorporated into a road traffic island, Dunfermline Eastern Expansion (DEX), Scotland (Urban Water Technology Centre, University of Abertay Dundee).

for the residents (Mochida and Lun, 2008), whilst also encouraging turbulent wind flow and, hence, dispersal of pollutants (Buccolieri *et al.*, 2009).

14.8.2 The Suburbs

Suburbs are less dense than the city centre, allowing for greater flexibility as to the types and coverage of vegetation that can be used. Obviously, green roofs and walls can also be used in this area, as can vegetated porous paving systems, but larger devices such as swales and ponds can also be integrated into roadways utilising roundabouts or road traffic islands, such as the Dunfermline Eastern Expansion Roundabout Detention Basin (Figure 14.6) to incorporate ponds and wetlands without having to use any extra space.

There are likely to be more gardens associated with individual households in the suburbs. Unfortunately, it has become common for these to become 'sealed' or covered in impermeable material in order to provide extra parking for family vehicles (Perry and Nawaz, 2008). This activity has led to the loss of up to two thirds of London's front gardens (London Assembly, 2005), which could amount to a total area of $12\,mi^2$, or $32\,km^2$ of vegetation removed, potential habitat lost, as well as the loss of permeable surface which would have infiltrated excess rainfall in the event of storms. The Royal Horticultural Society (RHS; 2005) quote a value of 10 litres of rainfall per minute as the capacity of an average suburban garden, or 10% of the incident rainfall absorbed. Cumulatively across a city, this could represent thousands of litres of water that would not subsequently contribute to flooding.

14.8.3 Urban Periphery

Less dense than the suburbs, the urban periphery offers the greatest opportunities for incorporating vegetation at a large scale. New housing estates, industrial estates, out-of-town shopping areas and distribution centres are often located here and incorporation of devices such as large vegetated porous paving areas can be planned and designed into any new build, possibly even using some of the 'sacrificial areas' mentioned by Water UK (2008). Here the opportunity should be taken to undertake the 'smart landscaping' and 'smart design' suggested by Antonelli (2008) to make the most use of the 'Ecosystem Services' provided by green areas.

A study in Greater Manchester (UK) by Gill *et al.* (2007) calculated that if towns and cities increased their green cover by just 10%, surface temperatures would remain the same in spite of climate change. There are many strands to the reasons for climate change and, therefore, approaches to its mitigation. There will, therefore, not be a single technique that can be used to solve the problem as a whole, but rather a suite of approaches, tailor-made depending on the situation (e.g. geographical location, local climate and city structure, amongst others), which can be applied (Yamamoto, 2006).

14.9 Conclusions

It would seem that 'Green Infrastructure' (e.g. green roof and walls or pavement planting) has a significant role to play in any strategy implemented to adapt to (and mitigate) climate change. Incorporating vegetation into new builds is relatively straightforward by planning them in at the design stage; however, it is the residential, commercial and industrial estates already built that present the most difficulties in terms of retrofitting. However, rather than be dismissive, there is a need to identify those buildings and particular areas of towns and cities that lend themselves to these devices.

Vegetation provides 'ecosystem services' that can provide the means to regulate climate, intercept stormwater and sequester or capture carbon, leading to economic impacts of increased house prices and lowered energy costs (Tratalos *et al.*, 2007). Greening and cooling the urban environment can reverse the negative impacts on human health due to global climate change (Maas *et al.*, 2006). Therefore, the value of vegetated devices is not confined to a single aspect, such as aesthetics, but instead there are multiple benefits in their utilisation.

References

Akbari, H., Rose, L.S. & Taha, H. (1999) *Characterizing the Fabric of the Urban Environment: a Case Study of Sacramento, California*. LBNL-44688. Lawrence Berkeley National Laboratory, Berkeley, CA.

Akbari, H., Rose, L.S. & Taha, H. (2003) Analyzing the land cover of an urban environment using high-resolution orthophotos. *Landscape and Urban Planning*, 63, 1–14.

AMICA (2007) *Adaptation and Mitigation: An Integrated Climate Policy Approach*. Climate Alliance, Frankfurt am Main.

Antonelli, L. (2008) Alive and well: bringing nature back into building design. *Construct Ireland*, **4**, 4–6.

Asaeda, T. & Ca, V.T. (2000) Characteristics of permeable pavement during hot summer weather and impact on the thermal environment. *Building and Environment*, **35**, 363–375.

Bird, W. (2004) *Natural fit: can green space and biodiversity increase levels of physical activity?* Royal Society for the Protection of Birds (RSPB), Sandy, UK.

Booth, C.A., Fullen, M.A., Jankauskas, B., Jankauskiene, G. & Slepetiene, A. (2008) Field case studies of soil organic matter sequestration in Lithuania and the U.K. *International Journal of Design, Nature and Ecodynamics*, **3**, 203–216.

Brack, C.L. (2002) Pollution mitigation and carbon sequestration by an urban forest. *Environmental Pollution*, **116**, S195–S200.

Buccolieri, R., Gromke, C., Di Sabatino, S. & Ruck, B. (2009) Aerodynamic effects of trees on pollutant concentration in street canyons. *Science of the Total Environment*, **407**, 5247–5256.

Butler, D. & Davies, J.W. (2004) *Urban Drainage*. E & FN Spon, London.

Ca, V.T., Asaeda, T. & Abu, E.M. (1998) Reductions in air-conditioning energy caused by a nearby park. *Energy and Buildings*, **29**, 83–92.

Campana, N.A. & Tucci, C.E.M. (2001) Predicting floods from urban development scenarios: case study of the Dilúvio Basin, Porto Alegre, Brazil. *Urban Water*, **2**, 113–124.

Charlesworth, S.M. (2010.) A review of the adaptation and mitigation of Global Climate Change using Sustainable Drainage in cities. *Journal of Water and Climate Change*, **1**(3), 165–180.

Charlesworth, S.M., Harker, E. & Rickard, S. (2003) A review of sustainable drainage systems (SUDS): a soft option for hard drainage questions? *Geography*, **88**, 99–107.

City of Los Angeles Environmental Affairs Department (2006) *Green Roofs – Cooling Los Angeles*. City of Los Angeles Environmental Affairs Department, Los Angeles.

Coutts, A.M., Beringer, J. & Tapper, N.J. (2007) Characteristics influencing the variability of urban CO_2 fluxes in Melbourne, Australia. *Atmospheric Environment*, **41**, 51–62.

de Vries, S., Verheij, R.A., Groenewegen, P.P. & Spreeuwenberg, P. (2003) Natural environments – healthy environments? An exploratory analysis of the relationship between greenspace and health. *Environment and Planning A*, **35**, 1717–1731.

DTI (2006) Sustainable drainage systems: a mission to the USA. Global Watch Mission Report. Available from http://www.britishwater.co.uk/Document/Download.aspx?uid=b0e0f905-9ffc-4a99-a057-38e7feff93a1.

Early, P., Gedge, D., Newton, J. & Wilson, S. (2007) *Building Greener: Guidance on the use of Green Roofs, Green Walls and Complementary Features on Buildings (C644)*. Construction Industry Research and Information Association (CIRIA), London.

Environmental Protection Agency (EPA) (2000) *National Menu of Stormwater Best Management Practices*. Available from http://www.epa.gov/npdes/stormwater/menuofbmps.

European Environment Agency (EEA) (2005) *Environment and Health*. European Environment Agency Report No 10/2005. European Environment Agency, Copenhagen.

Getter, K.L. & Rowe, D.B. (2009) Carbon sequestration potential of extensive green roofs. In: *Greening Rooftops for Sustainable Communities Conference*, Atlanta,

GA, USA, Session 3.1: Unravelling the Energy/Water/Carbon Sequestration Equation.

Gill, S.E., Handley, J.F., Ennos, A.R. & Pauleit, S. (2007) Adapting cities for climate change: the role of the green infrastructure. *Built Environment*, **33**, 115–133.

Greater London Authority (GLA) (2006) *London's Urban Heat Island: A Summary for Decision Makers*. Greater London Authority, London.

Groenewegen, P.P., van den Berg, A.E., de Vries, S. & Verheij, R.A. (2006) Vitamin G: effects of green space on health, well-being and social safety. *BioMed Central: Public Health*, **6–149**, 1–9.

Ip, K., M. Lam & A. Miller (2010) Shading performance of a vertical deciduous climbing plant canopy. *Building and Environment*, **45**, 81–88.

Jones, P. & Macdonald, N. (2007) Making space for unruly water: Sustainable drainage systems and the disciplining of surface runoff. *Geoforum*, **38**, 534–544.

Lafortezza, R., Carrus, G., Sanesi, G. & Davies, C. (2009) Benefits and well-being perceived by people visiting green spaces in periods of heat stress. *Urban Forestry and Urban Greening*, **8**, 97–108.

London Assembly (2005) *Crazy Paving: The Environmental Importance of London's Front Gardens*. Greater London Authority, London.

Maas, J., Verheij, R.A., Groenewegen, P.P., de Vries, S. & Spreeuwenberg, P. (2006) Green space, urbanity and health: how strong is the relation? *Journal of Epidemiology and Community Health*, **60**, 587–592.

Memon, R.A., Leung, D.Y.C. & Chunho, L. (2008) A review on the generation, determination and mitigation of Urban Heat Island. *Journal of Environmental Sciences*, **20**, 120–128.

Mentens, J., Raes, D. & Hermy, M. (2006) Green roofs as a tool for solving the rainwater runoff problem in the urbanised 21st century? *Landscape and Urban Planning*, **77**, 217–226.

Meteorological Office (2009) Floods 2008. Available from http://www.metoffice.gov.uk/corporate/verification/case_studies_08.html.

Mochida, A., & Lun, I.Y.F. (2008) Prediction of wind environment and thermal comfort at pedestrian level in urban area. *Journal of Wind Engineering and Industrial Aerodynamics*, **96**, 1498–1527.

Nicholls, N. & Alexander, L. (2007) Has the climate become more variable or extreme? Progress 1992–2006. *Progress in Physical Geography*, **31**, 77–87.

Nowak, D.J. & Crane, D.E. (2002) Carbon storage and sequestration by urban trees in the USA. *Environmental Pollution*, **116**, 381–389.

Oberndorfer, E., Lundholm, J., Bass, B., Coffman, R.R., Doshi, H., Bunnett, N., Gaffin, S., Kohler, M., Liu, K.K.Y. & Rowe, B. (2007) Green roofs as urban ecosystems: ecological structures, functions and services. *BioScience*, **57**, 823–833.

Onmura, S., Matsumoto, M. & Hokoi, S. (2001) Study on evaporative cooling effect of roof lawn gardens. *Energy and Buildings*, **33**, 653–666.

Pataki, D.E., Alig, R.J., Fung, A.S., Golubiewski, N.E., Kennedy, C.A., McPherson, E.G., Nowak, D.J., Pouyat, R.V. & Romero-Lamkaoss, P. (2006) Urban ecosystems and the North American carbon cycle. *Global Change Biology*, **12**, 1–11.

Perry, T. & Nawaz, R. (2008) An investigation into the extent and impacts of hard surfacing of domestic gardens in an area of Leeds, United Kingdom. *Landscape and Urban Planning*, **86**, 1–13.

Pouyat, P.V., Yesilonis, I. & Nowak, D.J. (2006) Carbon storage by urban soils in the USA. *Journal of Environmental Quality*, **35**, 1566–1575.

Robitu, M., Musy, M., Inard, C. & Groleau, D, (2006) Modeling the influence of vegetation and water pond on urban microclimate. *Solar Energy*, **80**, 435–447.

Rose, L.S., Akbari, H. & Taha, H. (1999) *Characterizing the fabric of an urban environment: a case study of Houston, Texas.* LBNL-51448. Lawrence Berkeley National Laboratory, Berkeley, CA.

Rosenzweig, C., Gaffin, S. & Parshall, L. (2006) *Green Roofs in the New York Metropolitan Region. Executive Summary.* Columbia University Centre for Climate Systems Research. NASA Goddard Institute for Space Studies, Greenbelt, MD.

Royal Horticultural Society (2005) Gardening matters: Front gardens. Are we parking on our gardens? Do driveways matter? Royal Horticultural Society, London.

Sailor, D.J. (1998) Simulations of annual degree-day impacts of urban vegetative augmentation. *Atmospheric Environment*, **21**(1), 43–52.

Schlesinger, W.H. (1999) Carbon sequestration in soils. *Science*, **284**, 725–730.

Tratalos, J., Fuller, R.A., Warren, P.H., Davies, R.G. & Gaston, K.J. (2007) Urban form, biodiversity potential and ecosystem services. *Landscape and Urban Planning*, **83**, 308–317.

Tzoulas, K., Korpela, K., Venn, S., Yli-Pelkonen, V., Kazmierczak, A., Niemela, J. & James, P. (2007) Promoting ecosystem and human health in urban areas using green infrastructure: a literature review. *Landscape and Urban Planning*, **81**, 167–178.

Wanphen, S. & Nagano, K. (2009) Experimental study of the performance of porous materials to moderate the roof surface temperature by its evaporative cooling effect. *Building and Environment*, **44**, 338–351.

Warwick, H. (2007) The garden up above. *Geographical Magazine*, **7**, 38–42.

Water UK (2008) *Lessons learnt from Summer Floods 2007: Phase 2 report – Long-term Issues.* Water UK's Review Group on Flooding, London.

Wheater, H. & Evans, E. (2009) Land use, water management and future flood risk. *Land Use Policy*, **26S**, S251–S264.

Whitford, V., Ennos, A.R. & Handley, J.F. (2001) 'City form and natural process' – indicators for the ecological performance of urban areas and their application to Merseyside, UK. *Landscape and Urban Planning*, **57**, 91–103.

Wilby, R.L. (2007) A review of climate change impacts on the built environment. *Built Environment*, **33**, 31–45.

Wilson, J.S., Clay, M., Martin, E., Stuckey, D. & Vedder-Risch, K. (2003) Evaluating environmental influences of zoning in urban ecosystems with remote sensing. *Remote Sensing of Environment*, **85**, 303–321.

Woods-Ballard, B., Kellagher, R., Martin, P., Jefferies, C., Bray, R. & Shaffer, P. (2007) *The SUDS Manual (C697).* Construction Industry Research and Information Association (CIRIA), London.

World Meteorological Organization (2008) Urban flood risk management: a tool for integrated flood management. Associated Programme on Flood Management. Available from: http://www.apfm.info/pdf/ifm_tools/Tools_Urban_Flood_Risk_Management.pdf.

Yamamoto, Y. (2006) Measures to mitigate urban heat islands. *Quarterly Review, Science and Technology Trends*, **18**, 65–83.

Yu, C. & Hien, W.N. (2006) Thermal benefits of city parks. *Energy and Buildings*, **38**, 105–120.

15

Particulate-Induced Soiling on Historic Limestone Buildings: Insights and the Effects of Climate Change

David E. Searle

15.1 Introduction

'Soiling', in the context of buildings and monuments, generally refers to the wet and dry deposition of atmospheric particulate material, both natural and anthropogenic, which results in a loss of reflectance or a 'darkening' of the stone surface (Hamilton et al., 1994). It could be argued that such darkening of buildings represents one of the earliest examples of human activities affecting the wider environment. In ancient Rome it was identified that poets complained about damage to temples caused by smoke (Brimblecombe, 1992). Such observations grew in number as history progressed, particularly in the great cities of the world where industry, powered by coal, dominated. Under Royal command from Charles II, John Evelyn produced an essay in 1661 entitled *Fumifugium, or The Inconvenience of the Aer and the Smoak of London Dissipated* in which he spoke of 'horrid smoke which obscures our churches and makes our palaces look old' (Mansfield et al., 1991) written in response to smoke obscuring his view of the palace of Charles II at Whitehall (Brimblecombe, 1992). This very early work on air pollution identified many of the effects and remedies for air pollution control that are still relevant today (Brimblecombe, 1987; Boubel et al., 1994; Stern, 1968; Lynn, 1976). In addition it also formed one of the starting points for the observation and recording of long-term climate change.

Solutions to Climate Change Challenges in the Built Environment, First Edition.
Edited by Colin A. Booth, Felix N. Hammond, Jessica E. Lamond and David G. Proverbs.
© 2012 Blackwell Publishing Ltd. Published 2012 by Blackwell Publishing Ltd.

15.2 Urban Particulate Pollution

Combustion processes have been, and still are, the most significant source of urban air pollution in both gaseous and particulate form. The nature and type of particulate pollution has varied over history but is strongly related to fuel source changes. In medieval Britain around the latter half of the thirteenth century, coal had just started to replace wood as the prime energy source; it initially had minimal effect upon the atmosphere of nucleated settlements, since many of the industries present at that time, such as the iron industry, were scattered and located in rural areas. This changed over time with widespread domestic use by the sixteenth century, culminating in coal becoming the most prevalent fuel used in many major European cities by the end of the nineteenth century (Grossi and Brimblecombe, 2007). The steady reduction in coal usage, observed in the first part of the twentieth century, due in part to control through legislation, continued in the latter half and contributed to a period in which particulate matter from vehicle derived sources has come to be the dominant source particularly in urban areas (Mansfield et al., 1991; Quality of Urban Air Review Group [QUARG], 1996; Grossi and Brimblecombe, 2007).

Particulate pollution originates from a number of sources and is a complex combination of substances that can be broadly classified into three main groups: (1) carbonaceous material, (2) hygroscopic inorganic salts and (3) insoluble minerals and crustal material (Watt and Hamilton, 2003). However, in the context of soiling on buildings at the current time, it is primarily the presence of particulate elemental carbon (PEC or EC) within particulate pollution, which has been responsible for the darkening of building surfaces over the last 20–25 years (Horvath, 1993). In this same period of time it was identified that emissions from diesel engines were responsible for the majority of PEC in urban atmospheres (Mansfield et al., 1991; Miguel et al., 1998). Diesel particulate can be described as combustion generated carbon soot, with condensed and/or adsorbed hydrocarbons and some inorganic species (Burtscher et al., 1998). They consist of spherical particles with an approximate diameter of 10–20 nm, which agglomerate to form larger structures. On a mass basis, diesel has a soiling factor three times greater when compared to particulate resulting from coal combustion (Hamilton and Mansfield, 1991), so it was considered that regardless of the reductions in coal particulates, historically responsible for building soiling, that significant potential still remains for such soiling to continue.

15.3 Soiling of Buildings

Effects of soiling on stone building surfaces can be considered in two ways. Firstly, as a degradation of the visual aesthetic of the building and how that aesthetic is perceived and secondly as a factor in the accelerated weathering of buildings stone, primarily through the formation of black crusts or scabs. On carbonate stone these crusts consist primarily of a matrix of gypsum ($CaSO_4 \cdot 2H_2O$) and calcite crystals, in which particulates, soil dust and pollen are imbedded (Schiavon and Zhou, 1996; Camuffo et al., 1983). There has been extensive study of the formation and effect of black crusts formed during times where coal pollution dominated urban centres (Sabbioni et al., 1996; Warke et al., 1996; Zappia et al., 1993; Bonazza et al., 2005). However the corresponding role of traffic derived particulates is less understood and investigated on only a comparatively small number of research studies (Rodriguez-Navarro and Sebastian, 1996; Ausset et al., 1996; Johnson et al., 1996; Viles et al., 2002; Searle and Mitchell, 2006). It could be argued that while traffic derived particulate pollution (primarily diesel based) has been identified as a potential factor in the accelerated weathering of carbonate stone, it is inconclusive as to whether it will be as significant as coal pollution has been over the last 200 years. Arguably of more widespread concern, particularly for the purposes of urban planning, is the visual degradation as a result of soiling on building surfaces and the rate at which this is currently occurring or may occur in the future.

The rate at which a stone surface soils is a combination of a number of complex factors. These include physical parameters of the particulate and the stone surface, concentration of the particulate in the urban atmosphere, orientation of building surface and the prevailing meteorological conditions (Grossi et al., 2003). These in turn control the deposition rate and the removal of previously deposited material through rain washing (Creighton et al., 1990). Soiling rate studies usually involve the exposure of small samples of materials to different atmospheric conditions, measuring the loss in reflectance over time and developing models relating soiling rate to time and particulate concentration (Berloin and Haynie, 1975; Pio et al., 1998, Grossi and Brimblecombe, 2007, 2008).

The study by Pio et al. (1998), arguably one of the most extensive on stone, exposed Portland Limestone tablets, in both sheltered and unsheltered conditions, for 2.4 years (one of the longest exposure periods for a soiling study in the literature) in Oporto, Portugal. Using a square root soiling model, originally proposed by Berloin and Haynie (1975), they calculated a 30% decrease in reflectance for sheltered stone samples would occur within 5.5–8.8 years. The same model described well, the most sheltered samples areas of an eight year study on the resoiling of the façade of the Victoria and Albert Museum, following cleaning in 1989 (Pretzel, 2004). The efficacy of the model was reduced when used with data from the more exposed sample areas, which reflects the variability of soiling when the stone surface is subject to direct rainfall or rain-wash. However, the scaling up of results of laboratory studies and small sample sizes to full size buildings can be problematic (Inkpen, 2003) and, arguably, studies looking at whole-building effects form a relatively diminished research area. This is reflected in the area of soiling patterns on complete buildings where very few studies are present in the literature (McLaughlin, 1992; Davidson et al, 2000; Ball and Young, 2004).

In an attempt to further explore the effect of contemporary traffic levels and composition on building soiling, this chapter presents the results of a number of surveys on building façades undertaken over a 17-year period in the city of Bath, Somerset, United Kingdom.

15.4 The Bath Study

Bath is a city of considerable architectural importance, containing 636 Grade 1, 61 Grade II* and 4283 Grade II listed buildings (McLaughlin, 1996). Some are structures remaining from the first-century Roman town of Aquae Sulis. However, the majority are Georgian terrace houses from the eighteenth century. Like many urban areas from the Georgian era, the City of Bath has suffered from extensive burning of high sulphur coals during the eighteenth and nineteenth centuries, causing severe soiling of stone surfaces, turning the 'Bath stone', an oolitic limestone from the Great Oolite of the Middle Jurassic (Leary, 1983), from a light buff colouration to black.

Over the past 25 years, Bath City Council has undertaken a number of qualitative condition surveys of listed structures. Although these were primarily external visual condition surveys, soiling levels were evaluated and recorded using the criteria given in Table 15.1 (McLaughlin, 1992, 1996).

In 1996, an appeal was made against a Council decision to refuse planning permission for the erection of a $2323\,m^2$ food retail store by the A4 London Road, Bath (McLaughlin, 1996). The Council put forward the argument that increased traffic levels along the A4 London Road (one of the main routes into and through the City of Bath) and hence increased pollution, would contribute significantly to the degradation of the façades of listed structures adjacent to the London Road. The 1996 survey was undertaken to provide supporting evidence.

Since the same basic criteria were used in both the 1989–1993 and 1996 surveys, it presented the opportunity to repeat the surveys of 355 facades and structures along the London Road, using the same criteria. This was undertaken in 2000 and again in 2009 which provides a visual perspective and evidence on soiling, primarily resulting from traffic, over a 17-year period along the A4 London Road (Figure 15.1).

The two recent surveys were undertaken when the building surfaces were ostensibly dry. Data collection consisted of marking the results of observations onto a large-scale map of the London Road. Façades were

Table 15.1 Previous soiling surveys on listed buildings in Bath, United Kingdom (McLaughlin, 1992, 1996).

Survey year	Title	Number of buildings	Soiling classification
1975	1975: Listed Building Condition Survey	4658	4658 listed buildings, soiling levels classified as 'black' or 'clean'
1989–1993	1989–1993: Buildings at Risk Survey,	4771	3 stages, 4771 listed buildings, soiling levels classified as 'clean', 'resoiling', 'grimy' or 'black'
1996	1996: Soiling Survey of London Road	355	355 listed buildings in London Road Area, soiling levels as for 1989–1993 with an additional 'badly resoiled' category

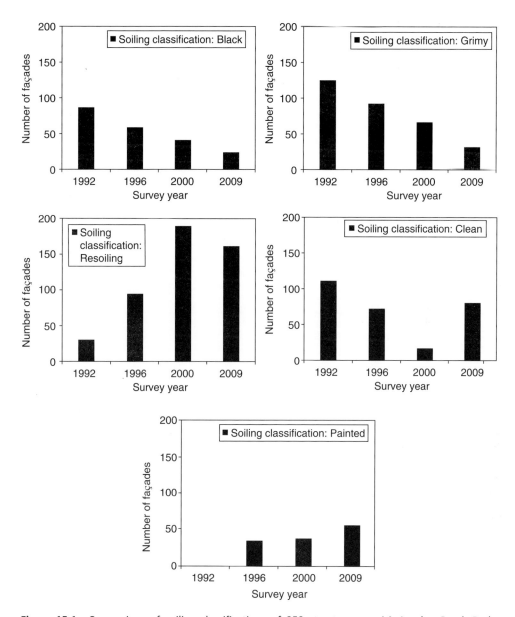

Figure 15.1 Comparison of soiling classifications of 352 structures on A4, London Road, Bath (1992–2009).

examined by two operators and then classified to one of the following criteria as used in the previous two surveys (Table 15.2). To facilitate consistency between surveys reference was also made to photo examples for each of the soiling classifications. Table 15.3 provides the results of the surveys carried out between 1992 and 2009 using the criteria given in Table 15.2.

There appears to be a change in the patterns of soiling observed between the two decades. From 1992 to 2000, buildings appear to be steadily resoiling as indicated by the 25% decrease in the number of 'clean buildings

Table 15.2 Soiling classifications used in surveys.

Classification	Description
Clean	Where the stonework had been cleaned and there was little or no evidence of discolouration due to resoiling around architectural features, and no evidence of resoiling on flat surfaces of stonework.
Resoiling	Where the stonework has been cleaned and there is some evidence of discolouration due to resoiling around architectural details and/or discolouration of the flat surfaces of stonework.
Grimy	Where there was no evidence of the original stonework of the building having ever been cleaned, but where the soiling was confined to specific, relatively small, areas such as below windowsills, band courses, cornices and similar details.
Black	Where there was no evidence of the original stonework of the building having ever been cleaned, but where the soiling extended over a substantial area in the form of the sooty deposits.
Painted	On building where the stonework of the ground floor and/or basement walls was painted, only the soiling or resoiling condition of the unpainted stonework of the upper floors was recorded. If the façade was totally painted, the building was recorded as being painted.

Table 15.3 Soiling classification of 352 structures on London Road, Bath, from 1992 to 2009.

	Classification									
	Black		Grimy		Resoiling		Clean		Painted	
Survey Year	Number	%	Number	%	Number	%	Number	%	Number	%
1992	86	24.4	127	36.1	28	8.0	111	31.5	n/a	n/a
1996	59	16.8	94	26.7	93	26.4	71	20.2	35	9.9
2000	41	11.7	68	19.3	187	53.1	17	4.8	39	11.1
2009	23	6.5	33	9.4	160	45.5	80	22.7	56	15.9

Note: The category 'Painted' was not used in the 1992 survey.

and corresponding rise in buildings 'resoiling'. Reductions in 'black' or 'grimy' buildings over this period is almost certainly due to cleaning and conservation work that has been undertaken (some which is grant aided), effectively removing 'historic' soiling from building facades. In the second half of the study period, this trend does not continue. Further examination of the data showed that of the 17 buildings classified as 'clean' in 2000, 15 were still classified as 'clean' in 2009. Again, there is evidence of considerable building improvement works during this period with decreases in 'black' and 'grimy' and increases in 'clean' and 'painted'.

15.5 Insights from the Bath Study

Monitoring of the levels of particulate matter with a diameter $\leq 10\,\mu m$ (PM_{10}) along the London Road commenced in 2009 (Spalding, 2010) and therefore provides minimal supporting explanation of the results obtained. However, records of traffic data identified a drop in traffic flows from 30,000 vehicles per day in 1997 to 27 000 in 2001 (Bath & North East Somerset Council, 2002) and then a further drop to 21,000 in 2002, which remained fairly constant over the next four years (Bath & North East Somerset Council, 2006). It could be argued that this reduction in traffic flows tentatively mirrors the overall patterns of soiling observed in the study.

In a national context, the overall trend for PM_{10} emissions has been decreasing with a 47% drop between 1970 to 2000 (AQEG, 2005). It has been identified as remaining static during the first decade of the twenty-first century (Harrison et al., 2008). Diesel particulate, containing a high percentage of elemental carbon or black smoke (the primary darkening agent in relation to building soiling), is a major proportion of this total. Although this would seem to present a case supporting cleaner buildings in the future, the following points should be taken into consideration:

- National PM_{10} emissions represent mean concentrations across a national network of monitoring stations, where deposition of particulates onto buildings is arguably dominated by point sources in their immediate vicinity and the effect of the local environment upon particulate deposition (e.g. street canyons). Roadside measurements of PM_{10} are often double that of background levels (AQEG 2005) with corresponding higher levels of elemental carbon, the primary darkening agent on building surfaces (Harrison and Yin, 2008).
- The strong downward trend observed in urban PM_{10} in the 1990's appears to be levelling-off over the period 2000–2010. This could indicate that predicted future reductions in urban PM_{10} are not robust.
- Uncertainty in the effectiveness of current and proposed European vehicle emission standards (e.g. Euro 5 and 6 for cars and light-delivery vehicles and Euro V and VI for heavy-delivery vehicles) to control emissions arising from the projected traffic growth of 67% for freight vehicles and 40% for cars by 2030 from 2005 levels (Kousoulidou et al., 2008).
- A significant proportion of traffic derived urban PM_{10} emissions do not come from combustion alone. This includes brake, clutch and tyre wear emissions, which would not be affected by any current or proposed emission controls but yet will grow with increases in vehicle numbers (Harrison et al., 2008).

While it would appear that over the study period the soiling of buildings displayed an increasing trend followed by a decreasing one, the results should be considered as indicative only, as the method used is qualitative and insensitive to small changes in soiling levels and is heavily operator dependent. However, considering the large sample size and duration of the surveys on the London Road (the largest and longest to date in the United Kingdom), the method could be argued to be sensitive enough to correctly identify the mass changes observed in soiling levels. To further explore this area, an improved method based on digital image analysis has been developed to quantify soiling coverage and intensity (Searle and Mitchell, 2008) and is currently being used on a long-term study of soiling levels on listed buildings in Bath.

15.6 Effects of Climate Change on the Soiling of Buildings

Latest projections for climate change in the United Kingdom under the UKCP09 medium emissions scenario show central estimated increases of 9–17% in winter precipitation and decreases of 10–22% in summer precipitation (UKCP09, 2009). In the context of falling PM_{10} levels, changes in rainfall, combined with increased temperatures and the potential for more wind driven rain, means there could be notable differences in the manner to which buildings respond to particulate deposition. This may manifest itself in two ways: (1) a reduction or even reversal of overall soiling levels (Grossi and Brimblecombe, 2007) and (2) a change in the soiling patterns produced and the depth of contrast of these patterns between heavily rain-washed and 'shaded' areas. Furthermore, soiling changes to the surface of buildings have also been postulated to occur though increased growth of micro-organisms, as a result of increased periods of wetness (Viles, 2002).

It is probable that future strategies to protect built heritage from the effects of traffic-derived soiling will not change significantly and will still consist of two approaches. The first of these is the cleaning or repainting of façades (as indicated by the surveys) once it is perceived by the owners that the soiled condition has reached an unacceptable level. Although the local council does have a measure of control over such cleaning activities through the listed building consent process (Bath & North East Somerset Council, 2010), it could be argued that the effective long-term control of particulate levels, and hence soiling, will occur though future traffic management policies and procedures implemented by Bath & North East Somerset Council.

15.7 Conclusions

In order to fully inform decisions on conservation management practices, it is clear that there is still much research to be undertaken in this area to fully elucidate the mechanisms and potential impacts of contemporary particulate pollution levels on monuments and stone buildings in our urban centres. Future research in this area will also need to identify how these impacts will evolve in the context of climate change and further pollution controls.

References

AQEG (2005) *Air Quality Expert Group. Particulate matter in the United Kingdom.* Department of Environment, Food and Rural Affairs, London.

Ausset, P., Crovisier, J.L., Del Monte, M., Furlan, V., Giradet, F., Hammecker, C., Jeannette, D. & Lefevre, R.A. (1996) Experimental study of limestone and sandstone sulphation in polluted realistic conditions: the Lausanne Atmospheric Simulation Chamber (LASC). *Atmospheric Environment*, 30(18), 3197–3207.

Ball, J. & Young, M.E. (2004) Comparative assessment of decay and soiling of masonry: Methodology and analysis of surveyor variability. In: *Stone Deterioration in Polluted Urban Environments*, ed. Mitchell, D.J. & Searle, D.E. Science Publishers, Plymouth, UK, 191–202.

Bath & North East Somerset Council (2002) *Review of Road Layout and Bus Lanes on London Road, Bath.* Planning, Transportation and Environment Committee, Bath. Available from http://www.bathnes.gov.uk/Committee_Papers/PTandE/PTE020321/11ReviewLondonRd.htm.

Bath & North East Somerset Council (2006) *Local Transport Plan 2001–6 Annual Progress Report July 2005.* Planning, Transportation and Environment Committee, Bath. Available from http://www.bathnes.gov.uk/NR/rdonlyres/31EB0E40-A416-44D5-AFA7-50618F54CB6B/0/APR05report.pdf.

Bath & North East Somerset Council (2010) *The Cleaning of Bath Stone.* Bath Preservation Trust Planning, Bath. Available from http://www.bathnes.gov.uk/NR/rdonlyres/956F28ED-5E26-483E-85F6-0F9E1BAB5413/0/CleanBathStone.pdf.

Berloin, N.J. & Haynie, F.H. (1975) Soiling of building surfaces. *Journal of Air Pollution Control Association*, 25, 393–403.

Bonazza, A., Sabbioni, C. & Ghedini, N. (2005) Quantitative data on carbon fractions in interpretation of black crusts and soiling on European built heritage. *Atmospheric Environment*, 39, 2607–2618.

Boubel, R.W, Fox, D.L., Turner, D.B. & Stern, A.C. (1994) *Fundamentals of Air Pollution*, 3rd ed. Academic Press, San Diego, CA.

Brimblecombe, P. (1987) *The Big Smoke: A History of Air Pollution in London since Medieval Times.* Methuen, London.

Brimblecombe, P. (1992) A brief history of grime: accumulation and removal of soot deposits on buildings since the 17th century. In: *Stone Cleaning and the Nature, Soiling and Decay Mechanisms of Stone*, ed. Webster, G.M., Donhead Publishing, London.

Burtscher, H., Kunzel, S. & Huglin, C. (1998) Characterisation of particles in combustion engine exhaust. *Journal of Aerosol Science*, 29(4), 389–396.

Camuffo, D., Del Monte, M. & Sabbioni, C. (1983) Origin and growth mechanisms of the sulfated crusts on urban limestone. *Water, Air and Soil Pollution*, 19, 351–359.

Creighton, P.J., Lloy, P.J, Haynie, F.H., Lemmons, T.J., Miller, J.L. & Gerhart, J. (1990) Soiling by atmospheric aerosols in an urban industrial area. *Journal of Air Waste Management Association*, 40(9), 1285–1289.

Davidson, C.I., Tang, W., Finger, S., Etyemezian, V., Striegel, M.F. & Sherwood, S.I. (2000) Soiling patterns in a tall limestone building: changes over 60 years. *Environmental Science & Technology*, 34(4), 560–565.

Grossi, C.M., Esbert, R.M., Díaz-Pache, F. & Alonso, F.J. (2003) Soiling of building stones in urban environments. *Building and Environment*, 3–8, 147–159.

Grossi, C.M. & Brimblecombe, B. (2007) Effect of long-term changes in air pollution and climate on the decay and blackening of European stone buildings. In: *Building Stone Decay: From Diagnosis to Conservation*, ed. Přikryl, R. and Smith, B.J. Geological Society Special Publications, 271. Geological Society, London, 117–130.

Grossi, C.M. & Brimblecombe, P. (2008) Past and future colouring patterns of historic stone buildings. *Materiales de Construcción*, 58, 289–290.

Hamilton, R.S. & Mansfield, T.A. (1991) Airborne particulate elemental carbon: its sources, transport and contribution to dark smoke and soiling. *Atmospheric environment*, 25A(3–4), 715–723.

Hamilton, R.S., Kershaw, P.R., Segarra, F., Spears, C.J. & Watt, J.M. (1994) Detection of airborne carbonaceous particulate matter by scanning electron microscopy. *The Science of the Total Environment*, 146–147, 303–308.

Harrison, R.M. & Yin, J. (2008) Sources and processes affecting carbonaceous aerosol in central England. *Atmospheric Environment*, 42, 1413–1423.

Harrison, R.M., Stedman, J. & Derwent, D. (2008) New directions: why are PM10 concentrations in Europe not falling? *Atmospheric Environment*, **42**, 603–606.

Horvath, H. (1993) Atmospheric light absorption – a review. *Atmospheric Environment Part A General Topics*, **27**, 293–317.

Inkpen, R. (2003) The whole building and patterns of degradation. In: *The effects of air pollution on the built Environment*, ed. P. Brimblecombe. Imperial College Press, London, 393–422.

Johnson, J.B., Montgomery, M., Thompson, G.E., Wood, G.C., Sage, P.W., & Cooke, M.J. (1996) The influence of combustion-derived pollutants on limestone deterioration: 2, the wet deposition of pollutant species. *Corrosion Science*, **38**(2), 267–278.

Kousoulidou, M., Ntziachristos, L. Mellios, G. & Samaras, Z. (2008) Road-transport emission projections to 2020 in European urban environments. *Atmospheric Environment*, **42**, 7465–7475.

Leary, E. (1983) *The Building Limestones of the British Isles*. HMSO, London.

Lynn, D.A. (1976) *Air Pollution – Threat and Response*. Addison Wesley Publishing, Reading, UK.

Mansfield, T., Hamilton, R., Ellis, B. & Newby, P. (1991) Diesel particulate emissions and the implications for the soiling of buildings. *The Environmentalist*, **11**(4), 243–254.

McLaughlin, D. (1992) 'Acid rain': the cleaning and conservation of stonework in Bath. In: *Stone Cleaning and the Nature, Soiling and Decay Mechanisms of Stone*, ed. Webster, R.G.M., Donhead, London, 183–226.

McLaughlin, D. (1996) *Appeal by Safeway Stores PLC against the Decision of Bath City Council to Refuse Permission for the Erection of a 2323 Square Metres Retail Foodstore, Together with Customer Facilities, Offices, Storage, Plant Rooms and Car Parking on Land at Kensington Bus Depot, London Road, Bath – Proof of Evidence by John David McLaughlin, Acid Rain and Bath Stone*. Doe Reference APP/PO105/A/95/855444. Bath City Council, Bath.

Miguel, A.H., Kirchstetter, T.W., Harley, R.A. & Herring S.V. (1998) On-road emissions of particulate polycyclic aromatic hydrocarbons and black carbon from gasoline acid diesel vehicles. *Environmental Science and Technology*, **32**, 450–455.

Pio, C.A., Ramos, M.M. & Duarte, A.C. (1998) Atmospheric aerosol and soiling of external surfaces in an urban environment. *Atmospheric Environment*, **32**(11), 1979–1989.

Pretzel, B. (2004) Colour changes of Portland stone: a study of the Victoria and Albert Museum façade 1989–1998. In: *Stone Deterioration in Polluted Urban Environments*, ed. Mitchell, D.J. & Searle, D.E. Science Publishers, Plymouth, UK, 191–202.

Quality of Urban Air Review Group (QUARG) (1996) *Airborne Particulate Matter in the United Kingdom*. Third report of the Quality of Urban Air Review Group. Institute of Public and Environmental Health, University of Birmingham, UK.

Rodriguez-Navarro, C. & Sebastian, E. (1996) Role of particulate matter from vehicle exhaust on porous building stones (limestone) sulfation. *The Science of the Total Environment*, **187**(2), 79–91.

Sabbioni, C., Zappia, G. & Gobbi, G. (1996) Carbonaceous particles and stone damage in a laboratory exposure system. *Journal of Geophysical Research*, **101**(D14), 19621–19627.

Schiavon, N. & Zhou, L. (1996) Magnetic, chemical and microscopical characterization of urban soiling on historical monuments. *Environmental Science and Technology*, **30**(12), 3624–3629.

Searle, D.E. & Mitchell, D.J. (2006) The effect of coal and diesel particulates on the weathering loss of Portland limestone in an urban environment. *Science of the Total Environment*, **370**, 207–233.

Searle, D.E. & Mitchell, D.J. (2008) A Methodology for the semi-quantitative analysis of soiling patterns on building façades In Bath, U.K. using digital image processing. In *11th International. Congress on Deterioration and Conservation of Stone*, ed. Lukaszewicz, J. & Niemcewicz, P. Nicolaus Copernicus University Press, Torun, Poland, 497–504.

Spalding, R. (2010) Personal correspondence, email to D. Searle, 16 March.

Stern, A.C. (1968) *Air Pollution*, 2nd ed. Academic Press, New York.

UKCP09 (2009) *UK Climate Projections Science Report: Climate Change Projections*. Available from http://www.defra.gov.uk/environment/climate/documents/uk-climate-projections.pdf.

Viles, H.A. (2002) Implications of future climate change for stone deterioration. In *Natural Stone, Weathering Phenomena, Conservation Strategies and Case Studies*, ed. Siegesmund, S. Weiss, T. & Vollbrecht, A. Geological Society Special Publications, 205. Geological Society, London, 407–418.

Viles, H.A., Taylor, M.P, Yates, T.J.S. & Massey, S.W. (2002) Soiling and decay of NMEP limestone tablets. *Science of the Total Environment*, **292**(3), 215–229.

Warke, P.A., Smith, B.J. & Magee, R.W. (1996) Thermal response characteristics of stone: implications for weathering of soiled surfaces in urban environments. *Earth Surface Processes and Landforms*, **21**, 295–306.

Watt, J. & Hamilton, R. (2003) The soiling of buildings by air pollution. In *The Effects of Air Pollution on the Built Environment*, ed. P. Brimblecombe. Imperial College Press, London.

Zappia, G., Sabbioni, C. & Gobbi, G. (1993) Non-carbonate carbon content on black and white areas of damaged stone monuments. *Atmospheric Environment*, **27A**(7), 1117–1121.

Sustainable Transportation

Panagiotis Georgakis and Christopher Nwagboso

16.1 Introduction

Transportation of people and goods is an important function for social and economic prosperity, since it enables the mobility of resources for a wide range of commercial activities, together with the mobility of people for various types of trips, including business, leisure and tourism. Nevertheless, the high demand of transport services has not been met without a cost, in the form of the near saturation of the transport network in many cities around the world. This has resulted in a number of negative effects on transport networks including high levels of road congestion, increased operational costs, increasing number of accidents and excessive energy consumption with subsequent environmental pollution.

Sustainable transportation is a concept that was introduced to emphasise the need for alternative solutions to the prevailing practices of the users of the transport networks, in order to address the existing negative effects and limitations of the modern transport networks. Different organisations and scholars have specified a plethora of meanings on what sustainable transportation is. Some of these definitions are as follows:

> …satisfying current transport and mobility needs without compromising the ability of future generations to meet these needs. (Black, 1996)

> …transportation that does not endanger public health or ecosystems and meets mobility needs consistent with (a) use of renewable resources at below their rates of regeneration and (b) use of non-renewable resources at below the rates of development of renewable substitutes. (OECD, 1996)

Solutions to Climate Change Challenges in the Built Environment, First Edition.
Edited by Colin A. Booth, Felix N. Hammond, Jessica E. Lamond and David G. Proverbs.
© 2012 Blackwell Publishing Ltd. Published 2012 by Blackwell Publishing Ltd.

...a sustainable transportation system is one that:

1. allows the basic access and development needs of individuals, companies and society to be met safely and in a manner consistent with human and ecosystem health, and promotes equity within and between successive generations
2. is affordable, operates fairly and efficiently, offers a choice of transport mode, supports a competitive economy, as well as balanced regional development
3. limits emissions and waste within the planet's ability to absorb them, uses renewable resources at or below their rates of generation, and uses non-renewable resources at or below the rates of development of renewable substitutes, while minimizing the impact on the use of land and the generation of noise. (EU Council of Ministers of Transport, 2004)

It is evident from the above definitions that a number of characteristics of the existing transportation systems and services hinder their sustainability. These include, amongst others:

- High dependence on nonrenewable resources as the primary source of fuel
- Reliance on private cars, which results in highly congested networks, as the primary mode for most transportation activities
- Development of new infrastructure, with its negative impact on land availability, to address the oversaturation of the transport networks

The aim of this chapter is to enable the reader to develop an understanding for different aspects related to sustainable transportation. The following sections explain the impact of transport on climate change, present the different perspectives and dimensions of sustainable transportation and describe policy making and measures for the development of sustainable transportation systems.

16.2 Climate Change and Sustainable Transportation

One of the adverse effects that have resulted from the high-mobility demands on the existing transportation systems is the excessive production of greenhouse gases. In Europe, transport accounts for 30.7% of the total final energy consumption, with road having the major share for both passenger (83%) and freight (45.6%) transport (EUROSTAT, 2009). In terms of greenhouse gas emissions, road transport accounts for 17% of the total emissions in Europe (EEA, 2009). In the United States, road transport accounts for approximately 81% of the total greenhouse gases generated from the transport sector (EPA, 2006). Combined with the fact that the United States is the largest producer of greenhouse gases in the world, and that transportation sources account for 31% of the total greenhouse gas emissions (USEIA, 2009), it is evident that road transport is one of the major contributors to the global amount of greenhouse gas emissions and thus the climate change crisis.

However, passenger cars are not solely responsible for the ramifications that transportation has to climate change. Air transportation is another major contributor of greenhouse emissions to the environment. The effect of non-CO_2 emissions from aviation varies according to the different altitudes of operation, resulting to adverse short-term impacts on climate change (Cairns and Newson, 2006). These types of effects are non-observable with emissions coming from other modes of transport.

The main contributor to the environmental impacts of transport is the immense use of fossil fuels. The products of the incomplete and complete combustion of engine fuels include 'carbon monoxide, volatile organic compounds (VOC), various oxides of nitrogen (precursors to ozone pollution) and fine particulates', together with 'carbon dioxide and water vapour' (Greene and Wegener, 1997). Some diachronic environmental impacts of transportation are summarised in Table 16.1.

16.3 Perspectives of Sustainable Transportation

Although the environmental impacts of transportation are the ones directly related to climate change, sustainable transportation encompasses a variety of dimensions beyond that. Black (2000) has identified the following dimensions of sustainable transportation:

- The first dimension reflects the fact that petroleum (97% of the transport systems use it as a primary fuel source) is an energy source that is depleting rapidly.
- The second dimension is the environmental degradation due to the pollution that results from excessive production of greenhouse gases.
- The third dimension is related to the highly congested transport systems due to the saturation of the transport networks.
- The fourth dimension is related to safety and the effects of accidents on the society.
- The fifth dimension is the fair allocation of land for transportation services.

In parallel with the above, Basler (1998) reflected three perspectives of sustainable transportation: the 'economic perspective, where sustainability deals with 'optimal use of resources, efficient markets and improved general well-being', the ecological perspective, which 'derives rules for the use of natural resources from their limited availability', and the 'societal perspective', which encompasses aspects of equal opportunities together with social cohesiveness and equity.

Figure 16.1, presents an adaptation for transportation of a diagrammatic representation of the trade-offs between the three pillars of sustainable development. Sustainable transportation strategies should aim to have a positive impact on all three perspectives.

Table 16.1 Diachronic environmental impacts of different transport modes.

	Air	Water resources	Land resources	Solid waste	Noise
Marine and inland water transport		Modification of water systems during port construction and canal cutting and dredging	Land taken for infrastructures; dereliction of obsolete port facilities and canals	Vessels and craft withdrawn from service	
Rail transport			Land taken for rights of way and terminals; dereliction of obsolete facilities	Abandoned lines, equipment and rolling stock	Noise and vibration around terminals and along railway lines
Road transport	Air pollution (CO, HC, NO_x, particulates and fuel additives such as lead) Global Pollution (CO_2, CFC)	Pollution of surface water and groundwater by surface runoff, modification of water systems by road building	Land taken for infrastructures; extraction of road building materials	Abandoned spoil tips and rubble from road works; road vehicles withdrawn from service; waste oil	Noise and vibration from cars, motorcycles and lorries in cities, and along main roads
Air transport	Air pollution	Modification of water tables, river courses, and field drainage in airport construction	Land taken for infrastructures; dereliction of obsolete facilities	Aircraft withdrawn from service	Noise around airports

Source: Linster (1990).

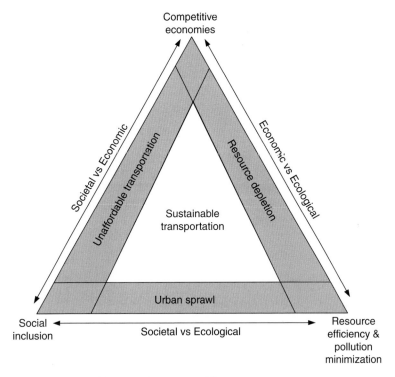

Figure 16.1 A trade-off approach to sustainable transportation.
Adapted from Feitelson (2002).

For example, a congestion charging scheme in an urban area may generate economic returns (competitive economies) and result in a modal shift towards public transport (pollution minimisation). However, if it is not supported by the provision of effective public transportation, accessibility to transportation for the wider public will be compromised. This will result in the social exclusion of passengers that cannot afford the use of private cars.

16.4 Development of Sustainable Transportation Systems

The development of sustainable transportation systems requires a step approach that incorporates policy making together with project implementation and appraisal. The following sections of this chapter aim to highlight different approaches towards the planning and development of such systems. A number of underlying issues, including appraisal indicators, organisational barriers, alternative strategies and the importance of public participation, are analysed and presented.

16.4.1 Policy Making

A number of different strategies can be adopted for the promotion of sustainable transport. Every strategy that promotes sustainable transportation must be supported by polices embraced by local authorities at a regional level, and by governments at a national level. There are three main groups of policies that can be taken to mitigate the environmental effects of transport and promote sustainability. These include (Greene and Wegener, 1997):

- 'Transport technology' that will improve the efficiency and reduce the impact of vehicles through the development of radical new engine and fuel technologies
- 'Transport supply', which will be supported by infrastructure development including highways networks, roads and streets, and also improvements to the existing provision of public transport
- 'Transport demand', which will include policies that will promote collective means of transport in order to minimize the use of private cars, and reduce the necessity to travel

In the context of sustainable transport, every new policy must take into account the effects this will have on the quality of life of the individual users that it affects. Policies that have a negative impact on the quality of life of people may face resistance from the wider public. That will in turn create pressure on the local authorities, or national governments and may alter their political will towards adopting such a policy. An example was the proposition for a congestion charging scheme in Edinburgh which was rejected by the public through a referendum. In some cases policy plans that hinder freedom of choice may have other outcomes than those envisaged (Tertoolen et al., 1998). For example, an attempt to introduce a policy that aims to restrict the use of private cars might instigate excessive driving, if it is perceived that the policy limits their freedom of choice (Steg and Gifford, 2005).

Policy making for environmental sustainable transport (EST) differs from policy making for conventional transport projects in three areas (OECD, 2002). These include the fact that EST objectives are consistent with 'specific requirements of sustainable transportation', EST policies aim to 'address the totality of transport's environmental impacts' and finally EST includes considerations to 'restrain growth in the most environmentally damaging forms of transport activity'.

16.4.2 Appraisal of Sustainable Transportation Projects

The development of any transport policy, plan, or system must be accompanied by some appraisal exercise, which will aim to assess the extent to which the objectives that were set at the inception of the project have been met. It is common practice for planning and project appraisal exercises to use a number of indicators, as required from national guidelines or laws (Johnston, 2008).

For sustainable transportation project appraisal different economic, social and environmental indicators have to be used. Analysis and use of such indicators is not always straightforward for decision and policy makers. The complexity in the use of sustainable transportation evaluation indicators can be summarised in the following points (Litman and Burwell, 2006):

Table 16.2 Levels of Analysis for Sustainable Transprotation Project Evaluation.

Level	Examples
External trends	Changes in population, income, economic activity, political pressures etc.
Decision making process	Planning process, pricing policies, stakeholder involvement etc.
Policies	Facility design and operations, transport services, prices, user information etc.
Response	Travel activity (VMT, mode choice etc.), pollution emissions, crashes, land development patterns etc.
Cumulative impacts	Changes in ambient pollution, traffic risk levels, overall accessibility, transportation costs etc.
Human and environmental effects	Changes in pollution exposure, health, traffic injuries and fatalities, ecological productivity etc.
Economic impacts	Property damages, medical expenses, productivity losses, mitigation and compensation costs
Performance evaluation	Ability to achieve specified targets

Source: Transport Research Board (2008).

- Some indicators may require analysis in terms of demographics, geographical factors and transport-related attributes (time, mode of transport and type of trip).
- In most cases, data collection requires investment in time and money, and therefore a balance should be kept in terms of the amount of data used for the appraisal. However, omission of data due to costs may reduce the accuracy of the evaluation results.
- Some indicators are sometimes difficult to measure. For example, 'low per capita vehicle mileage' may result due to an effective policy that encourages abandonment of private cars, or may result due to poverty or other economic constraints that prohibit certain users to drive.

Different indicators can reflect a range of levels of analysis as illustrated in Table 16.2. The fact that some of the indicators used for the assessment of sustainable transportation strategies are qualitative makes the adoption of traditional cost–benefit analysis (CBA) techniques inappropriate for the integrated appraisal of such projects. A study with local authorities (May et al., 2008) has identified the principal organisational, technical and external barriers to the development of sustainable transport strategies. These barriers are shown in Table 16.3.

16.4.3 Sustainable Infrastructure Development

Despite the fact that land resources have been consumed considerably, new transport infrastructure, or improvements on the existing transport network will continue to be a favourable measure for improving the existing

Table 16.3 Barriers for the Development of Sustainable Transport Strategies.

Organisational	Technical	External
Lack of interdepartmental working	Number of indicators	Inconsistency in national, regional and local priorities
Pursuit of pet schemes	Inappropriate indicators	Lack of control of bus and rail operators
Stakeholder numbers and diversity	Indicators which are difficult to measure	Poor public acceptance of certain instrument
Divided responsibilities	Failure to use indicators in the policy process	Short termism in decision making
Spatial boundaries	Poor management of data	
Funding too focused on specific solutions	Lack of understanding of certain impacts	
Lack of revenue funding	Inability to model certain instruments	
Lack of option generation staff and skills	Inability to appraise certain instruments	
Lack of modelling staff and skills	Lack of trust and transparency in models	
Organisational change	Incomplete appraisal of certain objectives	
	Inconsistency between appraisal and targets	
	Limited understanding of strategic environmental assessment	
	Risk-averse interpretation of national guidance	

Source: May et. al (2008).

transportation systems. Therefore sustainable development principles should be adopted in such measures. The Department for Transport in the United Kingdom (DFT, 2008) has defined a sustainability checklist of good practice for the developers of new infrastructure projects. The checklist includes five elements that should be taken into consideration during the planning process. These are:

- The provision of employment opportunities and community facilities as part of the new development
- The need for the new development to be well integrated with the existing infrastructure and land-use
- Evaluation of the existence of links between the location of the new development and surrounding urban areas, employment hubs and existing transportation supply
- Integration of facilities (flexible work/office space) and communication technologies (internet access for housing developments) that will minimize the need to travel
- The need for consultation sessions with stakeholders and the general public as part of the planning process for the new development

16.4.4 Sustainable Mobility Measures

Sustainable mobility refers to the adoption of innovative mobility concepts and policies that can promote sustainable transportation. These may include the promotion of sustainable modes of transport such as walking and cycling, better public transport systems and intermodalism, road pricing and taxing schemes and the reduction of single-occupant vehicles (SOV) with concepts such as carpooling and innovative car ownership schemes.

Promotion of walking and cycling requires improvement by ensuring that these modes are fully integrated into the development and monitoring of urban mobility policies. This necessitates diversification from traditional planning which assumes that transport progress requires replacement of slower modes with faster ones. In contrast, sustainable mobility requires the 'parallel model' paradigm, which assumes that each mode has to be utilised based on its performance and functional characteristics in order to achieve a balanced use of all modes (Litman and Burwell, 2006). A good example of the latter is the Vélib bike rental scheme in Paris. The developers identified that bicycle trips should last approximately 20 minutes and therefore adjusted the costing structure of the scheme based on that characteristic.

Congestion charging schemes can reduce the use of private cars and instigate a modal shift towards public transport. Early findings, from the implementation of congestion charging in London have demonstrated reductions of 16% in emissions of carbon dioxide (CO_2), and a reduction of between 40 and 70 collisions involving personal injury per year (Transport for London, 2008). In some cases road pricing or taxing schemes meet intense resistance from the public, which may result in the rejection of the schemes. In the case of the London congestion charging scheme, a number of conditions were already in place prior to its implementation (Banister, 2003). These were:

- 'The technical support in the form of a suitable technology, an area for implementation and realistic timescales.
- Legislative support through the Greater London Authority Act (1999) and the Transport Act 2000.
- Political support from central government and a Mayor committed to congestion charging.
- Management support through a dedicated team set up within Transport for London.
- Public support through extensive consultation and engagement.' (Banister, 2003)

Better public transport systems require easy switching and good travel integration between modes. This can be achieved by the implementation of real-time traveller information systems using intelligent transport systems, integrated fare schemes and increased and more efficient public transport management systems. A Norwegian study has documented that travel information services encourage people to change their habits towards more extensive use of public transport (Lodden and Brechan, 2003). In addition, a study with 69 Italian public transport authorities showed that integrated fare schemes had a positive impact on passenger demand (Abrate et al., 2009). In order to address the expected increase of public transport due to the introduction of the congestion charging, Transport for London developed the London Bus Priority Network, which used management and enforcement systems to improve the operational effectiveness of bus services (Goldman and Gorham, 2006).

Carpooling is a collective transportation system which aims to improve the average level of car occupancy by matching a vehicle (driver) with potential passengers that have common destinations or orientations. Effective carpooling implementation has the potential to reduce the number of SOVs on the roads. The United States was one of the pioneering countries in the development and analysis of carpooling schemes as a means of reducing SOVs (Ben-Akiva and Atherton, 1977). Statistics from the United Kingdom illustrate that average car occupancy for commuting trips is 1.2 persons per car. Similarly, in Germany the average commuter occupancy reaches 1.05 persons per car. Research carried out for the UK DFT has demonstrated that effective carpooling schemes can reduce the number of single occupant vehicles considerably (DFT, 2004). Fellows and Pitfield (2000) claimed that carpooling can benefit individuals by 'as much as halving their journey costs and the economy as whole benefits greatly in reduced vehicle kilometers, increased average speeds and saving in fuel, accidents and emissions'. The study revealed that the benefits of carpooling were analogous to those of major road schemes implemented in the area where the analysis was done (West Midlands, United Kingdom).

16.4.5 Public Participation

One of the requirements in the development of sustainable transportation strategies is consultation with the public prior to its implementation. However, this task is often challenging due to three distinct factors (Szyliowicz, 2003). Firstly, it not always clear what public participation entails, since it is a complicated exercise that can be implemented in different ways and produce various outputs. Secondly, public participation may be hampered by the lack of knowledge, or interest. Lastly, wide participation in public consultation requires an investment in resources that may not always be available. In order to overcome these limitations, a well-defined approach for public consultation and engagement, supported by the necessary resources, needs to be in place prior to the implementation of the project. As stated above, this was one of the conditions that enabled the successful implementation of the London congestion charging scheme.

16.5 Solutions for Sustainable Transportation

Historic trends of transportation point to the fact that the use of private cars will continue to grow for years to come. In conjunction with the fact that private cars account for the majority of road transport, sustainable transportation strategies should promote alternative means of transport. In the past we, as a society, have tried to 'build' our way out of transport problems, such as congestion, by the development of new road infrastructure. Such approaches will not pose as feasible alternatives in the future due to the exhaustion of available land and the inherent generation of additional traffic that characterises them. Therefore, more efficient use of the existing transport infrastructure is an important condition for sustainable transportation. Future projections suggest that oil prices will double by 2030 (EC, 2008). With this in mind, solutions that encompass new engine and vehicle technologies together with the use of collective means of transport could be the way forward for sustainable transportation in the future.

Implementation of different technologies, such as intelligent transport systems, has demonstrated that efficient use of the existing road networks is possible. Political determination will also be required for the implementation of strategies that may not be popular to all members of the public, but necessary for the survival of the planet. Road pricing and taxing schemes may offer desirable results for the minimisation of road build in the future. It is estimated that in the United Kingdom, a national road pricing scheme could 'reduce the case for inter-urban road build beyond 2015 by some 80 per cent' (DFT, 2006). Social equity considerations in the development of sustainable transportation could be a factor for the alleviation of public resistance. Studies have shown that there is a link between environmentally friendly mobility attitudes and knowledge related to environmental concerns (Nilsson and Küller, 2000). Therefore, educating future motorists may have a positive effect on the sustainability of the transportation systems. Finally, the existing undesirable situation can be attributed to the fact that transport problems have been viewed in isolation and therefore the development of sustainable transport in the future must be integrated with other policy areas such as 'health, social services and housing' (Cahill, 2007).

References

Abrate, G., Piacenza, M. & Vannoni, D. (2009) The impact of Integrated Tariff Systems on public transport demand: evidence from Italy. *Regional Science and Urban Economics*, **39**, 120–127.

Banister, D. (2003) Critical pragmatism and congestion charging in London. *International Social Science Journal*, **55**, 249–264.

Basler, E. (1998) Measuring the sustainability of transport. In: *Materials of NRP 41 'Transport and Environment'*, vol. M3. Programme Management NRP 41, Bern.

Ben-Akiva, M. & Atherton, T. J. (1977) Methodology for short-range travel demand predictions: analysis of carpooling incentives. *Journal of Transport Economics and Policy*, **11**, 224–261.

Black, W.R. (1996) Sustainable transportation: a US perspective. *Journal of Transport Geography*, **4**, 151–159.

Black, W.R. (2000) Socio-economic barriers to sustainable transport. *Journal of Transport Geography*, **8**, 141–147.

Cahill, M. (2007) Why the U-turn on sustainable transport? *Capitalism, Nature, Socialism*, **18**, 90–103.

Cairns, S. & Newson, C. (2006) *Predict and Decide – Aviation, Climate Change and UK Policy*. Environmental Change Institute, Oxford.

Department for Transport (DFT) (2004) *Making Car Sharing and Car Clubs Work – Final Report*. Department for Transport, London.

Department for Transport (DFT) (2006) *The Eddington Transport Study: The Case for Action: Sir Rod Eddington's Advice to Government*. Department for Transport, London.

Department for Transport (DFT) (2008) *Building Sustainable Transport into New Developments: A menu of options for Growth Points and Eco-towns*. Department for Transport, London.

European Commission (EC) (2008) *European Energy and Transport – Trends to 2030*. European Commission, Brussels.

European Conference of Ministers of Transportation (ECMT) (2004) *Assessment and decision making for sustainable transport*. European Conference of Ministers of Transportation, Paris.

European Environment Agency (EEA) (2009) *Greenhouse Gas Emission Trends and Projections in Europe 2009: Tracking Progress towards Kyoto Targets*. European Environment Agency Report No. 9/2009. European Environment Agency, Copenhagen.

Environmental Protection Agency (EPA) (2006) *Greenhouse gas emissions from the US Transportation Sector, 1990–2003*. Environmental Protection Agency, Office of Transportation and Air Quality, Washington, DC.

EUROSTAT (2009) *Panorama of Transport*. Eurostat Statistical Books, Brusssels.

Feitelson, E. (2002) Introducing environmental equity dimensions into the sustainable transport discourse: issues and pitfalls. *Transportation Research Part D: Transport and Environment*, **7**, 99–118.

Fellows, N.T. & Pitfield, D.E. (2000) An economic and operational evaluation of urban car-sharing. *Transportation Research Part D: Transport and Environment*, **5**, 1–10.

Goldman, T. & Gorham, R. (2006) Sustainable urban transport: Four innovative directions. *Technology in Society*, 28, 261–273.

Greene, D.L. & Wegener, M. (1997) Sustainable transport. *Journal of Transport Geography*, 5, 177–190.

Johnston, R.A. (2008) Indicators for sustainable transportation planning. *Transportation Research Record*, 146–154.

Linster, M. (1990) Background facts and figures. In: *Transport Policy and the Environment European Conference of Ministers of Transport*. European Conference of Ministers of Transport – OECD, Paris.

Litman, T. & Burwell, D. (2006) Issues in sustainable transportation. *International Journal of Global Environmental Issues*, 6, 331–347.

Lodden, U.B. & Brechan, I. (2003) Reiseinformasjonens betydning for bruk av kollektivtrafikk, Effekten av tjenestetilbudet til Trafikanten – The importance of travel information for the use of public transport – The effects of the travel information services of Trafikanten. Transportøkonomisk Institutt, 684/2003. Transportøkonomisk Institutt, Stockholm.

May, A.D., Page, M. & Hull, A. (2008) Developing a set of decision-support tools for sustainable urban transport in the UK. *Transport Policy*, 15, 328–340.

Nilsson, M. & Küller, R. (2000) Travel behaviour and environmental concern. *Transportation Research Part D: Transport and Environment*, 5, 211–234.

OECD (1996) *Pollution Prevention and Control: Environmental Criteria for Sustainable Transport*. Document OECD/GD(96)136. OECD, Paris.

OECD (2002) *OECD Guidelines towards Environmentally Sustainable Transport*. OECD, Paris.

Steg, L. & Gifford, R. (2005) Sustainable transportation and quality of life. *Journal of Transport Geography*, 13, 59–69.

Szyliowicz, J.S. (2003) Decision-making, intermodal transportation, and sustainable mobility: towards a new paradigm. *International Social Science Journal*, 55, 185–197.

Tertoolen, G., Van Kreveld, D. & Verstraten, B. (1998) Psychological resistance against attempts to reduce private car use. *Transportation Research Part A: Policy and Practice*, 32, 171–181.

Transport for London (2008) *Central London Congestion Charging – Impacts Monitoring, Sixth Annual Report*. Transport for London, London.

Transport Research Board (2008) *Sustainable Transportation Indicators – A Recommended Research Program for Developing Sustainable Transportation Indicators and Data*. Sustainable Transportation Indicators Subcommittee of the Transportation Research Board, Washington, DC. Available from http://www.vtpi.org/sustain/sti.pdf.

U.S. Energy Information Administration (USEIA) (2009) Emissions of Greenhouse Gases Report. U.S. Energy Information Administration Report DOE/EIA-0573(2008). U.S. Energy Information Administration, Washington, DC.

17

Linkages of Waste Management Strategies and Climate Change Issues

Kim Tannahill and Colin A. Booth

17.1 Introduction

The waste industry has seen increasing pressure in recent years due to a steady rise in waste production (Burnley, 2007), fuelled by increasing population growth, rapid urbanisation (Agdag, 2009) and the need for more environmentally acceptable waste management strategies (Hazra and Goel, 2009). As a result, waste management practices have evolved and there has been a paradigm shift from waste management to a more resource management philosophy (International Solid Waste Association [ISWA], 2007). With much of today's environmental focus on climate change and the potential effects of global warming, it has become necessary for both policy makers and the waste management industry to consider waste management strategies within this context (Department for Environment, Food and Rural Affairs [DEFRA], 2007a; Chen and Lin, 2008).

When considering waste management's contribution to climate change, it is necessary to take an holistic approach that encompasses the product lifecycle from gathering of raw materials through to the product's 'end of life' (Bebb and Kersey, 2003; DEFRA, 2007a). All of these processes contribute in some way to global warming through either direct or indirect production of greenhouse gases (GHG), particularly methane (CH_4), carbon dioxide (CO_2) (Bebb and Kersey, 2003; DEFRA, 2007a; Ramanathan and Feng, 2009), to a lesser extent, nitrous oxide (N_2O) (Bebb and Kersey, 2003; DEFRA, 2007a; Chen and Lin, 2008).

The most damaging greenhouse gas is CH_4 (23 times as damaging as CO_2 and produced in much larger quantities than N_2O) (DEFRA, 2007a) and the waste industry contributes to emissions primarily through CH_4 emissions

---

Solutions to Climate Change Challenges in the Built Environment, First Edition.
Edited by Colin A. Booth, Felix N. Hammond, Jessica E. Lamond and David G. Proverbs.
© 2012 Blackwell Publishing Ltd. Published 2012 by Blackwell Publishing Ltd.

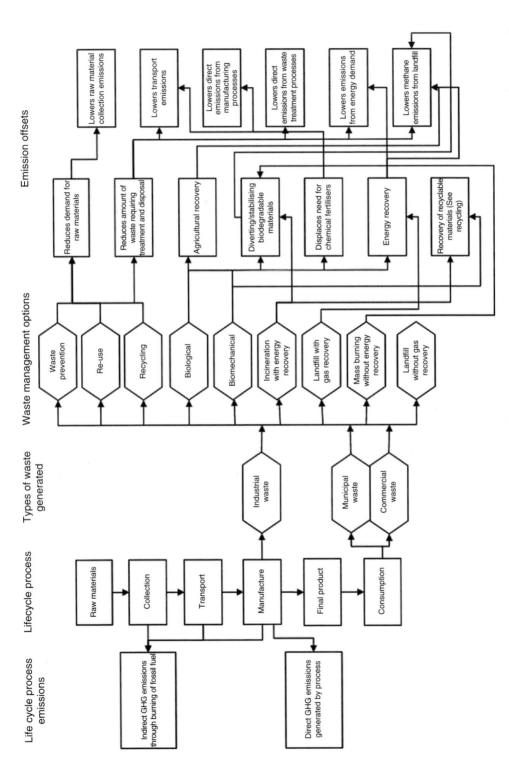

Figure 17.1 An example of an integrated solutions approach to waste management and the many stages involved in product lifecycle, which may contribute directly or indirectly to GHG emissions.

from landfill. It is thought that as much as one third of anthropogenic CH_4 emissions within the European Commission (EC) can be attributed to municipal solid waste (MSW) landfilling (DEFRA, 2007b; De Gioannis et al., 2009). Biodegradation of organic material in UK landfills is thought to account for 40% of all UK CH_4 emissions and 3% of all natural GHG emissions (DEFRA, 2007b). This is similarly reflected in the United States, where the Environmental Protection Agency (EPA) reports ~4% of anthropogenically generated natural GHG emissions are from the waste management sector and ~90% of GHG from the waste sector is due to CH_4 emissions from landfill (Batool and Chaudhry, 2009).

An integrated solutions approach is necessary to ascertain the areas of direct GHG production (i.e. direct emission to atmosphere from the landfilling processes) and indirect GHG contributions (i.e. through vehicle exhaust emissions during waste transport) throughout the product lifecycle. Therefore, interactions between these processes must be understood so that adjustments can be made to various components of the system that will, in turn, favourably impact other processes and allow the system, as a whole, to function in a much 'cleaner' manner where GHG emissions are concerned (Chen and Lin, 2008; De Gioannis et al., 2009).

17.2 Integrated Solutions Approach

17.2.1 Product Lifecycle and Lifecycle Assessment

Cherubini et al. (2009) describe the lifecycle assessment (LCA) study as 'a tool able to evaluate environmental burdens associated with a product, process, or service by identifying energy and materials used and emissions released to the environment; moreover, it allows also an identification of opportunities for environmental improvements'. LCA has become an important management tool when considering the environmental impact of a product or system and has been developed into an internationally standardised system (Arena et al., 2003).

Product lifecycle encompasses the creation of a product from the gathering of raw materials through to the end result that is available for consumption (Bebb and Kersey, 2003; DEFRA, 2007b; Rigamonti et al., 2009). There are many stages involved in the product lifecycle, all of which may contribute directly or indirectly to GHG emissions, such as fuel burning during collection and transport of raw materials, direct production during the manufacturing process and indirect production through electricity and energy usage (Figure 17.1). There may also be indirect effects on the GHG balance by the removal or reduction of carbon sequestration sources, such as forests if the required raw material is wood (Bebb and Kersey, 2003; Chen and Lin, 2008). This was illustrated effectively by Pickin et al. (2002) during lifecycle analysis of paper production and associated GHG emissions (summarised in Figure 17.2). The study was conducted to assess the emissions of the main

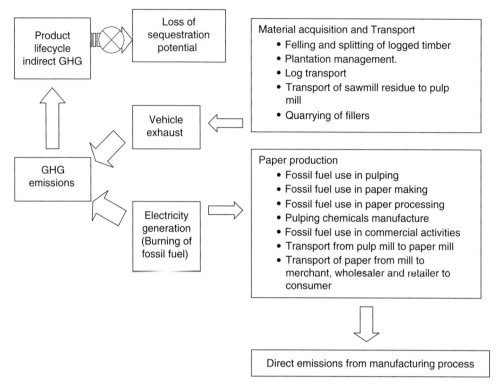

Figure 17.2 Lifecycle analysis of paper production and associated GHG emissions.

GHG (CO_2, CH_4 and N_2O) only and included all types of paper and paperboard and fossil fuel energy from production emissions, encompassing mining, refining, transport, storage and marketing. As Figure 17.2 shows, there are many considerations when trying to effectively monitor GHG emissions from product manufacture, and this is further complicated when waste management of the end product must be taken into account.

17.2.2 Waste Management Hierarchy

It is impossible to separate the waste management hierarchy and product lifecycle when tackling GHG emissions as they are each directly impacted by strategies implemented in the other. There are many potential options when considering waste management with some options being more preferable than others (DEFRA, 2007b).

There has been a notable shift in policy from the traditional collection and storage strategy resulting in a new waste management hierarchy that encourages waste prevention, minimisation, recycling and recovery (Banar *et al.*, 2008).

Waste Management Hierarchy

- Waste prevention
- Re-use
- Recycling
- Biological treatment
- Energy recovery
- Landfill

Landfilling has become a less preferable option as it requires large amounts of space, poses a potential threat to the environment and human health through leaching and gaseous emissions and recovers only a limited amount of energy (Messineo and Panno, 2008).

17.2.2.1 Waste Prevention

According to EC guidelines, the most effective option is to reduce waste generation, which will have cumulative effect in GHG emission reduction through the product lifecycle and management process (Marchettini *et al.*, 2007; Calabro, 2009). This reduces the demand for raw materials, positively impacting the first stage of the lifecycle, and has a knock-on effect reducing the need for collection and transportation of materials and subsequent manufacturing (ISWA, 2009).

17.2.2.2 Recycling and Re-use

Recycling and re-use have a similar impact by reducing the amount of virgin material being processed and, thus, reducing lifecycle emissions accordingly (Liamsanguan and Gheewala, 2008; Batool and Chaudhry, 2009). Re-use of waste delays the need for production of new materials and reduces the amount of waste requiring treatment and disposal (ISWA, 2009). Recycling requires collection and transport of waste to a manufacturing site, where it can be processed and converted to a new product (Calabro, 2009). This requires the burning of fossil fuel and energy use, which results in GHG emissions. However, the net gain through reduction in demand for raw materials and more intensive manufacturing processes results in this methodology contributing, ultimately, to a reduction in GHG emissions (ISWA, 2009). Studies carried out by Chen and Lin (2008) and Calabro (2009) support this fact and suggest recycling as the most favourable option when considering waste management strategies, because it produces lower GHG emissions, reduces the demand for energy, reduces manufacturing emissions and, when considering paper products, results in a lower demand for wood, which is beneficial in terms of forest carbon sequestration. Further studies conducted by Calabro (2009) indicate separate waste collections can maximise the quantity and quality of recyclable materials and can be more effective than alternative methods, such as waste-to-energy (WtE), landfilling or biological treatment that are very dependant on the level of technology utilised.

17.2.2.3 Mechanical Biological Treatment

Separation and treatment of organic waste from the waste stream are vital when trying to reduce CH_4 emissions from landfill (DEFRA, 2007c). Mechanical biological treatment (MBT) is commonly utilised to separate the biological and nonbiological components for further treatment and can be defined as 'a conversion of municipal solid waste (MSW) or residual MSW via a combination of mechanical with biological processes, aiming mainly at the stabilisation of biologically degradable components'. This technology can be characterised by low energy input, high process flexibility, low-cost facilities and possible material recovery. The mechanical phase, typically, involves the shredding and sieving of waste to produce two distinct streams: (1) the oversize fraction, which can be incinerated or partially recycled prior to disposal, and (2) the undersized, which typically comprises the mechanically sorted organic residues (MSOR). This requires further treatment to ensure that it is biologically stable prior to being landfilled or re-used.

Composting is a simple yet effective method of stabilising the organic fraction and resulting in a by-product with agricultural applications (De Gioannis *et al.*, 2009) and is being used with increasing frequency in developing countries, where the organic content of waste is generally high (occasionally above 50%) and investment in expensive technology is not an option. It is, however, also a popular option in some developed countries, such as the Netherlands, where 97% of organic waste is treated in composting facilities (ISWA, 2009). The process is aerobic, meaning the biodegradable material is broken down in the presence of oxygen resulting in the production of CO_2 and H_2O. Methane is typically produced in the absence of oxygen (i.e. during an anaerobic process) and it is thought that a limited amount of CH_4 may form in small oxygen deprived pockets. However, the majority of this CH_4 is eventually oxidised during the composting process resulting in negligible levels being emitted to atmosphere. There has also been conflicting results produced during studies to analyse the amount of N_2O released to atmosphere during the process through either incomplete de-nitrification or incomplete ammonium oxidation. One study reported high emissions levels at the start of the process with levels becoming negligible over time (He *et al.*, 2000); whereas, Beck-Friis *et al.* (2000) found emissions grew over time. Given N_2O is 310 times more effective as a GHG, this issue needs to be investigated further and quantified where possible.

Anaerobic digestion is, typically, employed as an integrated waste management solution, as during the breakdown of organic matter by microorganisms in the absence of oxygen, CH_4 and CO_2 are produced and this biogas can be captured and utilised as an energy fuel. Furthermore, the resulting liquor is nutrient rich and can be used a fertiliser (Lou and Nair, 2009). This method of biological stabilisation requires a much higher level of pre-sorting and does not accept as many different types of organic waste as composting. However, the potential for energy recovery, with an efficiency rate of 35% for the biogas energy content is valuable when offsetting against larger alternative emissions that may arise if this treatment was omitted from the management strategy (ISWA, 2009).

There have been many studies conducted that support the aptness of MBT as a tool to substantially reduce the CH_4 emissions from landfill (Cossu et al., 2003; Robinson et al., 2005; De Gioannis and Mutoni, 2007). Reductions of gas production of between 80% (De Gioannis et al., 2009) and 98% (Fricke et al., 2005) have been reported.

Aside from massive reductions in landfill gas generation, composting and anaerobic digestion can also contribute to the reduction of GHG emissions by displacing the use of chemical fertilisers that are more carbon intensive due to the large amounts of energy required to produce and transport. Furthermore, studies have shown the use of the resulting liquor as a fertiliser encourage more rapid plant growth, which results in increased carbon sequestration potential within the plant (Lou and Nair, 2009). However, this methodology should be used in conjunction with another, as once the inorganic fraction has been removed from the waste stream, it will require further treatment in the form of recycling where possible and then disposal. Often, this fraction can be sent to a WtE incinerator plant where it can be pre-treated prior to landfill.

17.2.2.4 Waste-to-Energy Incineration

Waste-to-energy incineration plants offer an alternative to landfilling unprocessed waste (DEFRA, 2007c). Modern combustion and pollution control technologies allow minimisation of emissions, and the process is advantageous in that it reduces the volume, weight and hazardousness of MSW and produces potentially re-usable materials (Messineo and Panno, 2008). Although widespread throughout other western and northern European countries, where national waste strategies employ WtE as a key component, the process is not widely utilised in the United Kingdom (Papageorgiou et al., 2009). At present, there are 19 operational incinerator plants throughout the United Kingdom, processing ~300 tonnes of waste per annum (DEFRA, 2007c). The process involves the controlled combustion of raw MSW with an aim to reduce the waste volume, stabilise any organic components and recover energy from the process. There are two main types of WtE plants operating in the United Kingdom, incinerators and advanced thermal treatment (ATT) plants.

ATT (pyrolysis and/or gasification) is a fairly new technology and is still in its infancy in the United Kingdom. There are two types of ATT treatment, pyrolysis and gasification. Pyrolysis involves the combustion of waste in the absence of oxygen in temperatures much lower than those used during incineration (~300°C to 850°C) using an external heat source. This produces a synthetic gas, which includes carbon monoxide, hydrogen and CH_4, and a solid residue, both of which can be used as fuels. Gasification involves the combustion of waste in a small amount of oxygen at temperatures around 650°C, resulting in partial oxidation of the waste. This results in synthetic gas, comparable to that produced during pyrolysis, along with a relatively low-level carbon containing solid residue.

Incinerator plants (with energy recovery) are far more common in the United Kingdom though much less common than throughout Europe. However, DEFRA's waste strategy for England (2007b) suggests by 2010, 25% of MSW will be processed at incinerators as opposed to only 11% in 2006. Incinerators operate using a mass burn technique in the presence of oxygen in a single-stage chamber (Papageorgiou *et al.*, 2009) at temperatures >850°C (DEFRA, 2007c). The calorific content of the waste is harnessed during the combustion process and used to produce electricity via the creation of superheated steam to drive the turbines. In some cases, residual heat is recovered and re-used by combined heat and power (CHP) incinerators (Papageorgiou *et al.*, 2009). The primary concerns with these WtE techologies within a climate change context are the emission of CO_2 and N_2O, through direct combustion emissions, and CO_2 emissions, through energy use (Chen and Lin, 2008). However, it should be noted N_2O emissions account for <1% of total GHG emissions during combustion. It is vital incinerators are equipped with an efficient flue gas treatment system able to cope with legislative limitations on emissions (Calabro, 2009). There has, traditionally, been a strong opposition to the use of incineration as a waste management strategy due to the high level of emissions generated by the old-style incinerators (Papageorgiou *et al.*, 2009). However, modern incinerators are characterised by efficient gas scrubbing systems and are subject to strict monitoring protocols that ensure emissions are below legal limits (Marchettini *et al.*, 2007). It has been shown that the GHG emission impacts associated with incineration can significantly reduce the amount of CH_4 generated in landfill by stabilising the organic fraction of the waste stream input (Papageorgiou *et al.*, 2009). Furthermore, emissions are offset by the electricity generation, thus reducing the consumption of fossil fuel (ISWA, 2009).

17.2.2.5 Landfill

The majority of the methods mentioned above are implemented in conjunction with landfilling, as this remains the final solution for waste that cannot be treated, separated or recovered any further (Calabro, 2009). Decomposition of biodegradable organic materials within a landfill facilitate its function as a bioreactor, producing CH_4, CO_2, H_2O and small amounts of N_2O (Chen *et al.*, 2008) and this equates to ~60% CH_4 and ~40% CO_2 (along with other trace gases) (Jha *et al.*, 2008). Furthermore, the Intergovernmental Panel on Climate Change (IPCC; 2007) reported, in 2001, 10–20% of all anthropogenic CH_4 emissions were a result of landfill gases (Einola *et al.*, 2008). Demographics and local environmental factors play an important role in the variation of these percentages (Chen *et al.*,

2008; Calabro, 2009) because waste input varies from country to country and area to area. For instance, developing countries will tend to have a higher ratio of biodegradable food materials within the waste stream, while waste in developed countries will contain more hardboard and wood (Jha et al., 2008). This, coupled with variations in soil composition, pressure, temperature, moisture content and pH, can lead to temporal and spatial variations in emission ratios (Chen et al., 2008).

In attempt to greatly reduce the amount of anthropogenic emissions to atmosphere from landfill, most waste management strategies are employed to facilitate waste stabilisation (Calabro, 2009). While these measures are implemented today, the release of CH_4 from landfill to atmosphere occurs over a period of many decades and, historically, there has been minimal or no landfill management with respect to waste input or landfill design. Recent years have seen the introduction of tighter controls over what waste types can be sent to landfill and has required design to consist of impermeable liners, a cap and a gas and leachate collection system (Marchettini et al., 2007). The gas extraction system is perhaps the most important mitigating factor when reducing emissions. Studies conducted by the IPCC (2007) suggest well managed landfills with good gas recovery systems can capture >90% of CH_4 generated. In sites where the technology is lacking, or the extraction system only partial, this results in fugitive emissions and gives a much lower recovery (~20%). It is important that systems are regularly checked and maintained to keep them running at peak capacity (IPCC, 2007). In some countries the law requires installation of gas recovery systems, particularly at those sites receiving any organic waste (ISWA 2009). All EU member states fall into this category and a similar system is in place in the United States, where both liquid effluents and gas must be collected for at least 30 years post closure of any site (Hao et al., 2008). In some cases, gas can be collected and utilised as a fuel to generate electricity or fuel industrial boilers (IPCC, 2007) but often it is flared, which converts the CH_4 content to CO_2 and vapour, reducing the overall greenhouse effect as CO_2 has a much lower global warming potential than CH_4 (Hao et al., 2008).

Aside from the collection of landfill gases, it is possible to further control CH_4 emissions through microbial oxidation to CO_2 in cover soils (IPCC, 2007). Studies conducted by Einola et al. (2008) suggest composted waste and MBT residuals may be a suitable medium in the cover layer to facilitate the process. For the medium to be effective, it requires high porosity, appropriate nutrient levels and water holding capacity and must be biochemically stable. Both composted waste and MBT residuals fit these criteria but, although MBT is successful at facilitating oxidation, the leachability of some heavy metals exceeded legal limits and this requires further study. It should also be noted that oxidation rates may vary as they are dependant on the physical properties of the soil, such as thickness, moisture content and temperature to allow for easy transport of CH_4 upwards through the strata from the anaerobic zones to the cover soil (IPCC, 2007).

In situ aeration methods have also proved successful in reducing emissions by changing the anaerobic conditions to aerobic and allowing for the oxidisation of CH_4 to CO_2 (Ritzkowski and Stegmann, 2010). Ritzkowski and

Stegmann (2010) claim accelerated biodegradation leads to biostabilisaton of the biodegradable wastes and the resulting stabilised landfills contribute only marginally to GHG emissions in the long term. Furthermore, in the case of CO_2 emissions from this process, they are climate-neutral as they originate from a biogenic source and are part of the natural carbon cycle.

It should be noted also, while landfills generate large amount of CH_4, they are also a long-term carbon sink. A recent report by the IPCC (2007) concluded: 'Since lignin is recalcitrant and cellulosic fractions decompose slowly, a minimum of 50% of the organic carbon landfilled is not typically converted to biogas carbon but remains in the landfill'. This fraction of carbon storage is variable as the input is not homogenous, but coupled with new technologies and management practices designed to reduce CH_4 emissions, this changes the face of landfilling and makes it a more environmentally friendly disposal method, within a climate change context.

17.3 Key Policy Drivers: A European Perspective

There has been a noticeable shift in waste management policy over recent decades as environmental issues have become more prominent and, as such, waste management solutions have required review to comply with the need for more environmentally friendly waste management practices. Until the early 1990s, policy focussed on diverting waste from landfill and promoting recycling, but at that time established no binding targets for member states (ISWA, 2009). Nowadays, one of the most prominent policy drivers at present is the EC Landfill Directive (Council Directive 1993/31/EC on the Landfilling of Waste; EC, 1993), which has been transposed in the United Kingdom through landfill regulations.

17.3.1 Landfill Directive (1993/31/EC)

Introduction of the EC landfill directive (EC, 1993) aimed to facilitate the new waste management hierarchy by limiting the use of landfill as a disposal option and ensuring waste sent to landfill was properly treated, both physically and chemically, to minimise potential environmental damage (Messineo and Panno, 2008). It has implemented standards for design and construction and states all member states must have liners and gas capture in new landfills, and the amount of biodegradable waste sent to landfill should be reduced to 75% (of the total biodegradable municipal waste by weight produced in 1995), 50% by 2013 and 35% by 2020 (Bebb and Kersey, 2003; ISWA, 2009). It further required the re-classification of landfills into: (1) hazardous waste; (2) non-inert, nonhazardous and (3) inert waste. Furthermore, landfilling of tyres has been banned and all waste must now receive pre-treatment where it poses any risk to human health or the environment (Bebb and Kersey, 2003). To date, studies show there has been a drop in waste sent to landfill in the European Union from 62% in 1995 to 41% in 2007 (ISWA, 2009).

17.3.2 Directive on Packaging and Packaging Waste (94/62/EC)

This directive (EC, 1994) has been transposed in law in the United Kingdom through the Producer Responsibility Obligations (Packaging Waste) Regulation 1997 (UK Government, 1997) and the Packaging (Essential Requirements) Regulations 1998 (UK Government, 1998). These aim to encourage producer responsibility (ISWA, 2009), and the Producer Responsibility Obligations Regulation sets targets for recycling of packaging waste and enforcing the producer to arrange for recycling or recovery of a proportion of the waste its handles.

17.3.3 EC Waste Incineration Directive (2000/67/EC)

The waste incineration directive (EC, 2000) is designed to implement high standards of cost effective emission control. This directive addresses municipal and hazardous wastes and co-incineration of wastes that had been excluded from the scope of the Hazardous Waste Incineration Directive (94/67/EC). It further imposes the requirement for all plants to obtain a Pollution Prevention and Control Permit that will set limitations on atmospheric emissions (Bebb and Kersey, 2003).

17.3.4 EC Directive on Waste (75/442/EEC) Amended 91/156/EEC AND 91/692/EEC), Articles 3, 4 and 5

Bebb and Kersey (2003) define this directive as placing 'a requirement on Member States to have a regard to the need to minimise waste, encourage recycling and waste recovery. There must also be a regard given to the need to protect the environment and human health in the context of potentially polluting developments'.

17.3.5 Clean Development Mechanism

Although not legislation, the Clean Development Mechanism (CDM) encourages investment and development of new technologies in developing countries and allows those contributing to offset their own emissions targets by claiming reduction credits for projects invested in, to supplement their own domestic targets (ISWA, 2009).

17.3.6 Waste Strategy for England 2007

This documents the implementation of European policy with short and long term goals and outlines the government's policy on the promotion of sustainable waste management. The strategy focuses on diverting waste from landfill and increasing recycling rates through the implementation of

Table 17.1 Waste management strategies.

Method	Solution
Waste prevention	• Reduce demand for raw materials. • Reduce indirect GHG emissions associated with material collection and transport. • Reduce direct GHG emissions from the manufacturing process. • Less waste requiring treatment and disposal. • Fewer emissions from waste collection and transportation. • Fewer emissions from waste management processes (ISWA, 2009).
Re-use	• Reduce amount of virgin material being processed by delaying the need for production of new materials. • Reduces the amount of waste requiring treatment and disposal. • Emission reductions are similar to those described for waste prevention, but less effective (ISWA, 2009).
Recycling	• Reduce demand for raw materials. • Reduces the amount of waste requiring treatment and disposal. • Requires collection, transport and manufacture to produce new product; however, the net gain in emission reduction makes this option an effective solution (Calabro, 2009).
Mechanical biological treatment	• Removal and treatment of organic waste. • Reduction in Ch_4 emission from landfill (DEFRA, 2007c). • Composting reduces Ch_4 emissions from biodegradation as aerobic process produces carbon dioxide, which has a much lower global warming potential. • Anaerobic digestion allows for harnessing of Ch_4 as fuel source. • This reduces Ch_4 emissions directly and GHG emissions indirectly through lower demand for energy. • Results in a by-product with agricultural applications (De Gioannis et al., 2009). • Displaces use of chemical fertilisers and associated GHG emissions through manufacture and transport. • Encourages rapid plant growth resulting in an increased carbon sequestration potential within plant (Lou and Nair, 2009). • Separates out inorganic fraction for re-use, recycling and/or incineration (De Gioannis et al., 2009).
Advanced thermal treatment	• Stabilises (or partially stabilises) organic fraction through combustion. • Reduces Ch_4 emissions from landfill. • Produces synthetic gas and solid residue, both of which can be used as fuels. • Lowers demand for energy and associated GHG emissions (DEFRA, 2007c).
Waste-to-energy incineration	• Complete oxidation (stabilisation) of the organic fraction. • Reduces Ch_4 emissions from landfill. • Produces electricity reducing the demand for energy and associated GHG emissions. • Highly effective gas scrubber systems and legislative emissions limitations greatly reduce any direct GHG emissions to atmosphere from the combustion process (DEFRA, 2007c).

Table 17.1 (Cont'd)

Method	Solution

Landfill	• Reduction of CH_4 emissions through:
	  • Stabilisation of organic fraction
	  • Introduction of gas collection systems
	• Flaring of CH_4 to produce carbon dioxide (lower global warming potential).
	• Tighter controls over what types of waste can be sent to landfill.
	• Introduction of microbial oxidation of CH_4 to carbon dioxide in cover soils.
	• In situ aeration to facilitate oxidation of CH_4 to carbon dioxide.
	• Landfills are a long-term carbon sink (IPCC, 2007).

legislation led by EU directives and placing more emphasis on business and consumer responsibility.

17.4 Solutions for the Waste Management Sector

The waste management sector contributes significantly to climate change through the global warming potential of GHG emissions. Of particular concern are CH_4 emissions from landfilling and recent years have seen a shift in policy and management strategy designed to tackle these emissions and encourage waste diversion through alternative technologies (DEFRA, 2007b and ISWA, 2009). The introduction of the Landfill Directive (1993/31/EC) along with corresponding waste policy drivers has put in place legislation that encourages compliance with the waste management hierarchy, promoting the reduction of waste production and recycling/reuse of products (Bebb and Kersey, 2003). A shift in strategy to analyse systems using an holistic approach encompassing the whole of the product lifecycle within the context of the new waste management hierarchy will significantly reduce the GHG emissions from the waste management sector (DEFRA, 2007b). The resulting waste management strategies are designed to include a combination of processes and treatments that will maximise recovery and provide solutions to the climate change problem through reduction of greenhouse gases. The processes utilised are dependant on varying factors such as economic viability and composition of waste input. The key methodologies that could be implemented have been discussed in this chapter and are summarised in Table 17.1

It is important to ensure that policy drivers are in place to enforce strategies that will encompass those solutions outlined in Table 17.1. These policies, coupled with incentives such as the Clean Development Mechanism, will offer a worldwide solution to waste management of GHG emissions and allow for a cleaner, more sustainable future (ISWA, 2009).

References

Agdag, O.N. (2009) Comparison of old and new municipal solid waste management systems in Denizli, Turkey. *Waste Management*, 29, 456–464.

Arena, U., Mastellone, M.L. & Perugini, F. (2003) The environmental performance of alternative solid waste management options: a lifecycle assessment study. *Chemical Engineering Journal*, 96, 207–222.

Banar, M., Cokaygil, Z. & Ozkan, A. (2008) Life cycle assessment of solid waste management options for Eskisehir, Turkey. *Waste Management*, 29, 54–62.

Batool, S.A., & Chaudhry, M.N. (2009) The impact of municipal solid waste treatment methods on greenhouse gas emissions in Lahore, Pakistan. *Waste Management*, 29, 63–69.

Bebb, J. & Kersey, J. (2003) *Potential impacts of climate change on waste management*. Environment Agency Technical Report X1-042, Bristol, UK.

Beck-Friis, B., Pell, M., Sonesson, U., Jonsson, H. & Kirchmann, H. (2000) Formation and emissions of N_2O and CH_4 from compost heaps of organic household waste. *Environmental Monitoring and Assessment*, 62, 317–331.

Burnley, S.J. (2007) A review of municipal solid waste composition in the United Kingdom. *Waste Management*, 27, 1274–1285.

Calabro, P. (2009) Greenhouse gases emission from municipal waste management: the role of separate collection. *Waste Management*, 29, 2178–2187.

Chen, T. & Lin, C. (2008) Greenhouse gases emissions from waste management practices using Life Cycle Inventory model. *Journal of Hazardous Materials*, 155, 23–31.

Chen, I., Hegde, U., Chang, C. & Yang, S. (2008) Methane and carbon dioxide emissions from closed landfill in Taiwan. *Chemosphere*, 70, 1484–1491.

Cherubini, F., Bargigli, S. & Ulgiati, S. (2009) Life cycle assessment (LCA) of waste management strategies: landfilling, sorting plant and incineration. *Energy*, 34, 2116–2123.

Cossu, R., Raga, R. & Rossetti, D. (2003) The PAF model: an integrated approach for landfill sustainability. *Waste Management*, 23, 37–44.

DEFRA (2007a) *Climate Change and Waste Management: The Link*. DEFRA, London.

DEFRA (2007b) *Waste Strategy for England 2007*. DEFRA, London.

DEFRA (2007c) *Incineration of Municipal Solid Waste*. DEFRA, London.

De Gioannis, G. & Muntoni, A. (2007) Dynamic transformations of nitrogen during mechanical–biological pre-treatment of municipal solid waste. *Waste Management*, 27, 1479–1485.

De Gioannis, G., Muntoni, A., Cappai, G. & Milia, S. (2009) Landfill gas generation after mechanical biological treatment of municipal solid waste: estimation of gas generation rate constants. *Waste Management*, 29, 1026–1034.

European Commission (EC) (1993) *Directive of the European Parliament and of the Council on the Landfilling of Waste*. Directive 1993/31/EC. European Commission, Brussels.

European Commission (EC) (1994) *Directive of the European Parliament and of the Council on Packaging and Packaging Waste*. Directive 94/62/EC. European Commission, Brussels.

European Commission (EC) (2000) *Directive of the European Parliament and of the Council on Waste Incineration*. Directive 2000/67/EC. European Commission, Brussels.

Einola, J.M., Karhu, A.E. & Rintala, J.A. (2008) Mechanically biologically treated municipal solid waste as a support medium for microbial methane oxidation to mitigate landfill greenhouse emissions. *Waste Management*, 28, 97–111.

Fricke, K., Santen, H. & Wallmann, R. (2005) Comparison of selected aerobic and anaerobic procedures for MSW treatment. *Waste Management*, **25**, 799–810.

Hao, X., Yang, H. & Zhang, G. (2008) Trigeneration: a new ways for landfill gas utilization and its feasibility in Hong Kong. *Energy Policy*, **36**, 3662–3673.

Hazra, T. & Goel, S. (2009) Solid waste management in Kolkata, India: practices and challenges. *Waste Management*, **29**, 470–478.

He, Y., Inamori, Y., Mizuochi, M., Kong, H., Iwami, N. & Sun, T. (2000) Measurement of N_2O and CH_4 from the aerated composting of food waste. *The Science of the Total Environment*, **254**, 65–74.

International Solid Waste Association (ISWA) (2007) *Waste and climate change*. White Paper. International Solid Waste Association, Vienna.

International Solid Waste Association (2009) *Waste and climate change*. White Paper. ISWA, Vienna.

Intergovernmental Panel on Climate Change, Bogner, J., Abdelrafie Ahmed, M., Diaz, C., Faaij, A., Gao, Q., Hashimoto, S., Mareckova, K., Pipatti, R. & Zhang, T. (2007) Waste management. In: *Climate Change 2007: Mitigation*, ed. Intergovernmental Panel on Climate Change. Contribution of working group III to the Fourth Assessment Report of the IPCC. Intergovernmental Panel on Climate Change, Geneva.

Jha, A.K., Sharma, C., Singh, N., Ramesh, R. & Purvaja, R. (2008) Greenhouse gas emissions from municpal solid waste management in Indian mega-cities: a case study of Chennai landfill sites. *Chemosphere*, **71**, 70–758.

Liamsanguan, C. & Gheewala, S.H. (2008) The holistic impact of integrated solid waste management on greenhouse gas emissions in Phuket. *Journal of Cleaner Production*, **16**, 1865–1871.

Lou, X.F. & Nair, J. (2009) The impact of landfilling and composting on greenhouse gas emissions – a review. *Bioresource Technology*, **100**, 3793–3798.

Marchettini, N., Ridolfi, R. & Rustici, M. (2007) An environmental analysis for comparing waste management options and strategies. *Waste Management*, **27**, 562–571.

Messineo, A. & Panno, D. (2008) Municipal waste management in Sicily: practices and challenges. *Waste Management*, **28**, 1201–1208.

Papageorgiou, A., Barton, J.R. & Karagiannidis, A. (2009) Assessment of the greenhouse effect impact of technologies used for energy recovery from municipal waste: a case for England *Journal of Environmental Management*, **90**, 2999–3012.

Pickin, J.G., Yuen, S.T.S. & Hennings, H. (2002) Waste Management options to reduce greenhouse gas emissions from paper in Australia *Atmospheric Environment*, **36**, 741–752.

Ramanathan, V. and Feng, Y. (2009) Air pollution, greenhouse gases and climate change: Global and regional perspectives. *Atmospheric Environment* 43, 37–50.

Rigamonti, L., Grosso, M. and Guigliano, M. (2009) Life cycle assessment for optimising the level of separated collection in integrated MSW management systems. *Waste Management* 29, 934–944.

Ritzkowski, M. & Stegmann, R. (2010) Generating CO2-credits through landfill in situ aeration. *Waste Management*, **30**, 702–706.

Robinson, H.D., Knox, K., Bone, B.D. & Picken, A. (2005) Leachate quality from landfilled MBT waste. *Waste Management*, **25**, 383–391.

UK Government (1997) Producer Responsibility Obligations (Packaging Waste) Regulations 1997. No. 648. Available from http://www.legislation.gov.uk/uksi/1997/648/contents/made.

UK Government (1998) Packaging (Essential Requirements) Regulations 1998. No. 1165. Available from http://www.coherent.com/Downloads/packagingrequirements.pdf.

Climate Change and the Geotechnical Stability of 'Engineered' Landfill Sites

Robert W. Sarsby

18.1 Introduction

It is widely acknowledged that the Earth is going through a period of climatic change. Currently, for many, the major concerns relating to climate change and infrastructure are flooding and sea level. However, more people are, potentially, at greater risk due to ground instability at landfill refuse disposal sites leading to enhanced mobility of pollution and negative impacts on ecosystems and drinking water supplies.

Higher surface temperatures are expected to cause an increased portion of precipitation to fall in the form of rain rather than snow. This is likely to increase both soil moisture and runoff. Increased heating will lead to increased evaporation and this will decrease the availability of soil moisture needed both for natural vegetation and agriculture in many places. The general concerns with regard to climate change and ground conditions can be summarised as:

- Saturation and instability of slopes;
- Drought conditions and loss of vegetation cover on slopes;
- Overland flows causing erosion, transport of solids, exposure of underlying materials;
- Saturation and detrimental effects on vegetation rooting systems; and
- 'Flooding' of waste masses leading to polluted seepage, wash-out of contaminants, excess pressure on containment systems.

Solutions to Climate Change Challenges in the Built Environment, First Edition.
Edited by Colin A. Booth, Felix N. Hammond, Jessica E. Lamond and David G. Proverbs.
© 2012 Blackwell Publishing Ltd. Published 2012 by Blackwell Publishing Ltd.

18.2 Ground Instability Effects

Climate change affects both shallow and deep groundwater and, therefore, affects the condition and stability of existing slopes, both natural and man-made. Whilst the traditional perception is that landslides are mainly natural hazards (Anon, 2004), this is far from the truth in a developed society with an industrial heritage. Within the infrastructure there will be numerous embankments and cuttings associated with roads and railways, spoil tips and heaps, engineered refuse disposal sites, covered deposits and lagoons of contaminated and polluted materials, repositories for sewage and dredging sludges, water reservoirs and drainage systems, amongst others. Furthermore, increased urbanisation and population growth have resulted in many people (and dwellings) living in close proximity to a variety of man-made slopes, many associated with environmental contaminants and pollutants.

Figure 18.1 is an aerial view of an old sewage treatment works. The lagoon in the centre of the picture is approximately 150 m in diameter and up to 8 m deep. The lagoon embankments were originally raised in the first part of the nineteenth century and were not designed or built to any engineering standard. The lagoon contains processed sewage, which is heavily contaminated with heavy metals, and if any of the embankments become unstable then residents in the houses adjacent to the site and the local groundwater system will be very detrimentally affected.

Figure 18.1 Old sewage sludge lagoon close to residential accommodation and alongside the M60 motorway and ship canal (Manchester in 1993).

Table 18.1 Threats from slope instability.

Case	Effects
Movement of geological materials from natural slopes	Changes in flow patterns, blockages, sedimentation, scour around structures, flooding. Damage to infrastructure, loss of land, damage to drainage systems. Loss of life.
Movement of surface materials on waste tips	Exposure of waste materials to precipitation – generation of polluted runoff water, penetration of water into wastes leading to greater instability, loss of vegetation and biodiversity, loss of land.
Movement of covers over landfilled waste	Exposure of waste to precipitation – generation of leachate, overload of leachate extraction and treatment systems, collapse of waste slopes and creation of flow slides.
Movement of the face of tailings 'dams'	Increase in egress of polluted water through the face, potential for instability of the whole facility (leading to widespread ground and water pollution).
Instability of dyke slopes	Reduction of capacity of flood banks, blockage of land drainage systems, increased sediment in water courses.

Any region with a long industrial heritage will have a variety of sites that can seriously affect the environment through slope instability. The potential threats arising from climate change are outlined in Table 18.1.

Natural slopes evolve as a result of processes such as erosion and movement over a long period of time, which result in a stable slope geometry. On the other hand, man-made slopes are imposed on nature and they are designed for a specific combination of geometry and ground strength at their most critical and vulnerable period. If the ambient conditions (such as climate) change after the construction works have been designed and built, this can drastically alter groundwater regimes from that anticipated. This alteration can negate the proper functioning of the works and may reduce factors of safety (FoS) by so much that slope failures should be expected.

18.3 Stability of Soil Slopes

In general, slope instability results from the in-situ shear stresses exceeding the available shearing resistance of the ground and the major factor influencing in situ ground strength is the groundwater regime and associated porewater pressures. These are affected by extreme rainfall and drainage pattern changes because they alter the weight of the potential sliding mass and decrease the shear strength of the slope material. The stability state of a slope may be assessed by calculating the ratio of the resisting force or moment (due to the shear strength of the slope material) to the disturbing force or moment (due to self-weight of the slope material). This defines the FoS and a value above unity is required for stability. The flatter a slope, the higher its FoS and the more stable it will be. Cost considerations (such as the amount of land occupied by a slope, the volume of material within the slope) determine that engineered man-made slopes are designed with low

FoS, typically in the range 1.15 to 1.50 (Perry et al., 2001). Furthermore, any slope that has undergone some movement and has then become stable will have an overall FoS very close to unity. Hence, it is evident that in any developed region there will be numerous ground slopes that can be destabilised by changes in the severity of rainfall patterns.

Ground movements can be put into three categories:

1. Surface erosion (i.e. loss of surface soil that leads to formation of rills and gullies and loss of vegetation)
2. Shallow planar slides wherein a relatively thin veneer of soil (up to ~2 m thick) slides down the slope parallel to the surface of the slope
3. Deep-seated rotational movement of a large volume of soil that travels as an integral mass

18.3.1 Surface Erosion

It has been estimated that soil erosion by water and wind is responsible for about 56% and 28%, respectively, of worldwide land degradation (Wibisono, 2000). Soil erosion involves a process of both particle detachment and transport by the disturbing agencies. Erosion is initiated by drag, impact or tractive forces acting on individual particles of soil. This erosion is controlled by a number of factors including intensity and duration of precipitation, ground roughness, length and steepness of slope, inherent soil strength, type and extent of cover (Rao and Balan, 2000). Surface erosion by water commences when the kinetic energy of rainfall is transferred to individual soil particles, breaking the bonding between particles and moving the particles upwards and laterally. If the energy of the falling rain can be absorbed or dissipated by vegetation or impacting on some other soil cover or surface obstruction, energy transfer to the soil particles will be reduced and there will be a consequent reduction in soil erosion.

If rainfall exceeds the infiltration capacity of a soil continued precipitation results in overland flow. The runoff collects in small rivulets that may erode very small channels (rills). These rills may eventually merge into larger and deeper channels (gullies). Man-made tips or heaps of waste materials and by-products of mineral processing are particularly prone to the foregoing type of erosion, because they are difficult media on which to establish good vegetation cover. Figure 18.2 shows gullies formed by water running down the surface of a coal-mining spoil tip. If the soil is permeable and has a favourable structure, infiltration will be enhanced and overland runoff will be reduced. Slowing the flow of water downslope reduces the soil transport capacity of the flow, thereby, minimising the displacement of dislodged soil particles.

Ground erosion is frequently the result of human activity, for example formation of cuttings and embankments when roads are built, creation of waste tips, removal of trees and other vegetation from slopes and placement

Figure 18.2 Surface erosion of a coal-mining spoil heap (Nottinghamshire in 2001).

of final caps over engineered landfill sites. In the latter case, the establishment of vegetation on the final caps over landfill sites is essential for long-term effectiveness of the cap as it settles and distorts and undergoes wetting–drying and heating–cooling cycles.

18.3.2 Shallow Slides

Shallow slides within slopes are usually more or less parallel to ground surface – with most failures occurring at depths between 0.5 m and 1.5 m. This type of failure frequently occurs in natural slopes with adverse geological history and in man-made slopes where ground saturation has occurred. The very form of construction in some man-made slopes (e.g. the presence of geomembranes in landfill caps) may promote translational sliding. For a long slope the steady-state groundwater regime consists of a water table approximately parallel to the ground surface; hence, water moves through the ground in a direction parallel to the ground surface. After failure the saturated materials of the slipped mass may easily degenerate into a mudslide that sloughs off the slope, leaving behind the same unstable geometry. The slide may also result in the exposure of man-made materials to adverse conditions (e.g. the exposure of polymeric geomembranes to UV attack, excessive heating and cooling).

Shallow slides are analysed on the basis of force equilibrium along a potential sliding plane. If a small element of the potential translational slide material has weight W, the ground surface has a slope angle of β and the resistance to sliding along the potential slip plane is T, then the FoS against sliding is given by:

$$\text{FoS} = \frac{T}{W \sin \beta} \qquad (18.1)$$

The estimation of W is relatively straightforward but, because of the thinness of the sliding mass, the value of W is relatively small. Thus, the value of FoS is rather sensitive to the absolute value of T – this term is dependent on a number of parameters that are very sensitive to groundwater regime (because of the low value of overburden stress) and ground movement.

18.3.3 Circular Rotation

Deep-seated failure usually involves a large mass of soil sliding along an approximately circular failure surface. Such failure has a variety of consequences; damage/disruption of systems and structures within and on the ground, exposure of underlying materials (both natural and man-made, loss of containment and retention functions, release of contaminated semifluid masses, amongst others. In this type of instability, a large volume of material is involved in the movement and this mass is very destructive by virtue of its size. Furthermore, the failed slope may be unable to resist forces from material within or behind it (such as water within a reservoir dam or slimes within a tailings facility) and a large-scale catastrophic collapse could ensue.

The stability of a given situation is assessed in terms of the moment equilibrium and the FoS is written as

$$\text{FoS} = \frac{M_R}{M_D} \quad (18.2)$$

where M_R is the total resisting moment due to the shear strength of the ground and M_D is the total disturbing moment due to the self-weight of the unstable mass. M_R and M_D are both large. M_D is a direct function of the unit weight of the sliding mass, which is relatively insensitive to likely changes in groundwater regime – most of the material will already have a high degree of saturation (around 80% or more) and a change to full saturation will only increase the unit weight by 4–5%. The greatest potential change is to M_R due to uplift pressures and reduction in frictional resistance.

Many structures composed essentially of waste, particularly those formed some time ago, were not subject to rigorous analysis or design, but because they were stable after construction and still appear stable today, there is tacit assumption they will remain stable regardless of environmental changes. However, the stability may be a result of fortuitous beneficial effects and the incorporation of hidden FoS by discounting effects, which are difficult to quantify in an engineering way. Unfortunately, variations in patterns of rainfall/precipitation, ambient air temperature and evapotranspiration activity due to global climate change will have significant effects on the groundwater regime within soil and the vegetation that it supports. These effects are highly likely to negate the fortuitous factors referred to earlier.

18.4 Soil Shear Strength

The shear strength of a soil (τ) is traditionally defined as arising from two components (i.e. friction and cohesion). The frictional resistance originates from the relative sliding of solid, rough grains within the soil mass and is proportional to the stress pressing the rough surfaces together. Cohesion is an inherent strength, which is due to the previous geological history of the soil and is independent of the current stress between solid particles. However, when considering the stability of the near surface region of soils there are two other effects that need to be taken into account, that is, soil suction and reinforcement by vegetation. Thus, for an examination of the shearing behaviour of soil at very shallow depths the shear strength (τ) should be written as:

$$\tau = f + c' + s + r \tag{18.3}$$

where:

- f = sliding resistance due to contact between hard surfaces which are pressed together;
- c' = inherent cohesion resulting from the geological history of the soil;
- s = sliding resistance due to suction between particles in a partially saturated soil; and
- r = sliding resistance due reinforcement by the roots of vegetation.

18.4.1 Friction

In soils the materials being pressed together are not solids, they are a collection of grains with a variety of shapes with spaces (voids) in between the grains – the volume of a soil, which comprises voids, is typically in the region of 35–50% of the total volume. Terzaghi (1966) accounted for the presence of water on the sliding interface between the solid materials by introducing the concept of effective stress (σ'_n). This is a notional stress that is directly proportional to the stress transmitted across the actual contact points between the solid materials that are sliding. Thus, in general, the total stress (σ_n) acting across a sliding surface has two components – the effective stress (which generates frictional shearing resistance) and the pressure (u) in the pore fluid, which cannot generate directly any shearing resistance.

Thus:

$$\sigma'_n = \sigma_n - u \tag{18.4}$$

and

$$f = \sigma'_n \tan \phi' \equiv \sigma'_n \mu' \tag{18.5}$$

The term μ' is the friction coefficient between solid materials on either side of the sliding surface. The solid materials on either side of the sliding surface may comprise particles of geological materials but in some cases (such as in engineered landfill sites) one side of the sliding surface may be a solid, nongeological material (e.g. polymeric lining material used in the base or cap of an engineered refuse disposal site). If a protective layer (such as a polymer geosynthetic) is used in conjunction with the foregoing type of liner, then the materials on either side of a potential sliding surface could be solid, nongeological material. For soils, μ' ranges from about 0.4 (for clays) to around 1.2 (for gravels). If changes in groundwater regime mean that a soil moves from an unsaturated state to a fully saturated state, then σ'_n may be effectively halved.

18.4.2 Cohesion

Effective cohesion (c') is an inherent shear strength, which is dependent on the current internal state or structure of the slope material, and it is determined by the stress and strain history of the soil and how this has affected the packing of the particles within the slope material. For instance, desiccation of a clay soil creates very high suctions in the water within the voids between the particles (due to surface tension in the water), which pull the particles together so that they interlock to form a dense mass. Subsequent wetting of the particulate mass will reduce the suction within the pore water, but since the particles are not smooth the friction between them will maintain the interlocked nature of the mass and the particles will not return to their original packing state. The interlocking is readily destroyed by physical distortion and straining of the soil so that particles are separated. Such distortion may be a result of the post-construction progressive development of that part of the shear strength of a soil mass, which is due to friction between individual soil grains. However, when a small, 'undisturbed' soil sample is tested in a laboratory significant values of c' (around 20kN/m²) may be measured and initial stability of engineered slopes may support the existence of such large values of c'. Nevertheless, it must be realised that this effective cohesion is rapidly lost due to soil deformation.

18.4.3 Pore Fluid Suctions

When soils are partially saturated the water between individual grains behaves as if it were in a capillary tube and manifests surface tension effects. This induces a negative pressure within the pore water. This pressure is inversely proportional to the equivalent diameter of the 'capillary tube' within the soil and the degree of saturation of the soil. The suction in the pore water pulls the solid particles together and, thereby, creates frictional resistance to sliding of the particles even without the application of an external stress pressing the particles together – it is precisely this partial saturation/surface tension effect that enables a sandcastle (damp sand) to be built with vertical faces but if the sandcastle dries out it collapses as the internal suctions are lost. Mitchell (1976) has quoted potential capillary

Table 18.2 Effect of vegetation on groundwater.

Effect	Comments
Rainfall interception	The volume of rainfall reaching the ground, and the rate at which it does so, is reduced (by the canopy action of plants) so that saturation of the upper soil layer is likely to be prevented.
Surface water runoff	Vegetation reduces the velocity of flow so that the potential for surface erosion is less. Filtering of soil particles out of runoff is performed by stems and organic litter.
Infiltration	Vegetation increases the permeability and infiltration of the upper soil because of the voids formed by roots and the loosening of the surface soil which is caused by their growth. Man-made inclusions divert the water from its downwards path.
Soil moisture depletion	Plant roots extract water from the soil and the effect can extend well beyond the physical extent of the roots – zones of greatest moisture deficiency are 2–4 m below ground surface.

heads within fine sand and silt, which would give rise to apparent effective cohesions, in the range 10–20 kN/m². Fine-grained soils contain very narrow capillary passages and have low permeability and so exhibit very slow moisture content changes and can maintain pore suctions equivalent to tens of metres of water (thereby generating apparent cohesions greater than 50 kN/m²) more or less indefinitely.

18.4.4 Vegetation

Vegetation is the major factor in the prevention of erosion of a soil slope through its influence on the effects of precipitation (Table 18.2). Vegetation also affects the stability of a slope directly by providing mechanical enhancement of the soil strength. The roots (which have a relatively high tensile strength) and the soil form a composite material akin to 'reinforced soil' so that soil and roots act as a united mass. The soil friction angle is unchanged, but there can be an apparent increase in cohesion of around 3–10 kN/m², which can extend for several metres depth in the case of tree roots (Coppin and Richards, 1990). In addition, the tap roots of many large shrubs and trees penetrate to great depths and, thus, into ground that is considerably stronger than the surface material. The trunks and principal roots then act as individual cantilever pile walls providing a buttressing effect to up-slope soil masses. If the trees are sufficiently closely spaced, arching may occur between the up-slope 'buttresses'. Thus, the presence of vegetation, and its type, can be crucial in determining the surface stability of both natural and man-made slopes.

18.5 Landfill Sites

Over the past 50 years there has been a huge increase in the complexity and level of sophistication of design and construction procedures for engineered waste disposal by landfill. In all countries, early landfilling operations involved the uncontrolled infilling of natural depressions and man-made excavations. This approach assumed that within the groundwater the concentrations of any contaminants derived from landfill would reduce to acceptable levels as they dispersed and were diluted under natural processes. Subsequently, landfill philosophy moved to the objective of total containment and isolation of wastes, thereby leading to the following developments in engineered waste disposal practice:

- Earthworks techniques employed by the civil engineering industry are employed to form low-permeability clay liners beneath the impounded refuse;
- Man-made materials (e.g. geomembranes and composite liners) were introduced into the ground to provide basal barriers to the movement of leachate from the base and sides of the refuse;
- Compacted clay caps, and subsequently composite caps incorporating geosynthetics and geomembranes, were placed over the top of refuse when a site was full; and
- The development of vegetation on completed landfills was encouraged to enhance the visual appearance of the site.

Landfilling has progressively given way to above-ground disposal, (i.e. 'landraising' or 'landforming'), and this means that the stability of the side slopes of the refuse mound assumes even greater importance than before.

Since the active life of waste within a landfill deposit is between 25 and 40 years, it is obvious that Europe contains many thousands of active and 'dormant' landfills of very different kinds, in terms of size, type, structural configuration and composition. These facilities could seriously pollute the environment and ecosystems throughout the whole of Europe if climate change caused their containment systems to fail. The design measures that have the potential to be significantly affected by climate change are the final cap over the refuse and the basal lining system.

The final cap has several vital functions:

- To minimise leachate production by preventing the ingress of rain and surface water into the underlying waste;
- To prevent uncontrolled escape of landfill gas from decomposing refuse; and
- To provide a growing medium for vegetation, which will restrict water ingress into the refuse and will also give an acceptable appearance to the completed landfill site.

Consequently, final caps need to be composed of several layers, each of which performs a specific function so that the complete system operates as required. Figure 18.3 illustrates a typical current arrangement of layers – most are relatively thin and create a number of interfaces, which are orientated parallel to the surface of the cap. Such an arrangement of thin sloping layers has great potential for instability through translational sliding.

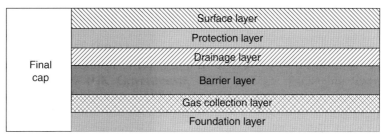

Figure 18.3 Multilayered landfill cap.

To illustrate the potential effects of groundwater regime alteration due to climate change, the following typical situation is analysed:

1. A cap with an overall slope of 1 vertical to 2 horizontal, i.e. a slope angle (β) of 26.5°;
2. Soil with a bulk density of 2 Mg/m^3 (corresponding to unit weight, γ, of 19.6 kN/m^3);
3. The water table is parallel to the surface of the slope, i.e. there is steady-state seepage down the slope; and
4. The upper surface of the soil is vegetated, partially saturated and has been subjected to desiccation due to surface evaporation and water extraction by vegetation.

For translational instability a conventional geotechnical analysis generates the following expression for FoS against downslope sliding (Sarsby, 2000):

$$\text{FoS} = \frac{c'}{\gamma z \sin\beta \cos\beta} + \left(1 - \frac{\gamma_w d}{\gamma z}\right)\frac{\tan\phi'}{\tan\beta} \qquad (18.6)$$

where:

z is the depth to the failure surface; and
d is the height of the water table above the failure surface.

If the apparent shear strength due to pore water suction effects and surface vegetation is included (as mentioned previously), then equation (18.6) becomes

$$\text{FoS} = \frac{c' + s + r}{\gamma z \sin\beta \cos\beta} + \left(1 - \frac{\gamma_w d}{\gamma z}\right)\frac{\tan\phi'}{\tan\beta} \qquad (18.7)$$

Consider now the stability of the cap at the base of the surface layer shown in Figure 18.3 (a summary of the outcome of the analyses is given in Table 18.3). The surface layer is intended to support vegetation, and z would typically be around 0.3 m. Since most vegetation cannot survive in permanently saturated ground the designer of the cap would assume that the surface layer was not saturated (hence, the presence of the drainage layer in Figure 18.4) so that d was equal to zero. Since the surface growing medium

Table 18.3 Effect of changing climate on landfill cap stability.

Precipitation pattern	Stage in life of the slope	Groundwater regime	Vegetation condition	Global Factor of Safety
Current	Cap constructed	Surface layer drained	No vegetation	1.8
	Shortly after construction	Surface layer drained	Healthy vegetation	11.0
Changed due to global warming	Long time after construction	Surface layer saturated	Healthy vegetation	4.3
			Vegetation dies	1.3 initially; slope creeps, factor of safety moves towards unity, and *slope fails*.

Figure 18.4 Landfill failure due to waste sliding on a geomembrane liner (near Oldham in 1997).

would need to be a fine-grained soil, even a conservative designer would be unlikely to assume an effective cohesion $<2\,kN/m^2$ and an effective friction angle $<25°$. For these conditions, the conventional estimation of FoS immediately after construction would give:

$$\text{FoS} = \frac{2}{19.6 \times 0.3 \times \sin 26.5 \cos 26.5} + \left(1 - \frac{9.8 \times 0}{19.6 \times 0.3}\right)\frac{\tan 25}{\tan 26.5} = 1.79 \quad (18.8)$$

The value of 1.79 is significantly greater than unity and would be deemed acceptable. In fact, once vegetation had thoroughly established itself the overall FoS would be considerably larger, due to inclusion of pore water suction effects and root reinforcement (as discussed in Sections 4.3 and 4.4 of this chapter, respectively). If typical values for the foregoing effects are included in the slope stability analyses then the FoS is around 11.0. Consequently, the slope would be very stable and quite acceptable.

Should the surface layer in the cap become saturated (due to increased rainfall and blockage of the drainage layer because of increased washout of fine particles), then the pore suction effect would be lost and d would become 0.3. Thus, the FoS would become 4.3. The new FoS is still significantly greater than unity and so the slope would still be stable and subject to minimal ground movement.

If the ground saturation persists then the vegetation will die and its reinforcing effect will be greatly reduced or even lost completely. In the latter case, the FoS will become 1.3 and the slope should still be stable. However, the contribution that frictional resistance needs to make to achieve an overall FoS of unity is ~32% of the total available and this will mean that significant deformation of the soil will be needed. This movement will cause some loss of the effective cohesion (for the reasons given earlier) and more frictional resistance will have to be developed – this means more ground movement and more loss of effective cohesion and so on, until the slope creeps to failure because all cohesion has been lost.

In the case of a deep-seated failure, the FoS is defined as in equation (18.2), and it is the M_R term that is most prone to decrease as a result of long-term conditions being different from that assumed by a designer. For general rotational failure:

$$M_R = \text{function } \{(\sigma_n - n)\alpha \tan\phi'\} \quad (18.9)$$

The term α is a function of the roughness of the sliding interface and for a particulate material, such as refuse, it is equal to unity. However, if the most critical sliding surface is a soil-on-solid surface (as could occur with a composite liner) the value of α could be around 0.5 (the early forms of geomembrane had very smooth surfaces). As a consequence, changes in water table position coupled with the presence of low-friction interfaces could reduce the assumed design shear strength on the failure surface by up to 75%. An example of a classical deep-seated rotational failure occurred at the Kettleman Hills hazardous waste repository (United States), which was essentially a very large, oval-shaped bowl excavated in the ground to a depth of about 30 m, into which the waste fill was dumped. It contained a highly designed, properly constructed composite double-liner system (consisting of impermeable membranes, compacted clay layers and drainage layers) to protect the ground and groundwater against transport of leachate out of the containment system. In March 1988, a slope failure occurred at the site with lateral displacements of up to 11 m and vertical settlements of up to 4 m being measured. There was significant escape of polluted water. The subsequent technical investigation revealed the failure resulted from reduction of the sliding resistance of the liner system due to the ingress of water (Mitchell et al., 1990). This type of instability is illustrated in Figure 18.4.

18.6 Insights and Solutions

The evidence of the detrimental effect of climate change on the stability of both natural and man-made slopes is clear to see. However, the severity of the after-effects of rainfall-induced slope instability is increasing with time and some ground engineering works, which have to been stable to date, are likely to fail in the future. Of particular concern is the potential for large-scale destabilisation of engineered landfills by future extreme precipitation events and the resultant pollution of groundwater and continued discharge of greenhouse gases into the atmosphere.

The dumps, tips and lagoons resulting from mineral extraction and processing are also potential sources of major disasters due to climate change. Not only does rainfall leach-out toxic substances but also it can cause slope failures and flowslides that lead to loss of life. The problem of colliery tip instability was brought into sharp focus by the disaster at Aberfan, United Kingdom, in 1966, when a slide involving some $100\,000\,m^3$ of colliery waste resulted in the death of 144 people. The flowslide travelled a distance of about 480 m to a junior school that it largely destroyed – 116 children were killed. Flowslides resulting from waste heap failure are not confined to tips of colliery spoil – notable failures have occurred within limestone waste, China clay spoil, fly ash, gold and copper tailings, amongst others. In all cases, the primary cause of failure was rapid saturation of dumped waste.

Consequently, there is a need to develop national and regional integrated risk assessment management plans for man-made slopes that will:

- Identify the causal mechanisms of slope instability;
- Take account of the influence of human activities;
- Develop an accessible method for rapid assessment of site vulnerability; and
- Incorporate indirect appraisal and monitoring of slope condition.

References

Anon (2004) *Research for erosion, compaction, floods and landslides*. European Commission, Directorate-General Environment, Soil Thematic Strategy, Task Group 1 on, Final Report of Working Group. European Commission, Brussels.

Coppin, N.J. & Richards, I.G. (1990) *Use of Vegetation in Civil Engineering*. CIRIA Butterworths, London.

Mitchell, J.K., Seed, R.B. & Seed, H.B. (1990) Kettleman Hills waste landfill slope failure. 1: liner-system properties. *ASCE Geotechnical Engineering*, **116**(4), 647–668.

Perry, J., Pedley, M. & Reid, M. (2001) *Infrastructure Embankments C550*. CIRIA, London.

Rao, G.V. & Balan, K. (2000) *Improving the Environment by Vegetation Growth, in Coir Geotextiles – Emerging Trends*. Kerala State Coir Corporation, Alappuzha, India.

Sarsby, R.W. (2000) *Environmental Geotechnics*. Thomas Telford, London.

Terzaghi, K. (1966) *Theoretical Soil Mechanics*. John Wiley & Sons, New York.

Wibisono, G. (2000) *The effect of water on the erodibility of fibre reinforced or stabilised kaolin*. MSc thesis, University of Birmingham, UK.

Water Resources Issues and Solutions for the Built Environment: Too Little versus Too Much

Susanne M. Charlesworth and Colin A. Booth

19.1 Introduction

Water is a fundamental requirement for an array of environmental, economic and social needs, from ecosystems, habitats, biodiversity and agricultural necessities through to industrial, commercial and recreational activities, plus it also offers wealth and political stability, but none is more important than its requirement for human quality of life (e.g. drinking, cooking and hygiene). Unfortunately, water is not universally abundant and for many nations the uneven distribution of water and human settlement continues to create growing problems of freshwater availability and accessibility. Moreover, based on UN medium population projections, >2.8 billion people will face water stress or scarcity conditions by 2025.

In extreme contrast to water shortages, it is also a fact that the number of global disasters attributable to floods (e.g. Pakistan and China in 2010) is increasing and, with ~82% of the global population living on lands prone to flooding, it is a problem that will undoubtedly worsen with climate change. The built environment has become more prone to flooding because urbanisation has meant that landscapes, which were once porous and allowed surface water to infiltrate, have been stripped of vegetation and soil and have been covered with impermeable roads, pavements and buildings. As a consequence, this has meant precipitation, which would have once soaked away, now increasingly collects as surface runoff, enhancing the storm peak (on a hydrograph), reducing the lag time (the time between peak rainfall and peak river discharge) and promoting flooding of immediate neighbourhoods and/or those downstream.

Unfortunately, those circumstances portrayed above will further deteriorate when they are coupled with predicted climate change because it is anticipated

---

Solutions to Climate Change Challenges in the Built Environment, First Edition.
Edited by Colin A. Booth, Felix N. Hammond, Jessica E. Lamond and David G. Proverbs.
© 2012 Blackwell Publishing Ltd. Published 2012 by Blackwell Publishing Ltd.

that global rainfall regimes will dramatically alter, some areas will experience wetter winters and drier summers but may also experience changes in the patterns of rainfall frequency and intensity (Intergovernmental Panel on Climate Change, 2007). Therefore, more than ever before, water supply and urban water management have become necessary requirements for the attention of governments, local authorities, relevant organisations and property owners.

This chapter addresses the main water resources challenges surrounding both water supply shortages and property flooding issues, by outlining appropriate solutions to cope with present and future climate changes. For the focus of the reader, this article is directed at the UK scenario but its content accords with similar issues and solutions attributable to the built environments of many global towns and cities.

19.2 Too Little Water: Water Supply Shortages

Most media coverage often implies water shortages to be an almost exclusive issue for developing countries, so it may surprise many people to know that the United Kingdom has water shortages. For instance, in England, particularly in the southeast (where ~70% of the public supply is from groundwater), water demand outweighs natural water availability and groundwater recharge, such that there is less water available per capita than in many Mediterranean countries (Environment Agency, 2007a). Coupled with an increasing population opting to live in this region and a concomitant demand for new homes to be built, these circumstances are clearly unsustainable and are the focus of government attention.

Legislative guidance is provided from the European Commission, through the EU Water Framework Directive (2000/60/EC). This was introduced for the protection, improvement and sustainable use of water across Europe because the existing water policy was fragmented and there was a need for an integrated approach to water management. Member states implemented the ruling through a Common Implementation Strategy, which (in the United Kingdom) resulted in the Water Act 2003 (England and Wales), the Water Environment and Water Services Act 2003 (Scotland) and the Water Environment (WFD) Regulations 2003 (Northern Ireland).

These legislations have four wide-reaching intentions: (1) achieve the sustainable use of water resources, (2) strengthen the voice of consumers, (3) permit a measured increase in competition and (4) promote water conservation (http://www.opsi.gov.uk).

As a consequence, the Department for Farming and Rural Affairs (DEFRA) produced an outline of the government's water strategy for England through a report entitled *Future Water* (DEFRA, 2008a) and, in 2010, the Department for Communities and Local Government (DCLG) produced a revised *Building Regulations G* document (http://www.planningportal.gov.uk).

19.2.1 Household Water Usage and Savings

There has been a dramatic increase in water demand since the 1940s, which is largely attributed to an increase in the number of households and a decrease in occupancy numbers (e.g. single-person households). Furthermore, changes in lifestyles have meant there has been a dramatic increase in personal water use (70%) since the 1970s (DCLG, 2007), which is mainly attributable to the number and frequency of use of household appliances (e.g. washing machines and dishwashers). As a consequence, nowadays, average UK water use per person per day is 150 litres (DEFRA, 2008b), which is sizeably more than the European neighbours of Germany (127 litres), Netherlands (124 litres) and Belgium (108 litres).

The UK government has set ambitious water usage targets of an average of 130 litres per person per day for 2030 (DEFRA, 2008a). It is anticipated this will largely be achieved by improving standards in newly constructed housing stock through compliance with category 2 of the Code for Sustainable Homes (DCLG, 2008), which offers higher ratings for homes built with water efficiency measures installed (e.g. the highest target, level 6, aims for 80 litres per person per day), and through amendments to the Building Regulations. Further achievements will improve water efficiency in existing housing stock through refurbishment and end-of-life replacement of fittings that are innovative and improve performance.

For further details on water efficiency, the reader is referred to the literature of the Department of Communities and Local Government (DCLG; 2006, 2007), the Environment Agency (2007b) and the National House Building Council Foundation (Griggs and Burns, 2009), plus the visionary report of Thompson *et al.* (2007), which outlines the feasibility of approaches to achieve water neutrality (i.e. for every development, total water use in the region after the development must be equal to or less than the total water use in the region before the new development).

19.2.2 Domestic Water Savings

A breakdown of UK water usage reveals: toilet flushing 30%, personal washing from baths and taps 21%, clothes washing 13%, personal washing from showers 12%, washing up 8%, outdoors 7%, other 5% and drinking 4% (http://www.waterwise.org.uk). These figures illustrate the potential for where the simplest changes in household water use could have the greatest influence on future water demand. Therefore, many exemplars of domestic water saving devices typically focus on toilets, taps, showers, reuse and outdoor aspects (Table 19.1). Other recognized water saving approaches include fixing leaks and dripping taps, updating household 'white-goods' appliances and installing water meters (Waterwise, 2008), plus the development of water-efficient landscaping (see the SUDS section later in this chapter 19.3.1).

Some of the most noteworthy examples of domestic water savings include: changing a toilet from an old (9 litre) to new type (4.5 litre) can halve the water used per flush, fitting an aerated showerhead can reduce the flow rate

Table 19.1 Exemplars of domestic water saving devices (adapted from Waterwise, 2008).

Water savings use	Exemplar devices
Toilets	Cistern displacement devices Retrofit dual flush devices Retrofit interruptible devices Replacement dual-flush toilets Replacement low-flush toilets
Taps	Tap insert aerators Low-flow restrictors Push taps Infrared taps
Showers	Low-flow showerheads Aerated showerheads Low-flow restrictors Shower timers Bath measures
Reuse	Small-scale rainwater harvesting water butts Large-scale rainwater harvesting systems Greywater recycling
Outdoors	Hosepipe flow restrictors Hosepipe siphons Hose guns Drip irrigation systems

by 28% (~3 litres per minute), fixing a tap that is leaking only two drops per second can save 9500 litres per year and updating to a modern water efficient dishwasher that uses only 12–18 litres of water to wash 12 place settings is notably lower than the equivalent 40 litres of hot water used to wash the same crockery by hand (Environment Agency, 2007b).

Water meters have been compulsorily installed in all new UK homes since 1997 and, as a consequence, it is now widely acknowledged that metering reduces water usage because it raises awareness of the financial incentive of using less. Moreover, it is generally regarded by many that paying for the quantity of water consumed is the fairest way to pay (similar to other utility services). At present, ~30% of homes have water meters but it is expected that it will take many years before this will hugely alter patterns of water demand. Therefore, the Environment Agency would like the majority of homes (80%) in seriously stressed areas of the England and Wales to be metered by 2015 (Thompson et al., 2007), which will also allow more complex tariff structures to be formulated. This accords with the recommendations of the housing review of Barker (2004) for changes in the water charging system to encourage customers to use water wisely but also highlights that it should be affordable for all.

For further details on water saving devices, the reader is referred to the literature of the Environment Agency (2003, 2007a,b), the Water Regulations Advisory Scheme (2005) and Waterwise (2008).

19.2.3 Rainwater Harvesting

In tandem with many homeowners becoming increasingly aware of the financial benefits of water conservation inside their homes, they are also becoming increasingly aware that precipitation falling outside their home offers a free supply – if harvested. Rainwater harvesting is the collection of water derived from roofs or hard standings, which, rather than draining directly offsite (normally connecting to a community drainage system), can be harvested and, after treatment, used for domestic purposes not requiring drinking water quality (e.g. toilet flushing, washing machines, garden watering or vehicle washing).

Rainwater harvesting systems can range from the straightforwardness of attaching a water butt to the down-pipe outside a building and collecting the gutter water draining from the roof area and then using it for garden watering, or it can stretch to the complexity of directing collected roof water through a filter (removing debris) and then transferring the water to an underground storage tank (available for individual or communal dwellings) buried beneath gardens, where unfiltered particles settle out, before the treated water is returned (pumped on demand) to the house for inside activities not requiring potable water (e.g. toilet flushing and washing machines) or to an external tap for outside activities (e.g. garden watering or vehicle washing). The latter systems can be expensive (costing ~£2000, depending on tank size) and the payback can take several years; by comparison, water butts can be purchased relatively cheaply (~£20), with rapid payback for those households with water meters.

Based on the domestic usage figures outlined earlier, those activities not requiring potable water (toilet flushing, clothes washing and outdoors), collectively accounting for ~50% of the daily water requirement, could potentially be supplied from a rainwater harvesting system. As an example, it is feasible, where rainwater-harvesting technology is installed, for a house offering a roof area of $120\,m^2$, with a 0.85 pitch tile roof collection co-efficient and a 0.6 rainwater pre-filter efficiency, situated in a region where the annual rainfall is 900 mm, could harvest and supply 55,080 litres of water per annum. Now, assuming the house had a two person occupancy requiring 54,750 litres per annum for nonpotable quality activities (based on 50% of 150 litres per day), rainwater harvesting would be suffice to supply these uses. However, where the occupancy is more and/or to avoid seasonal shortfalls (particularly from anticipated climate changes), it would be recommended that the system should be backed up from public supplies.

For further details on rainwater harvesting, the reader is referred to the literature of the Construction Industry Research and Information Association (Leggett, 2001a, b) and the Environment Agency (2003).

19.2.4 Greywater Harvesting

There is a greater public awareness of the waste hierarchy (see Chapter 17) to highlight opportunities to reduce, reuse and recycle; there is also a growing recognition that this can extend to water already used inside the home.

Greywater is the unwanted household wastewater, derived from showers, baths, washbasins, washing machines and kitchen sinks, which can be harvested and, after treatment, used for domestic purposes not requiring drinking water quality (e.g. toilet flushing, garden watering or vehicle washing).

Greywater systems can range from the simplicity of installing a valve (costing ~£30) to direct wastewater from the waste pipe, draining the bath or shower, to a water butt and then used for garden watering (since the water is untreated, it is important that it is not stored for long as the water quality deteriorates (Dixon and Fewkes, 1999). Alternatively, greywater technology can extend to the sophistication of biomechanical systems (costing ~£3500), which combine biological and physical treatments (using aeration, settlement, filtration and disinfectant) before storing the treated water in tanks enclosed within the housing of the system. The latter systems are not always appropriate for retrofitting so they need to be incorporated as part of new builds. It is also noteworthy, for house builders to meet the target (80 litres) for level 6 of the Code for Sustainable Homes, greywater and/or rainwater harvesting is most likely to be a necessity.

Again based on the domestic usage figures outlined earlier, it is immediately apparent that treated greywater collected from personal washing alone (33%) could easily be used to offset personal toilet usage requirements (30%) and the surplus could contribute to outdoor usage activities (7%). Unfortunately, greywater installations are not widespread in the United Kingdom because the initial purchase and maintenance of the systems can be expensive. However, those households with water meters could make financial savings. Elsewhere, in other countries, greywater recycling has been made mandatory for particular types of buildings (e.g. Japan), while other governments offer grants and subsidies for those households installing greywater systems (e.g. Cyprus and Australia).

For specific details on greywater technologies, the reader is referred to the literature of the Construction Industry Research and Information Association (Leggett, 2001a, b) and the Environment Agency (2008).

19.3 Too Much Water: Urban Flooding

There are many reasons for flooding in the built environment, which causes considerable mortality globally each year. Jonkman (2005) lists six: coastal flooding, flash floods, fluvial flooding, drainage problems, tsunamis and tidal waves or tidal bores. According to the Parliamentary Office of Science and Technology (2007), urban flooding is due to endogenic factors within the drainage basin. These include drainage infrastructure that is ageing and, therefore, unable to cope with excess storm water; additional construction that seals the surface, making it impermeable together with the addition of the building's drainage to a sewerage infrastructure, which is already at capacity; changes to river hydrology and the encroachment of impermeable surfaces from buildings already in place as householders pave over their front gardens (Perry and Nawaz, 2008). Finally, the impacts of global climate change are an exogenic factor that will change the characteristics of urban drainage basins according to their geographic location (Evans *et al.*, 2008; White and Howe, 2004).

It is feasible that engineers could build even larger and stronger permanent flood defences alongside river corridors to protect those properties built on floodplains but these can be costly and are not always aesthetically pleasing to riverside communities (National Audit Office, 2001, 2007). Alternative approaches include temporary demountable or inflatable flood defences, which have been utilised on the River Severn at Shrewsbury and Ironbridge, Shropshire. These can be less costly but do require considerable human resources to deploy, and also need storing between flood events. However, not allowing floodwater to spread over the natural floodplain simply means that floodwater is being redirected elsewhere and, in doing so, can cause greater flooding and devastation further downstream.

The factors outlined above contribute to an unsustainable state of affairs; whereby, even without the unpredictable nature of the impact of climate change on hydrological regimes, flooding in cities is becoming more frequent and more intense. The following sections review a sustainable approach to drainage, which mimics natural infiltration, and detention of stormwater and which is enshrined in the English Flood and Water Management Bill (2010).

19.3.1 Sustainable Drainage Systems (SUDS)

Conventional hard drainage tends to concentrate on managing water quantity by gathering all the runoff water from impervious streets and pavements into storm sewer systems, which pass via gullypots, pipes and water treatment facilities into the receiving watercourse. SUDS are a suite of measures that treats water in a different way by encouraging infiltration and detention of surface water on site.

The often-quoted SUDS triangle (e.g. Martin *et al.*, 2001) illustrates the multiple benefits of the SUDS approach, in which there is an equal balance between water quantity, water quality, and biodiversity or amenity, in contrast to that of conventional drainage. There is much evidence of water quality improvements in the literature (e.g. Charlesworth *et al.*, 2003; Stovin *et al.*, 2007), although less research has been carried out regarding biodiversity benefits and even less on amenity. For further details on the SUDS approach, the reader is referred to the literature of the Construction Industry Research and Information Association (Martin *et al.*, 2001; Woods-Ballard *et al.*, 2007).

19.3.2 SUDS Surface Water Management Train

Figure 19.1 shows the manner in which the SUDS Surface Water Management Train (after Martin *et al.*, 2001) begins with prevention for individual premises, involving good housekeeping measures, then controlling as much of the water at source as possible, followed by water management at site scale, and then regionally. At each stage water is allowed to percolate either to the receiving watercourse or to groundwater through devices, which provide water treatment capabilities and amenity and biodiversity benefits.

The water being conveyed through each part of the system is intentionally slowed down, so that each step offers an opportunity for runoff waters to be infiltrated to join groundwater stores, or evaporated, either by utilising vegetated SUDS devices such as swales, wetlands or green roofs, or from the hard surfacing and subsurface aggregates of a porous paving system. Therefore, any runoff water reaching a watercourse will have an increased lag time and will also have been lessened in quantity.

Many developed country's planning laws (e.g. England's Planning Policy Statement 25; PPS25) (http://www.communities.gov.uk), stipulate that a new build must render the site able to deal with surface water at greenfield runoff rates (i.e. the rate at which the site would have infiltrated or stored the water prior to development). It is, therefore, necessary to implement good site design from the initial planning of the site to be developed and to integrate measures to prevent runoff and pollution by minimizing the installation of impermeable paved areas, utilization of roof space by installing green roofs (e.g. Stovin, 2010; see Chapter 14) and optimizing the use of 'green infrastructure' such as street trees in the drainage design (e.g. Stovin et al., 2008).

The SUDS surface water management train (Figure 19.1) begins at the individual house level, whereby runoff should be restricted to the curtilage of the house and prevented as far as possible. This can be accomplished using rainwater harvesting, pervious pavements, green roofs or soakaways on individual plots. At a larger scale, source control manages surface water runoff from several dwellings using swales. For example, as used in the Upton development, Northamptonshire (Figure 19.2), where swales have been used to convey stormwater throughout the site and are directed into ponds and wetlands downstream of the development.

Site control involves the management of water from several subcatchments, or a larger area and can include routing water from roofs and car parks to one large soakaway or infiltration basin for the whole site. The Oxford and Hopwood motorway service areas (M6 and M42, respectively), in England, have SUDS trains managing vehicle parking, amenity and fuel supply areas and also some access roads, utilising porous paving, swales, ponds, filter strips and constructed wetlands. Hopwood covers 34 hectares in total, of

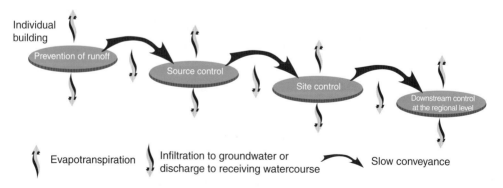

Figure 19.1 The SUDS management train (after Martin et al., 2001).

Figure 19.2 A swale installed to convey stormwater at the Upton development, Northamptonshire, UK.

Figure 19.3 An amenity feature that is part of the SUDS management train at Hopwood motorway services area, UK.

which 9 hectares is the motorway service area itself and 25 hectares has been designated a wildlife reserve (Figure 19.3). According to Heal *et al.* (2008), peak flows and pollutant concentrations in both sediment and water sampled from the management train at Hopwood decreased both spatially throughout the train and into the receiving watercourse, but also temporally over the eight study years.

The Dunfermline Eastern Expansion (DEX) is identified as being the largest UK site to use widespread sustainable drainage methods, with 550 hectares of industrial, commercial, residential and recreational areas. SUDS devices include soakaways to individual houses, filter drains and swales, detention ponds and wetlands as well as a set of regional ponds and wetlands, which are used to achieve the maximum attenuation of stormwater flow. Porous paving is also used to convey water into further attenuation basins and thence into the regional wetland. Whilst the quality of runoff initially drove the installation of the SUDS train by SEPA, other concerns included historical problems with flooding downstream, as well as the potentially high velocity and volume of water leaving the DEX site.

It has been shown that SUDS are flexible and have multiple benefits. However, circumstances may occur where a sustainable drainage approach cannot be employed or cannot be retrospectively adapted to an existing built-up area. The following sections explore what the individual householder can do to protect their home from floods using flood resilient infrastructure at the individual property scale.

19.4 Property-Level Flood Resistance versus Resilience Measures

Where properties are known to be in a flood risk area (or future risk), rather than ignore the situation and wait for a flood event to happen, there is an opportunity for property builders and/or owners to install flood resistance and/or flood resilience measures outside and/or inside buildings. Both of these approaches are outlined beneath but the reader is referred to the literature of the Department of Farming and Rural Affairs (2008a, b; and see Bowker, 2007) and the Construction Industry Research and Information Association (McBain *et al.*, 2010) for further reading.

19.4.1 Flood-Resistant Measures

Flood resistance attempts to halt floodwater entry into a property to prevent floodwater damaging its fabric (Bowker, 2007). These measures can be either permanent (e.g. earth bund walls, periphery fences with sealed gates, raising building thresholds, storm porches to external doors and anti-backflow valves on sewers) or temporary (e.g. air-brick covers, external door guards and flood skirts) features. Unfortunately, because of the nature of these features, some permanent measures do require planning permission and, furthermore, they are mostly disruptive to the property owner and can also be expensive to install. By comparison, temporary measures are low cost and the installation is simple, with minimal disruption for the property owner, but, because they are temporary features, a person must always be present at the time of a flood to install the devices.

Flood resistance features are known to function as designed (BSI kite mark) but, in some instances, they have not stopped properties being flooded (e.g. from groundwater ingress or from shared walls of adjoining properties).

Moreover, when floodwaters reach >600 mm outside a property it is advisable to allow water entry, or the owner will risk damaging the structural integrity of their building. Although flood resistance products do offer a viable solution for property owners to address flooding issues, these measures alone are not always the complete answer to the property flooding problem.

19.4.2 Flood Resilience Measures

In contrast to flood resistance, flood resilience permits floodwater entry because the property is designed and constructed in a manner that floodwater impact is minimised, with no permanent damage caused, structural integrity is maintained and subsequent drying and cleaning are easily facilitated (Bowker, 2007). Most resilience measures are permanent and expensive, and require professional craftsmanship such as internal tanking, concrete floors, raised electrical sockets, horizontal replacement plasterboard, resilient kitchens (plastic, stainless steel and freestanding removable units) with raised white goods (fridges, freezers and washing machines), resilient internal walls (rendered, tiled and/or coated) and plastic skirting board. However, flood resilience features do offer the advantage to property owners that they can return and continue normal activities, after drying and cleaning, within a matter of days, as apposed to several months or sometimes more than a year after a flood event. Furthermore, because internal property damage will be limited and replacement costs will be minimal, insurance company policy premiums are likely to be lessened.

Until recently, normal insurance company policy after a flood was repair with no betterment. However, changes are happening, insurance companies will now discuss resilience as part of the repair strategy, but the property owner must pay for any additional costs. Unfortunately, the majority of flooded property owners simply want their homes returned to normal and most opt not to install flood resistance and/or resilience measures. Therefore, there is a need for the flood repair industry to develop flood products that are affordable to property owners, cause minimal disruption, are easy to install, address all water entry routes into buildings and also allow property owners to continue occupancy soon or immediately after a flood event. That said, a flexible skirting system (patent GB2452423) has been proposed (Beddoes & Booth, 2010) but it seems the flood repair industry is reticent in considering alternatives, maybe because of the expense of new product testing for BSI kite marking and/or because many companies have already heavily invested in existing products.

19.5 Present and Future Water Resources Solutions

Various approaches have been outlined and solutions proposed above, which address the main water resources challenges surrounding both water supply shortages and property flooding challenges in the United Kingdom. For both issues, examination of relevant literature illustrates that change is almost exclusively initiated by European and national governments and seldom by a particular sector of the relevant industries. Furthermore, the fact that Government has to take actions to encourage (or force) property builders to act, raises the notion that the construction industry chooses not to introduce water-related measures voluntarily. That said, this may be attributed to ignorance or may be because they are opting not to actively address the issues or, perhaps mostly likely, because it is not in their own financial interests. Therefore, the perceived way forward is to feed-down further mandatory water policy and legislation, with minimal freedom of choice, through a programme of increasingly tighter controls on new housing design and water consumption. For instance, it is mandatory for all new social housing to be built to a minimum level 3 for the Code for Sustainable Homes but compliance remains voluntary for privately built housing. However, to encourage the participation of house builders and raise the awareness of house buyers, all new houses are required to have a code rating, similar to the former Home Information Pack (http://www.homeinformationpacks.gov.uk) and those houses not assessed against the Code must include a nil-rated certificate, which may affect the sale and/or profit they can make.

Rather than remaining reliant on the Government and relevant industries to address water resources challenges, it is apparent that society must also take responsibility to change its frivolous behaviour and attitudes to water and, where appropriate, embrace alternative water supply and/or property flooding technologies. For instance, as a nation where there are water shortages, and in a world where energy use and carbon footprints are measurable, there is no justifiable reason why society should use potable water for flushing toilets – people should embrace water saving and harvesting approaches. Similarly, in the current economic climate, there is no sizeable reason why the insurance industry should be subsidised by the government (or by the premiums of other policyholders) to pay for the repairs and replacements of properties that have been flooded, where the owner has chosen not to install flood resistance and/or resilience measures. Therefore, although no mandatory policy exists for builders or homeowners (yet), future flooding policy and legislation should have minimal freedom of choice and be intrinsically linked to re-educating people to change and promoting awareness campaigns (e.g. Environment Agency flood pamphlets).

19.6 Conclusions

Without doubt, the next generation of water resources managers have real and complex engineering, environmental, social, economic and political issues to consider in their decision making, which requires the forward-thinking integration of governments and industry with an informed and steadfast society. Ultimately, their roles will influence the sustainability of the built environment and lifestyle quality and, moreover, may also contribute to global security.

References

Barker, K. (2004) *Delivering Stability: Securing Our Future Housing Needs*. HM Treasury, London.

Beddoes, D.E. & Booth, C.A. (2010) Property level flood protection: a new effective and affordable solution. In: *Flood Recovery Innovation and Response II*, ed. de Wrachien, D., Proverbs, D.G., Brebbia, C.A. & Mambretti, S. WIT Press, Southampton, 271–280.

Bowker, P. (2007) *Flood Resistance and Resilience Solutions: an R&D Scoping Study*. Department of Environment, Food and Rural Affairs (DEFRA), London.

Charlesworth, S.M., Harker, E. & Rickard, S. (2003) A review of sustainable drainage systems (SUDS): a soft option for hard drainage questions? *Geography*, 88, 99–107.

Department of Communities and Local Government (DCLG) (2006) *Water Efficiency in New Buildings: A Consultation Document*. Communities and Local Government Publications, Wetherby.

Department of Communities and Local Government (DCLG) (2007) *Water Efficiency in New Buildings: A Joint DEFRA and DCLG Policy Statement*. Communities and Local Government Publications, Wetherby.

Department of Communities and Local Government (DCLG) (2008) *The Code for Sustainable Homes: Setting the Standard in Sustainability for New Homes*. Communities and Local Government Publications, Wetherby.

Department of Farming and Rural Affairs (2008a) *Future Water: The Government's Water Strategy for England*. The Stationery Office, London.

Department of Farming and Rural Affairs (2008b) *Action Taken by Government to Encourage the Conservation of Water*. The Stationery Office, London.

Dixon, A.M. & Fewkes, A. (1999) Guidelines for greywater re-use: health issues. *Water and Environment*, 13, 322–326.

Environment Agency (2003) *Harvesting Rainwater for Domestic Uses: An Information Guide*. Environment Agency Publications, Bristol, UK.

Environment Agency (2007a) *Making Better Use of Our Water Resources: Identifying Areas of Water Stress*. Environment Agency Publications, Bristol, UK.

Environment Agency (2007b) *Conserving Water in Buildings*. Environment Agency Publications, Bristol, UK.

Environment Agency (2008) *Greywater: An Information Guide*. Environment Agency Publications, Bristol, UK.

Evans, E.P., Simm, J.D., Thorne, C.R., Arnell, N.W., Ashley, R.M., Hess, T.M., Lane, S.N., Morris, J., Nicholls, R.J., Penning-Rowsell, E.C., Reynard, N.S., Saul, A.J., Tapsell, S.M., Watkinson, A.R. & Wheater, H.S. (2008) *An update of the Foresight Future Flooding 2004 Qualitative Risk Assessment*. Cabinet Office, London.

Griggs, J. & Burns, J. (2009) *Water Efficiency in New Homes: An Introductory Guide for Housebuilders*. BRE Press, National House Building Council Foundation, Amersham, UK.

Heal, K.V., Bray, R., Willingale, S.A.J., Briers, M., Napier, F., Jefferies, C. & Fogg, P. (2008) Medium-term performance and maintenance of SUDS: a case study of Hopwood Park Motorway Service Area, UK. 11th International Conference on Urban Drainage, Edinburgh.

Intergovernmental Panel on Climate Change (2007) *Climate Change 2007: Synthesis Report. Contribution of Working Groups I, II and III to the Fourth Assessment Report of the Intergovernmental Panel on Climate Change*, ed. Pachauri, R.K. & Reisinger, A. IPCC, Geneva.

Jonkman, S.N. (2005) Global perspectives on loss of human life caused by floods. *Natural Hazards*, **34**, 151–175.

Leggett, D., Brown, R., Stanfield, G., Brewer, D. & Holliday, E. (2001a) *Rainwater and Greywater Use in Buildings: Best Practice Guidance*. Construction Industry Research and Information Association, London.

Leggett, D., Brown, R., Stanfield, G., Brewer, D. & Holliday, E. (2001b) *Rainwater and Greywater Use in Buildings: Decision-Making for Water Conservation*. Construction Industry Research and Information Association, London.

Martin, P., Turner, B., Dell, J., Payne, J., Elliot, C. & Reed, B. (2001) *Sustainable Urban Drainage Systems – Best Practice Manual for England, Scotland, Wales and Northern Ireland (C523)*. Construction Industry Research and Information Association (CIRIA), London.

McBain, W., Wilkes, D. & Retter, M. (2010) *Flood Resilience and Resistance for Critical Infrastructure (C688)*. Construction Industry Research and Information Association (CIRIA), London.

National Audit Office (2001) *Inland Flood Defence*. The Stationery Office Limited, London.

National Audit Office (2007) *Building and Maintaining of River and Coastal Flood Defences in England*. The Stationery Office Limited, Norwich.

Perry, T. & Nawaz, R. (2008) An investigation into the extent and impacts of hard surfacing of domestic gardens in an area of Leeds, United Kingdom. *Landscape and Urban Planning*, **86**, 1–13.

Parliamentary Office of Science and Technology (2007) *Postnote 289: Urban Flooding*. Available from http://www.parliament.uk/documents/upload/postpn289.pdf.

Stovin, V. (2010) The potential of green roofs to manage urban stormwater. *Water and Environment Journal*, **24**, 192–199.

Stovin, V., Swan, A. & Moore, S. (2007) *Retrofit SUDS for Urban Water Quality Enhancement*. EA, BOC Foundation, London.

Stovin, V.R., Jorgensen, A. & Clayden, A. (2008) Street trees and stormwater management. *Arboricultural Journal*, **30**, 297–310.

Thompson, J., Langley, E., Ballantyne, J. & Perry, C. (2007) *Towards Water Neutrality in the Thames Gateway*. Environment Agency Publications, Bristol, UK.

Water Regulations Advisory Scheme (2005) *Conservation of Water: An IGN for Architects, Designers, Installers and Occupiers of Premises*. WRAS Information and Guidance Note, Issue 2. Water Regulations Advisory Scheme, Oakdale, UK.

Waterwise (2008) *Evidence Base for Large-Scale Water Efficiency in Homes*. Waterwise Publications, London.

White, I. & Howe, J. (2004) The mismanagement of surface water. *Applied Geography*, **24**, 261–280.

Woods-Ballard, B., Kellagher, R., Martin, P., Jefferies, C., Bray, R. & Shaffer, P. (2007) *The SUDS Manual (C697)*. Construction Industry Research and Information Association (CIRIA), London.

Organisational Culture and Climate Change Driven Construction

Nii A. Ankrah and Patrick A. Manu

20.1 Introduction

In trying to establish appropriate mechanisms for delivering sustainability solutions that address climate change needs, there is a need to understand how the radical processes, protocols, structures, products and tools associated with this development paradigm will be impacted by organisational value systems, beliefs and behaviour norms. This chapter provides insight into cultural issues to be considered in the drive to minimise emissions in construction, and to propose strategies for overcoming potential cultural inertia. A theoretical discussion on the role of organisational culture in climate change driven construction is provided, followed by an examination of the antecedents of cultural change. Key publications informing this review include the *Stern Review* (Stern, 2006a), commentary on the implications for construction (cf. Department of Business, Innovation and Skills [BIS], 2010; Innes and le Grand, 2006; Shipworth, 2007) and Knott *et al.* (2007), which provides a robust summary on achieving cultural change. The chapter concludes with a case study of an organisation's efforts to stimulate culture change. This is used to highlight key success factors that facilitate the cultural transformation required to support a climate change driven construction agenda.

20.2 Climate Change and Construction

The construction industry consumes a significant amount of scarce natural resources and generates a lot of waste. The extent to which such consumption can be sustained without significant impacts on future generations has been

---

Solutions to Climate Change Challenges in the Built Environment, First Edition.
Edited by Colin A. Booth, Felix N. Hammond, Jessica E. Lamond and David G. Proverbs.
© 2012 Blackwell Publishing Ltd. Published 2012 by Blackwell Publishing Ltd.

debated extensively. What is quite apparent is that there is a significant environmental cost associated with current patterns of consumption. In the United Kingdom, it is on record that construction consumes over 420 million tonnes of material resources (Environment Agency, 2003) and generates 90 million tonnes of waste per annum of which 19 million (ca. 20%) is material delivered to sites but never used. As much as 40% of all carbon emissions come from domestic and nondomestic buildings (Carbon Trust, 2010). It is widely believed that in the long term, these patterns of consumption and waste will translate into very significant economic costs, the cost of which will far outweigh any present costs of trying to take corrective action (Stern, 2006a, b). Consequently, climate change driven construction is being advocated to change the current patterns of consumption and waste in the construction industry and wider society (cf. BIS, 2010). It is within this context of stimulating change that the role of culture takes significance.

20.3 Climate Change Driven Construction

The construction industry is responsible for delivering the built environment which is the cause of many challenges with climate change. To reverse this pattern, much effort has gone into encouraging the industry through its professionals, developers and contractors to design, construct, operate and maintain low-carbon buildings and infrastructure (BIS, 2010). Consequently the face of construction is changing, driven in part by changing customer requirements, regulatory environment, increases in energy cost and advances in technology (cf. BIS, 2010). In this new environment, construction organisations must operate differently. Resistance to change due to cultural inertia will make construction organisations unresponsive to the needs of the market and will threaten their survival.

Climate change driven construction encompasses a wide range of interventions to minimise the environmental impact of construction. The challenges (and opportunities) for construction organisations in implementing these interventions have been explored by commentators such as Innes and le Grand (2006), Shipworth (2007) and Lowe (2007). Innes and le Grand (2006) for instance refer to the requirement for construction organisations to upgrade and develop skills, knowledge and understanding of resource efficiency, and adaptation and mitigation of climate change (see also BIS, 2010). Shipworth (2007) also identifies introduction of new schemes such as emissions trading (see also BIS, 2010). Diversification into new markets has also been identified as a likely development within industry as a result of increased market activity in particular segments of the market such as flood defence schemes and water reservoirs (Shipworth, 2007). Recent flood events in the United Kingdom suggest the inevitability of such development.

Climate change driven construction also implies regulatory interventions including changes in planning regimes and performance standards as defined by the Building Regulations (Shipworth, 2007). It is noted in Shipworth

(2007) that these are likely to remain the most dominant mechanisms for addressing mitigation and adaptation. For construction organisations this practically means changes in 'what we build, where we build, and how we build it' (Innes and le Grand, 2006). Changes in taxes such as the aggregates levy and landfill tax will mean that the industry will have to do more with less and waste less as government will increasingly seek to ensure that those causing more damage pay more to reflect the damage they do.

Insurance, particularly professional indemnity insurance, is another issue identified by Innes and le Grand (2006). It will come at extra cost to reflect the emerging risks. Professionals will also need to understand and exercise a new dimension of 'reasonable skill and care' and 'professional competence' in relation to climate change (Innes and le Grand, 2006).

It is also very probable that environmental impact assessments will become more widespread, comprehensive and onerous (Shipworth, 2007). Some of the more obvious challenges relate to energy use and transport (BIS, 2010).

Evidently, the challenges are many and inevitable. In accelerating and entrenching change, the culture of organisations within the construction industry will certainly play a significant role and hence requires attention.

20.4 The Role of Culture

The phenomenon of culture is a powerful force that regulates thinking and behaviour. It is this force that shapes society and creates an identity that is unique and recognisable (cf. Eldridge and Crombie, 1975). Regardless of whether society is viewed in its broadest sense (e.g. a country or region) or in a more narrow sense (e.g. an organisation or club), the power of culture to shape thinking and behaviour and the conflicts that this leads to are the same (Allaire and Firsirotu, 1984).

The phenomenon of culture as affects organisations and the manner in which they conduct their business is well documented, and some of the contradictions and key points in the culture debate are summarised in Figure 20.1. Distilling some of the key points from Figure 20.1 to put organisational culture as used in this chapter in perspective, it can be argued that organisational culture is a distinctive feature of organisations that defines how organisational problems are perceived and consequently what solutions are deployed in response. It often manifests in the form of common values and practices. Hofstede's (1997) definition of organisational culture as the collective mental programming that distinguishes the members of one organisation from another is also instructive. Collective programming suggests that organisational members perceive problems in an identical fashion, process information through similar 'algorithms' (beliefs and value systems) and produce similar solutions. As shown in Figure 20.1, values are often evident in the espoused philosophy, vision, mission or goals of the organisation. Alongside constraints in the business environment, these values set the tone for the organisational practices which emerge. This point is echoed by Knott *et al.* (2007) who also state that cultural attitudes and values are a key determinant of behaviour towards inter alia climate change.

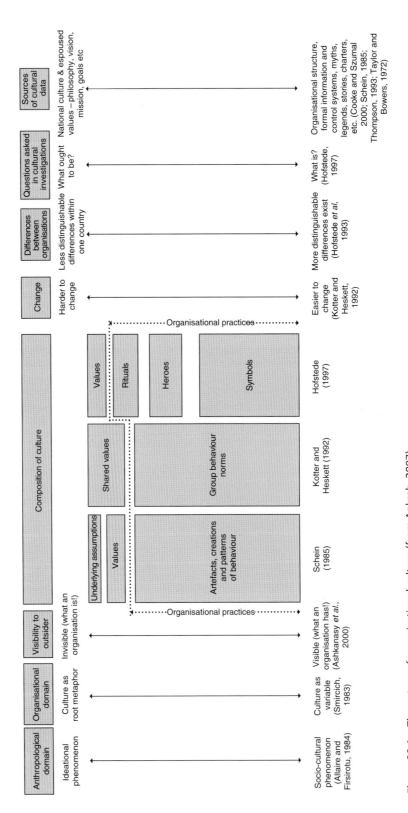

Figure 20.1 The nature of organisational culture (from Ankrah, 2007).

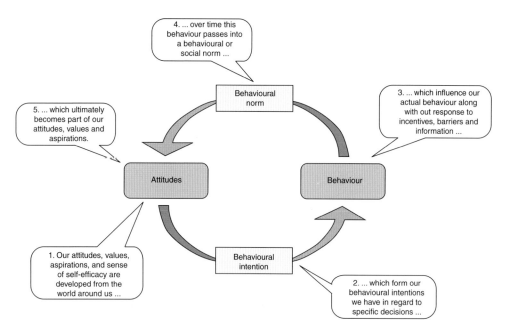

Figure 20.2 The cycle of culture change (from Knott et al., 2007).

Due to business constraints (such as cost), in reality there is often a gap between espoused values and actual behaviours, and it is this gap that also reflects the disparity between what should be done to achieve positive outcomes and what is actually done (Knott et al., 2007). Espoused values, however, remain an important element in the diagnosis of culture. Practices can be disaggregated into symbols, legends, rituals, organisational structure, leadership styles, communication patterns, information and control systems as well as individual behaviour patterns (cf. Schein, 1985). Practices are easier to change than values (Kotter and Heskett, 1992) and as illustrated in Figure 20.2, sustained behaviour (practices) in response to any emerging challenge over time becomes a behavioural norm, which ultimately becomes part of the value and belief system (Knott et al., 2007). This reciprocal determinism between values and behaviour has significant implications for stimulating culture change within construction organisations.

Importantly, culture has everything to do with the implementation of management ideas and how success is actually achieved (Schneider, 2000). It is very often most evident during times of change. This is highlighted by Serpell and Rodriguez (2002) who pointed to a number of sources suggesting that the main difficulties in organisational change implementation in construction are rooted in the cultural conditions of organisations. Further to this, they suggested that experience has shown that the study of culture has allowed the development of strategies for organisational change with higher probability of success. It is precisely this unique quality that makes culture's role in the paradigm shift that climate change driven construction heralds important.

20.5 The Culture of the UK Construction Industry

Taking the definition of culture as examined previously and transposing it into the domain of the construction industry, culture can be considered to be about the 'characteristics of the industry, approaches to construction, competence of craftsmen and people who work in the industry, and the goals, values and strategies of the organisations they work in' (Abeysekera, 2002). This definition provides a robust platform for operationalising and describing the culture of the construction industry.

The UK construction industry is still fairly traditional both in terms of technology and process. Although construction technology has grown increasingly complex over the years, much of construction activity still remains traditional in situ construction. In terms of the process, two fundamental characteristics of the construction industry are particularly pertinent in this discussion. The first characteristic is the highly fragmented nature of the industry both in terms of the process and structure (BIS, 2010). The process is typically divided into multiple phases each of which requires a wide range of participants whose involvement may be limited to just that particular phase and to specific aspects of the phase. Consequently any sense of ownership of the project overall is non-existent or minimal at best. Multiple participants also imply potential interface conflicts where activities overlap or are back to back. Inevitably mistrust among the participants is high particularly because margins are very small and everyone is struggling to avoid making a loss. Consequently they work towards objectives which are often in conflict with each other, and as a result relationships are very often adversarial (cf. Shammas-Toma *et al.*, 1998; Baiden *et al.*, 2006).

A second related characteristic is the extensive subcontracting that is required to cope with the fragmentation and maintain economic viability. As noted in Manu *et al.* (2009) there is a proliferation of small production units in the United Kingdom, which constitute over 90% of construction companies, and majority of them obtain work as subcontractors, therefore forming an important group in the supply chain. Earlier research such as Kheni *et al.* (2005 cited in Manu *et al.*, 2009) also indicates that 80% of construction work undertaken by UK main contractors is subcontracted. These small production units are characterised by fierce competition for contracts resulting in unreasonable cost minimization, lack of resources to invest in areas such as skills development and health and safety, ambiguities about responsibilities and work relationships due to complex subcontracting relationships, and inadequate regulatory control (cf. Manu *et al.*, 2009). These characteristics of the industry have significant implications for risk orientation, quality management, health and safety, cost and time management, competencies of personnel, training and concern for the environment (cf. Harvey and Ashworth, 1997; Hsieh, 1998; Barthorpe, 2002; Fellows *et al.*, 2002; Manu *et al.*, 2009). Attitudes and behaviours in these regards are what define the culture of individual organisations and the industry as a whole.

Construction organisations are generally very task-orientated, often to the detriment of team orientation, client orientation and people orientation (cf. Egan, 1998; Riley and Clare-Brown, 2001; Ankrah *et al.*, 2007). They are also characterised by litigiousness, casual approaches to recruitment, machismo, prejudice towards women and discrimination (cf. Latham, 1994; Egan, 1998; Dainty *et al.*, 2002; Duncan *et al.*, 2002; Loosemore, 2002; Serpell and Rodriguez, 2002; Rooke *et al.*, 2004). Training still remains two-pronged with vocational craft-based training providing the technical skills and degree-based training providing the professional and managerial skills.

Overall, the above discussion appears to suggest that construction organisations still remain fairly traditional in terms of attitudes and behaviours (see also Wolstenholme, 2009). Significantly and perhaps worryingly, not much of the literature on the culture of organisations in the industry refers to the environmental dimension. To gauge orientations on this dimension, it is necessary to interrogate the goals, values, strategies and practices of construction organisations in relation to the environment.

Table 20.1 summarises extracts of environmental philosophies from the websites of a selection of large and small contractors in the UK construction industry. Although it is by no means scientific, these extracts highlight the dichotomy between large and small construction organisations. As argued previously (see also Figure 20.2) behaviours flow out of the organisational philosophy and values, and consequently where the philosophy and values on the environment as captured in an environmental policy is clearly communicated this will translate into behavioural intentions with a strong likelihood of corresponding behaviours that reflect those values. Conversely where an environmental policy is non-existent or poorly communicated, the response in terms of relevant behaviours towards the environment is likely to be indifferent. In this regard, it is significant to observe the espoused values of the large construction organisations and examples of actions that have emerged from these values as captured in Table 20.1. Whilst the lack of presence on a website is not necessarily evidence of the absence of an environmental policy, it does signal that it is not really important to communicate this to (potential) employees, employers, suppliers, and the wider public. This seems to be the case with all except one of the small contractors in Table 20.1. Given the role of small contractors in the construction market as previously established, this trend is worrying. Potentially this could undermine efforts to achieve construction that is responsive to climate change imperatives.

20.6 Achieving and Sustaining a Culture of Sustainability

It is evident from the foregoing discussion that for the construction industry through its diverse agents to efficiently and effectively live up to its responsibilities of delivering sustainable construction practice, a framework for culture change founded on sound theoretical principles that recognises the fragmentation and complexity of construction must be in place. Such a framework will have profound significance particularly for small and medium-sized construction organisations, which appear to lack the organisational culture or structures through which they might fully and effectively respond to the challenges of climate change. In this regards Knott *et al.*

Table 20.1 Examples of sustainability philosophies of construction organizations in the United Kingdom.

Selected Large Construction Companies (Obtained from websites of top 150 UK contractors and house builders)

Company	Turnover (£000)	Policy on (statement of commitment to) climate change/sustainability	Sustainability case studies and efforts
Balfour Beatty	9 486 000	To contribute positively to the physical and social environments in which we operate.	96% of UK operations certified to ISO 14001 and 95% of non-UK operations certified to ISO 14001 as at 2008. Reduction in carbon emission by 19% since 2005. Recycle 79% of waste. M5 project, UK (e.g. using about 400 000 tonnes of recycled material, re-use of about 87% of waste generated on site, etc.). Biodiversity and local regeneration at Thames Barrier Park. Recycling at Harplands Hospital. Minimising environmental impacts at Snowdonia National Park. Energy recycling at Bristol.
Carillion	5 205 800	We will work to reduce our impact on climate change by minimizing all our uses of energy. We will work with our customers and suppliers to achieve 'more with less', through more efficient use of human and material resources, considering life cycle impacts and delivering sustainable, profitable and socially beneficial outcomes.	
Laing O'Rourke	3 603 100	By changing the way we work we can create a more sustainable future in both the long and short term.	Cardiff Central Library: Features a grass roof, which insulates the building, shields it from external noise and reduces the risk of local flooding and pollution. Novacem: Revolutionary cement which offsets global warming. It hardens by absorbing CO_2 from the atmosphere, permanently locking it into building materials.
Barratt	3 554 700	A responsibility to protect and care for the environment	Using recyclable materials in all our homes. Minimising the environmental impact of their business through systematic research. An example is the funding and construction of an 'Eco-House' in conjunction with the University of Nottingham. The house is used as a test bed for a host of innovations and new technologies including a heating system powered by a solar chimney and solar collectors that heat water.

Taylor Wimpey	3 467 700	Climate change policy. We aim to: Work with our customers to maximise the benefit of the increased energy efficiency of the homes we build. Where necessary to promote those benefits and inform our customers. Build our homes to the higher standards set out in the new Building Regulations, and other regulations, in the most effective way to maximise reductions in CO_2. Review with our suppliers the embodied energy within the materials we use. Work towards reduction of the embodied energy where feasible. Reduce CO_2 emissions from our general business operational activities and during construction. Reduce the amount of construction waste we produce and the percentage disposed of in landfill.	Academy Central, East London: When completed, 20% of total energy use will come from renewable sources; the site will generate 40% less carbon dioxide than is required under 2006 building regulations; 13 apartment buildings will have either a green or brown roof. Leybourne Grange, Kent: 702 new homes, all of which will be built to EcoHomes Excellent standard with the 30% affordable homes also achieving Code for Sustainable Homes level three. 65% of UK homes were built on brownfield land in 2009 Decrease in construction waste.
Morgan Sindall	2 548 100	'Delivering Today for Tomorrow' is Morgan Sindall's new sustainability statement. It sums up, in just four words, the Group's commitment to a more sustainable future. It embraces the things that characterise us as a Group.	£8.6 million sustainable extension to a Milton Keynes school which achieved a BREEAM Very Good rating. BREEAM is a way of measuring how environmentally friendly a building is at the time of construction or refurbishment. The Tarmac Homes project – a partnership between Lovell, Tarmac, ZEDfactory and University of Nottingham – aimed to develop a simple, affordable and repeatable blueprint for energy-efficient family homes.
Kier	2 374 200	Kier Group is committed to addressing sustainability through a responsible approach to economic, environmental and social issues.	Kier Group energy saving campaign. Energy-efficient homes at Queens road, Stourbridge; Roke Lane, Witley; and Richmond Lock.
Galliford Try	1 868 700	To integrate the assessment, management and control of environmental issues into the management of our business. To ensure the long-term success of the business by contributing economically, environmentally and socially (our corporate responsibilities) to the communities in which we operate.	26 out of 39 business units have now achieved ISO 14001:2004 certification. Over 1400 employees attending an in-house environmental course. 90% of projects completed environmental risk registers in 2009. Waste diverted from landfill for 2008 was 53% (excluding on-site recycling of inert materials).

(continued)

Table 20.1 (Cont'd).

Selected Large Construction Companies (Obtained from websites of top 150 UK contractors and house builders)

Company	Turnover (£000)	Policy on (statement of commitment to) climate change/sustainability	Sustainability case studies and efforts
Newarthill (Sir Robert Mc Alpine)	1 816 914	Sustainability is built into our projects from the outset and affects all aspects of operations on site from selection of materials to waste management. We are considerate of the impact of our activities on the environment and our stakeholders and aim to operate a successful business in a way that is safe, sustainable and socially responsible.	Watermark Place: Sustainability was integral to the design and construction of Watermark Place in London. A19 Carriageway Upgrade: All lighting on the 120km section of the A19 we maintain in the North East is powered by low carbon energy. In 2007/08 92% of construction, demolition and excavation waste diverted from landfill.
Interserve	1 800 000	We are committed to helping our clients meet their sustainability challenges and to providing a sustainable business for our employees, our communities and our shareholders.	Developed a guide in response to the Carbon Reduction Commitment Energy Efficiency Scheme to help clients understand how the Scheme may affect them, highlighting the key areas and considerations that businesses need to know. Delivering a major new health facility and minimising whole life energy usage through thermal modelling during design to achieve optimum utilisation of natural ventilation to create a comfortable environment for staff and patients.

Selected Small Construction Companies (Obtained from the Federation of Small Businesses)

Deane Public Works Ltd	N/A	None on company website	None on company website
Shera Construction Ltd	N/A	No company website	
Grand Alliance Ltd	N/A	None on company website	None on company website
KET Construction Ltd	N/A	None on company website	None on company website
Newtec Construction Ltd	N/A	None on company website	None on company website

Martin Blake Ltd	N/A	None on company website
G R V Groundworks	N/A	None on company website
G M Developments	N/A	None on company website
Simon Richard Parsons	N/A	No company website
Jim Knight Brickwork Ltd	N/A	None on company website
M K Wildin & Sons Ltd	N/A	Environment Agency Controlled Waste certification
Hirst Construction Ltd	N/A	None on company website
Lawson Surfacing Limited	N/A	None on company website
Mainlink Maintenance Ltd	N/A	Environmental policy – Mainlink recognises the importance of environmental protection and is committed to operating its business responsibly and in compliance with all legal requirements relating to the repair and maintenance of domestic and commercial properties
		ISO 14001 certified
Raisco Ltd	N/A	None on company website

N/A: Not available

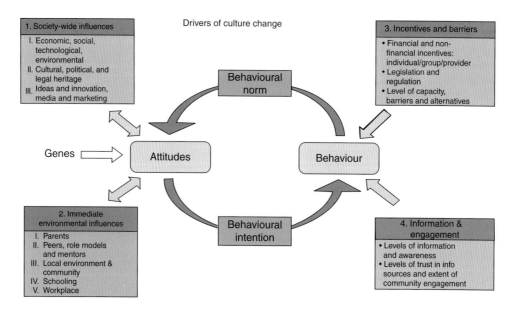

Figure 20.3 A culture change framework (from Knott et al., 2007).

(2007) provide a robust framework for conceptualising and operationalising the culture change process (Figure 20.3). Although Knott et al. (2007) construed and applied this framework within the broader context of policy making at the societal level, this framework can be adapted to reflect the needs of the construction industry. The antecedents of change captured in this framework are examined below within the context of construction organisations.

Society-wide influences are by definition beyond the control of individual construction organisations and encompass economic, social, technological, environmental, cultural, political, and legal forces. The theoretical basis of the influence of these antecedents of culture is well documented in Knott et al. (2007) and will not be revisited here. However to illustrate, high power distance in the wider societal culture might reflect in hierarchical organisational cultures (Hofstede, 1997). Another example might be the impact of new communication technologies (e.g. skyping and social media on communications culture within organisations). From a construction organisation perspective, significant shifts in the wider society in relation to, for example, the environment or political leadership are required to provide the stimulus for evolving more climate change sensitive values. Over this, organisations have no control.

Immediate influences encompass circles of socialisation including parents, peers, role models and mentors, local environment, neighbours and schooling. From a construction organisation perspective, higher and further education institutions, professional bodies and social networks could potentially provide stimulus for change. Here also there is very limited direct control by construction organisations. However there is an opportunity to bring indirect influence to bear on higher and further education institutions and professional bodies through existing education partnerships by helping to define appropriate competencies and training needs. Another relevant circle of socialisation is the workplace. Construction organisations have an opportunity to create an environment in the workplace that encourages an attitude of concern over climate change and an interest in making a difference. The *workplace* refers not only to the organisational 'space' but also to the project environment where a variety of relationships are evident between clients, main contractors, subcontractors, suppliers and professionals. To create a climate in this workplace that promotes concern for the environment, procurement arrangements and contract conditions potentially play a key role. For instance, long-term partnering arrangements could potentially justify investment in skills and technologies for climate change driven construction (particularly by smaller organisations), which will eventually become embedded in the organisations. Skills and technologies play a central role in stimulating cultural change because of their impact on self-efficacy. Also contracts could include specific targets and/or requirements geared towards energy efficiency and emissions reduction beyond what the regulations require.

To stimulate immediate behavioural change the framework shows that incentives, regulation, removal of barriers, information and engagement are the most critical drivers. The construction organisation is capable of exercising some control over these drivers. Very often they may also be introduced by external forces. For instance regulation by public authorities will compel behaviour change by directly or indirectly creating incentives or barriers. According to Knott *et al.* (2007), how people respond to these stimuli determines the behaviours exhibited. The responses to these stimuli will influence cultural change through the creation of behavioural norms over the long term. The theories underpinning this are also explored at length in Knott *et al.* (2007).

These drivers of culture provide an opportunity for specific mechanisms to be adopted by construction organisations, industry bodies and public institutions to stimulate changes in the culture of construction organisations to make them more responsive to the needs of climate change. An important factor to remember is that whilst underlying values and assumptions develop incrementally evolving gradually over time (Meudell and Gadd, 1994), behaviours on the other hand are relatively easier to change (see Figure 20.1). Ultimately behaviours lead to behavioural norms that contribute to the evolution of the culture. Consequently in trying to promote a culture that is congruent with the needs of climate change driven construction it seems logical to focus on the behaviour side of the culture change cycle to engender the needed cultural change.

20.7 Theory to Practice – Case Study

This brief case study brings to life some of the concepts discussed above as operationalised by a major UK construction and support services contractor. This contractor has annual revenue in excess of £5 billion and employs around 50 000 people. The culture of this organisation is described as based on values of openness, collaboration, mutual dependency, professional delivery, sustainable, profitable growth and innovation. Ultimately they seek to grow long-term relationships with customers.

The philosophy on the environment, and specifically on climate change, is clearly communicated on their website and in very comprehensive sustainability reports. Fundamentally it seeks to actively reduce the use of energy and emissions in transport, design, construction, maintenance and the operation of facilities. A content analysis of the 2009 sustainability report showed the most prominent words to be sustainability, business, employees, targets, people, customers, strategy, governance, projects and construction, signalling recognition of the close association between sustainability and various elements of the business.

To create a culture capable of responding effectively to the challenges of climate change, a number of practical measures were adopted by this company. Firstly there was an emphasis on gradual development starting with one or two initiatives and then developing from there. Leadership was considered very important, but significantly it was approached from both sides of the organisational hierarchy. Whilst leadership from the chief executive was essential because 'he says "do it!" and it gets done', specific effort was made to drive leadership from the bottom as well with graduates given free rein to work on specific initiatives and drive them through the organisation. This philosophy was not only to secure their buy-in but also to release their energy and enthusiasm in driving through change. For their efforts, there was recognition and reward at annual award ceremonies. Bonuses also proved important incentives for securing change.

It was important to establish the business case and demonstrate the benefits and business opportunities inherent in operating sustainably. This had to be communicated effectively throughout the organisation. To do this various management tools in the form of simple accessible sustainability models and diagrams, and appropriate excellence models were utilised. Special effort was made to exclude unnecessary jargon in communications. Raising awareness in this way and sharing ownership of ideas was recognition that people are the vehicles of change. Cultural training and development were provided to change mind-sets, promote understanding and share know-how. Mentors were also mentioned.

To secure commitment for the change initiatives and create a culture of collaborative working, the entire supply chain was involved by forming working groups comprising supply chain professionals and sustainability professionals to devise appropriate solutions for meeting sustainability targets. Targets were used to provide impetus for action. For instance, targets for individuals and the organisation as a whole on travel were used to promote a new culture of communication involving video conferencing.

To promote and secure cultural alignment, emphasis was also placed on embedding in the business local sourcing, avoidance of 'hard-nose' aggressive contractors, and building in firm commitments in contracts.

Ultimately there must be success quickly, no matter how small. In the case of this organisation, successes were shouted about and celebrated to reinforce those attitudes and behaviours that generated those results. An example of success that was achieved was the reduction in landfill waste from 100 tonnes per month to 0.5 tonnes per month. These principles are equally applicable to small and medium-sized construction organisations.

20.8 Emerging Solutions for a More Responsive Climate Change Culture in Construction

Ultimately congruence between culture and the business environment will determine the business performance and survival of an organisation (Thompson, 1993). As established already, change is inevitable and without the appropriate organisational culture or structures to respond to the challenges of climate change, there will be a gap between what ought to be done and what is actually done. This gap may result in failure to meet legal obligations, dissatisfied clients and a poor public image all of which will threaten the survival of the construction organisation.

What is needed is an organisational culture that is people and team orientated, concerned for the environment and is sensitive to the impact of organisational activities on climate change. A culture that is dynamic, willing to embrace new technologies and processes, a learning culture, and a culture that values doing the right thing.

To evolve these cultural attributes, interventions are required. From the arguments in the preceding sections, these interventions must focus on the behavioural dimensions and must encompass clear unambiguous communication using appropriate platforms of what needs to be done, training, incentives, appropriate targets, a sustainability czar (or consultant), providing figurehead leadership as well as giving all employees (and those downstream of the supply chain) ownership by promoting bottom-up leadership, the celebration of achievements (including recognition and reward) and the introduction of new technologies and processes.

At the industry level there is also a need for better enforcement of existing and future regulations. There are also implications for continuing professional development and curriculum design at higher and further education institutions to provide the necessary knowledge, skills and competencies required.

The results of behavioural changes will challenge existing value systems, which in the long-term will fundamentally alter long-held beliefs, and underlying assumptions about the way construction organisations should operate.

References

Abeysekera, V. (2002) Understanding 'culture' in an international construction context. In: *Perspectives on culture in construction*, ed. Fellows, R. and Seymour, D.E. CIB Report 275. CIB, Rotterdam.

Allaire, Y. & Firsirotu, M. E. (1984) Theories of organizational culture. *Organization Studies*, 5(3), 193–226.

Ankrah, N.A. (2007) *An investigation into the impact of culture on construction project performance*. PhD thesis, University of Wolverhampton, Wolverhampton, UK.

Ankrah, N.A., Proverbs, D. & Debrah, Y. (2007) A cultural profile of construction project organisations in the UK, 23rd Annual ARCOM Conference, Belfast. ARCOM, Belfast.

Baiden, B.K., Price, A.D.F. & Dainty, A.R.J. (2006) The extent of team integration within construction projects, *International Journal of Project Management*, **24**(1), 13–23.

Barthorpe, S. (2002) The origins and organisational perspectives of culture. *CIB Report*, 7–24.

Carbon Trust (2010) Buildings policy. Available from http://www.carbontrust.co.uk/policy-legislation/Business-Public-Sector/Pages/building-regulations.aspx.

Dainty, A.R.J., Bagilhole, B.M. & Neale, R.H. (2002) Coping with construction culture: a longitudinal case study of a woman's experiences of working on a British construction site. In: *Perspectives on Culture in Construction*, ed. Fellows, R. and Seymour, D.E. CIB Report 275. CIB, Rotterdam.

Department of Business, Innovation and Skills (BIS) (2010) *Low Carbon Construction: Emerging Findings*. Innovation & Growth Team, Department of Business, Innovation and Skills, HM Government, London.

Duncan, R., Neale, R. & Bagilhole, B. (2002) Equality of opportunity, family friendliness and UK construction industry culture. *Perspectives on Culture in Construction*, ed. Fellows, R. and Seymour, D.E. CIB Report 275. CIB, Rotterdam.

Egan, J. (1998) *Rethinking Construction*. HMSO, London.

Eldridge, J.E.T. & Crombie, A.D. (1975) *A Sociology of Organisations*. International Publications Service, New York.

Environment Agency (2003) Sustainable construction: position statement, environment agency. Available from http://www.environment-agency.gov.uk/static/documents/Research/ea_sustainable_908180.pdf.

Fellows, R., Langford, D., Newcombe, R. & Urry, S. (2002) *Construction Management in Practice*. Oxford, Blackwell Science.

Harvey, R.C. & Ashworth, A. (1997) *The Construction Industry of Great Britain*. Oxford, Laxton's.

Hofstede, G.H. (1997) *Cultures and Organizations: Software of the Mind*, rev. ed. McGraw-Hill, New York.

Hofstede, G. (2001) *Culture's Consequences: Comparing Values, Behaviors, Institutions, and Organizations across Nations*. Sage Publications, Thousand Oaks, CA.

Hsieh, T-Y. (1998) Impact of Subcontracting on Site Productivity: Lessons Learned in Taiwan. *Journal of Construction Engineering & Management*, **124**(2), 91–100.

Innes, S. & le Grand, Z. (2006) *Briefing Note on 'Stern Review: The Economics of Climate Change'*. Constructing Excellence, London.

Knott, D., Muers, S. & Aldridge, S. (2007) *Achieving Culture Change: A Policy Framework*. Prime Minister's Strategy Unit, Cabinet Office, London.

Kotter, J.P. & Heskett, J.L. (1992) *Corporate Culture and Performance*. New York, Maxwell Macmillan International.

Latham, M. (1994) *Constructing the Team: A Joint Review of Procurement and Contractual Arrangements in the UK Construction Industry*. HMSO, London.

Loosemore, M. (2002) Prejudice and racism in the construction industry. In: *Perspectives on Culture in Construction*, ed. Fellows, R. and Seymour, D.E. CIB Report 275. CIB, Rotterdam.

Lowe, R. (2007) Addressing the challenges of climate change for the built environment. *Building Research & Information*, **35**(4), 343–350.

Manu, P., Ankrah, N., Proverbs, D., Suresh, S. & Callaghan, E. (2009) Subcontracting versus health and safety: an inverse relationship, In: *Proceeding of CIB W099*

2009 Conference, 21–23 October 2009, Melbourne, Australia, ed. Lingard, H., Cooke, T., & Turner, M. CIB, Rotterdam.

Meudell, K. & Gadd, K. (1994) Culture and climate in short life organizations: sunny spells or thunderstorms. *International Journal of Contemporary Hospitality Management*, **6**(5), 27–32.

Riley, M.J. & Clare-Brown, D. (2001) Comparison of cultures in construction and manufacturing industries, *Journal of Management in Engineering*, **17**(3), 149–158.

Rooke, J., Seymour, D. & Fellows, R. (2004) Planning for claims: an ethnography of industry culture, *Construction Management and Economics*, **22**(6), 655–662.

Schein, E. (1985) *Organizational culture and leadership*, San Francisco, Jossey-Bass.

Schneider, B. (2000) The psychological life of organizations In: Handbook of Organizational Culture and Climate, ed. Ashkanasy, N. M., Wilderom, C. P. M. & Peterson, M. F. Sage, Thousand Oaks, CA.

Serpell, A.F. & Rodriguez, D. (2002) Studying the organisational culture of construction companies: a proposed methodology. In: *Perspectives on Culture in Construction*, ed. Fellows, R. and Seymour, D.E. CIB Report 275. CIB, Rotterdam.

Shammas-Toma, M., Seymour, D. & Clark, L. (1998) Obstacles to implementing total quality management in the UK construction industry. *Construction Management & Economics*, **16**(2), 177.

Shipworth, D. (2007) The Stern Review: implications for construction. *Building Research & Information*, **35**(4), 478–484.

Stern, N. (2006a) *The Economics of Climate Change*. Cambridge University Press, Cambridge. Available from http://www.hm-treasury.gov.uk.

Stern, N. (2006b) What is the economics of climate change? *World Economics*, **7**(2), 1–10.

Thompson, J. L. (1993) *Strategic Management: Awareness and Change*. London, Chapman & Hall, University and Professional Division.

Wolstenholme, A. (2009) Never waste a good crisis. Constructing Excellence. Available from www.constructingexcellence.org.uk.

Preparing for Extreme Weather Events: A Risk Assessment Approach

Keith Jones

21.1 Introduction

Extreme weather events, and their impact on communities, are not new. In the United Kingdom, wide-spread flooding in 2000 resulted in £500 million worth of insurance claims (RMS, 2000), whilst managing the 2004 floods cost £2.2 billion (OST, 2004). The Cumbria floods of 2009, at the time quoted as a 1 in 1000-year event, flooded 2239 properties, 80% of which were residential, with 470 local households requiring immediate relocation. In addition, damage to critical infrastructure saw the destruction of three major road bridges and the temporary closure of 20 minor bridges, causing significant disruption to business at a cost to the tourist industry alone of £12 million (Cumbria Intelligence Observatory, 2010). In addition to flooding, heat waves have also affected the United Kingdom and Europe. In August 2003 over 2000 premature deaths were attributed to the heat wave in southern England (Kovats et al., 2006) with this rising to 80 000 across Europe as a whole (Robine et al., 2007). Finally, the 'Hurricane' storm in October 1987 caused £1.4 billion of insured losses and killed approximately 20 people in the United Kingdom alone (RMS, 2007) whilst the 'Lothar and Martin' storms in December 1999 caused €13.5 million damage and killed 125 people across Europe (EQE International, 2010).

According to IPPC (2007), the frequency and severity of extreme weather events across the world are set to increase over the next century and the consequences for communities are likely to be profound. The ability to withstand and recover from an extreme weather event is dependent upon the resilience and adaptive capacity of the community under threat. Whilst there are procedures in place to assess national and regional level resilience and adaptive capacity, the spatial and temporal scales of the assessment

---

Solutions to Climate Change Challenges in the Built Environment, First Edition.
Edited by Colin A. Booth, Felix N. Hammond, Jessica E. Lamond and David G. Proverbs.
© 2012 Blackwell Publishing Ltd. Published 2012 by Blackwell Publishing Ltd.

tools do not translate well to the local or individual stakeholder level. Individual households, businesses and local policy makers respond to risks that they can understand. In this context risk needs to be articulated within a frame of reference that individuals within the community are familiar and can associate with. The challenge of downsizing the strategic-level assessment tools is making them relevant at the individual stakeholder level. This chapter describes the development of a risk assessment framework that addresses this challenge.

21.2 What Is an Extreme Weather Event?

From the climate science perspective, an *extreme weather event* is one that has a 'low probability' of occurring: from a meteorological perspective, it is one that is 'out of the ordinary' or 'rare' to the prevailing conditions at a particular geographical location. In both of these definitions problems arise when decision makers have to interpret 'low probability' and 'rare' in their particular context. Whilst strategic-level decision makers (e.g. regional planners an emergency responders) normally interpret the probability of occurrence at the extremes of the observed probability density function (IPCC, 2007), such interpretations are not well understood by local decision makers (e.g. households and businesses) responsible for risks assessments and contingency planning (Berkhout *et al.*, 2004). This group are not so concerned with the likelihood of an extreme weather event occurring but with the impact that the event will have if it does occur. In essence it is the 'hazards' associated with the extreme event and the impacts that these have on the physical and social system (e.g. loss of life or injury, property damage, social and economic disruption and environmental degradation) that need to be predicted and planned for. For those responsible for managing the built environment (building owners and users, households and local policy makers), such assessments normally involve a risk-based approach in which the likelihood of the impacts of an extreme weather event occurring and their consequences on the built environment system are balanced against the costs (financial and social) of protection and recovery. However, in this process, risk should not be mistaken or confused with uncertainty. Risk assumes that the impacts of a hazard can be defined, quantified, evaluated and managed. Uncertainty implies that whilst the impacts of a hazard may be known qualitatively, they cannot be predicted (or managed) quantitatively. It is this paradox that lies at the heart of developing a risk-based approach to managing the consequences of extreme weather events at a local level. Whilst uncertainty can be accommodated in strategic-level models of community resilience to extreme weather events it cannot be easily incorporated into operational risk assessment frameworks. The challenge to those advising local decision makers on preparing (and recovering from) extreme weather events is to find a way of converting the uncertainty surrounding the impacts of such events into risks that can be quantified and managed.

Assuming that uncertainty can be converted to risk, then developing a risk assessment framework to evaluate the impacts of extreme weather events on the built environment requires an assessment of:

- The impacts that the event will have on the physical, social and economic performance of the built environment system as a whole (reductionist approaches to risk assessment don't work);
- The probability of the impacts occurring; and
- The development of management strategies aimed at minimising the impacts and controlling their potential adverse effects on the built environment system and overseeing the recovery process.

It must be remembered at this point that risk management is not about preventing risk occurring, but being prepared for the risk and having contingency plans in place to manage the consequences associated with the risk. Indeed, fundamental to the concept of risk management is establishing thresholds or system tolerances that define acceptable damage levels to the system. In sociotechnical systems (such as the built environment), whilst risk assessments require quantitative metrics of system performance, risk management uses qualitative factors based on the value system of individuals within the system to establish thresholds and contextualise the risk management framework to their particular circumstances (thresholds for one community may not be considered acceptable by another community). As such, the qualitative values (e.g. psychological factors, sociological factors, economic factors and belief systems), whilst recognised as fundamental to the ability of communities to withstand and recover from an extreme weather event (Paton, 2007; Tierney & Bruneau, 2007; Cutter *et al.*, 2008) are by their nature difficult to measure in a robust, reliable and repeatable fashion. Thus, even if the uncertainty surrounding the physical impacts of an extreme weather event can be converted in measurable risks, it is unclear whether qualitative values can. It is also unclear whether a generic risk management framework model can be developed (or is even desirable) that can be applied to any given community.

The fundamental distinction between risk and uncertainty coupled with the need to define quantitative and qualitative metrics to traditional risk assessment models lie at the heart of the challenge facing those who seek to understand the impacts that extreme weather events might have on the performance of the built environment as a sociotechnical system. The remainder of this chapter will examine current thinking around these problems and will present the findings of an ongoing research project which addresses this challenge as it seeks to develop a risk framework model that can be used to assess the resilience of local communities to extreme weather events.

21.3 Relationship between Vulnerability, Resilience and Adaptive Capacity

Fundamental to the understanding of community resilience to extreme weather events, and as such to the development of a risk management framework, are the relationships between vulnerability, resilience and adaptive capacity.

Vulnerability, resilience and adaptive capacity are concepts that originated in the biophysical and social realms, but are now being applied to understanding the complex relationships between communities, the built environment, and the drivers that affect change. In this context, vulnerability relates to the characteristics and circumstances (factors) of the built environment (e.g. physical and technical, social, economic, cultural, political and environmental) system that determines how susceptible (combination of likelihood of exposure and the adverse consequences resulting from that exposure) it is to the impact of an extreme weather event hazard. The concept can be divided into physical vulnerability, which relates to the chances of being exposed because of location and social vulnerability, the chances of suffering impacts if exposed.

Resilience (in the context of the built environment) is slightly more difficult to define. Firstly, resilience should not be confused with resistance, which is primarily concerned with pre-disaster mitigation measures that enhance the performance of both physical elements (built environment) and social institutions in reducing loss. Resilience is a much broader concept than resistance, considering both the socio-environmental context (properties of social systems including communities and attributes of specific individuals, groups or organisations) and the sociotechnical context (properties of social systems and the technological environment within which they exist) as the basis for a more holistic approach toward managing risks and planning transitions through adaptation towards a more sustainable future. Resilience therefore relates to the ability of the built environment system to prevent, withstand, recover and learn from the impacts of an extreme weather event hazard.

Adaptive capacity relates to the ability of the built environment system to implement effective adaptation measures (physical, economic, temporal etc.) to change (adapt) to meet the new conditions brought about by the forces that fundamentally change the system. In essence, the adaptive capacity of the built environment system determines the degree to which it returns to a state similar to that which existed before the extreme weather event occurred (physical, social and economic) or re-organises to a changed state that reflects new, post-disaster community dynamics.

The concepts of vulnerability, resilience and adaptive capacity outlined above can be mapped quite closely to the risk assessment framework described earlier in this chapter. Vulnerability can be equated to the probability of the impacts occurring; whilst resilience and adaptive capacity relate to the management strategies for reducing impacts and aiding recovery. Thus a risk analysis framework for dealing with the impacts of extreme weather events could be informed by similar models that have applied the concepts of vulnerability, resilience and adaptive capacity to other extreme events.

There are a number of explanatory models (Paton, 2007; Tierney & Bruneau, 2007; Cutter et al., 2008) that consider the relationships between vulnerability, resilience (in its broadest sense) and adaptive capacity for different extreme event scenarios (although not limited to extreme weather events). Whilst each author examined different types of event, there was a high level of agreement on the factors, both prior to an event and following an event, that affected resilience. These include:

- The robustness of the system (in this case the built environment) to withstand disaster forces without significant degradation or loss of performance. This includes the physical components within a system, the organisations that manage these components, and the redundancy (the extent to which system components are substitutable) in the system.
- The ability to intervene in the system through effective diagnosis and prioritisation of problems and initiate solutions (by identifying and mobilising material, monetary, informational, technological and human resources) to restore functionality to the system as quickly as possible. This includes an understanding of population characteristics that may render different social groups more or less vulnerable to a hazard and the ability to initiate a range of coping strategies to deal with the hazards.
- The interrelationships between personal factors, community factors and institutional factors. These include critical awareness, self-efficacy, outcome expectancy, community values, the role of trusted advocates, social networks, faith groups, empowerment, trust, effective communication and a sense of community.

In addition to explanatory models, Cutter *et al.* (2008) also developed a predictive model of natural disaster resilience (Figure 21.1) supported by a set of candidate variables for measuring community resilience[1]. Cutter's disaster resilience of place (DROP) model considered the inherent (antecedent conditions) vulnerability and resilience of existing communities (combination of natural systems, social systems and the built environment) to an extreme event. The antecedent conditions interact with the hazard event characteristics to produce immediate effects. The event characteristics include frequency, duration, intensity, magnitude and rate of onset of the event. These vary depending on the type of hazard and geographical location. The immediate effects of an extreme event are either reduced or amplified by the presence (or lack of) mitigation actions and coping responses. After any coping strategies are implemented, the hazard impact is realised. The impact of the event is moderated by the absorptive capacity of the local community. If the absorptive capacity is not exceeded, then recovery is relatively quick. If the absorptive capacity is exceeded (because either the scale of the event is overwhelming or the coping responses are insufficient), then the system either adapts (through improvisation or social learning) and recovers relatively quickly, or does not adapt and recovers much more slowly (or, in extreme cases, doesn't recover at all). If social learning occurs, then there is a greater likelihood that mitigation and

---

[1] There is considerable debate at present regarding the relationship between vulnerability and resilience. For further consideration of the debate, see Cutter *et al.* (2008).

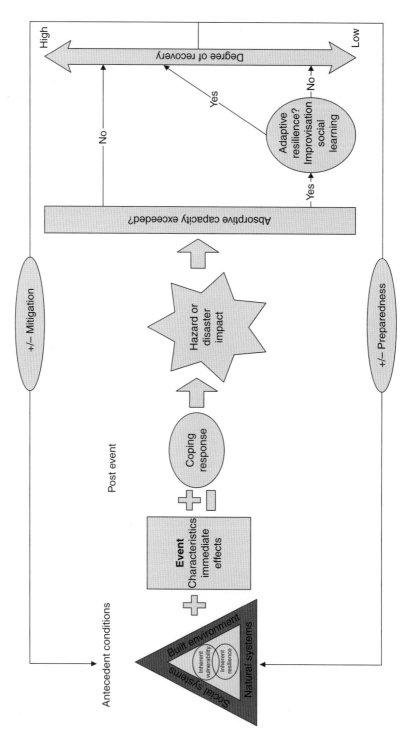

Figure 21.1 Schematic representation of the disaster resilience of place (DROP) model (from Cutter et al., 2008).

preparedness will be improved. The significance of this model to those responsible for managing the built environment should not be underestimated. Although many of the steps outlined by Cutter may seem commonsense, each contains a degree of complexity (both in the sense of the range of variables that need to be considered and of the inherent interrelationships and feedback loops that exist between variables – it could be argued that the true behaviour of the system is more nonlinear than Cutter's model implies and as such should be analysed as a complex system) that is currently not well understood.

Although concerns exist around operationalising the DROP model, a similar range of factors to those considered by Cutter form the basis of the UK government's draft strategic national framework on community resilience (Cabinet Office, 2010). Although the framework is not exclusively designed for extreme weather events, the guiding principles within it form the basis of current advice to communities preparing for an extreme weather event. The framework emphasises the need for communities to engage in action to prepare for an event and the importance of self-reliance and harnessing individual and local resources to support community action during and following an event. The strategy identifies a range of community characteristics and topologies (geographical, communities of interest, communities of circumstance, communities of practice etc.) that could be considered as the focus for community action and emphasises the importance of the inter linkages and interdependencies between stakeholders (in particular the need for a good working relationship between the communities themselves, those who support the communities on a professional and voluntary basis, and the agencies and organisations involved in community networks) as fundamental to ensuring community resilience to an extreme event. The framework identified the key characteristics of a resilient community as one

- in which people within the community use their skills, knowledge and resources to prepare for and deal with the consequence of an event, and they adapt everyday skills and use them in extraordinary circumstances;
- where people are aware of and understand the threats and hazards that may affect them and have a mature understanding of risk;
- in which people have taken steps to make their homes more resilient;
- that has a champion who facilitates effective communication to the wider community and motivates and encourages others to get involved and stay involved;
- that works in partnership with emergency services, the local authority and other relevant organisations before during and after an emergency; and
- that is empowered and actively involved in influencing and making decisions affecting them.

Finally, the framework identifies unrealistic expectations of stakeholders as a key factor that can lead to unresilient communities. Again however, whilst the strategy outlines the factors that need to be considered when assessing how resilient a community is to an extreme event, it doesn't provide quantitative interpretations (metrics) that would allow these factors to be integrated into a pre-event risk assessment framework.

21.4 A Risk Assessment Framework Model

There are a number of risk models (UKCIP, 2010; AXA, 2007a, b; BCI, 2007) currently available to assess vulnerability, resilience and adaptive capacity to extreme weather events. Whilst each model addresses the problems of risk in slightly different ways they all follow the same generic methodology. An initial scoping exercise contextualises the system being studied and identifies system boundaries. Once the system boundaries are established, the types of risk that can affect the system can be identified. This process involves identifying what is at risk, who is at risk, the causes of the risk, the impacts of the risk and the threshold levels at which the risk becomes unacceptable. For each identified risk, a risk appraisal is undertaken where the consequences of the threshold being exceeded are examined and strategies for managing the consequences considered. This process invariably involves the use of scenarios to both identify the potential consequences and evaluate alternative management strategies. Once the risk appraisal is complete, a risk evaluation takes place where the various options are prioritised. Finally the highest priority options are instigated and their performance is monitored. Unfortunately, whilst this generic approach to risk assessment is fairly well understood, its application, particularly in the United Kingdom, is patchy and its impact on community resilience to extreme weather events minimal.

The problem with the application of current risk assessment frameworks to assessing the resilience of communities to extreme weather events is one of scale. Whilst the approaches appear to work at a national or regional scale, they do not appear applicable to the local scale. Individual households find the prescriptive nature of the risk assessment methodology cumbersome and the interpretation of the hazard impact difficult to assess in terms that are meaningful to their particular circumstances. As such households tend to be disinclined to consider the potential consequences of exposure to extreme weather event unless they have prior experience of one (Rose *et al.*, 2009). Businesses, where preparing for the impact of extreme weather events should be a matter of extending current disaster recovery and business continuity plans, also find it difficult to relate extreme weather events to the risk models that they currently use to support business continuity planning (Berkhout *et al.*, 2004). The problems revolve around quantifying the impacts (either actual or perceived) that an extreme weather event may have on their operations in the context of their own, and their competitors' performance and of identifying solutions (adaptations) that are both cost-effective and have a measurable benefit to the performance of the business. In essence, both stakeholder groups have difficulty interpreting absolute measures of risk in the context of their own circumstances. The hazards need to be interpreted relative to a frame of reference that the individual and business can relate to (e.g. in the case of households the degree of disruption caused, and in the case of businesses the impact on business function). Indeed, when such relative metrics are used, businesses are more inclined to develop business continuity and disaster recovery plans to deal with extreme weather events (Climate South East, 2008).

21.5 Solutions: A New Risk Framework Model

A risk assessment framework for community resilience to extreme weather events has been developed that integrates current risk assessment models with Cutter's DROP model to produce an approach to risk assessment in which extreme weather event hazards are presented as relative impacts against a known frame of reference (e.g. at the household or individual business level) rather than absolute impacts against national thresholds (e.g. 1 in 100 return period). The risk assessment framework is show in Figure 21.2.

The first phase in assessing resilience to extreme weather events involves establishing the system boundaries and identifying the physical topology that constitutes the focus of the study. This topology can be spatial, organisational, continuous or discreet. Once the system boundaries have been established, antecedent conditions are examined to establish inherent vulnerability and resilience to current (and historic) extreme weather events. The analysis of the antecedent conditions is undertaken with reference to local histories (e.g. conversations with community members, examination of local newspapers, reference to local library archives etc.) and existing strategic risk assessment frameworks (e.g. national and regionally generated extreme weather events impact maps, local impact assessments, regional resilience plans etc.). Once the generic antecedent conditions are established, individuals within each stakeholder group can assess their vulnerability and resilience to the impacts of the extreme weather events and local policy makers can test their emergency response plans. At this stage of the model community buy in should be fairly high, as the extreme weather events being considered have a history of occurring within the locality.

The next phase of the model extends the range of extreme weather events to take into account the impact of future climate change on the type, nature and intensity of events. This phase inevitably involves the use of future scenarios, (e.g. the UK Climate Projections)[2] to develop a range of weather patterns that can be superimposed onto the system topology (area of interest) to allow specific hazard impacts to be developed for each scenario (e.g. flooding, overheating, subsidence, storm etc.). These impacts can then be related in relative terms to the antecedent assessments carried out in the previous phase. In this way stakeholders can assess the relative significance

---

[2] In the United Kingdom, a toolkit exists that allows the impact of climate change on a range of weather factors to be calculated for a 25 km grid. See http://ukclimateprojections.defra.gov.uk.

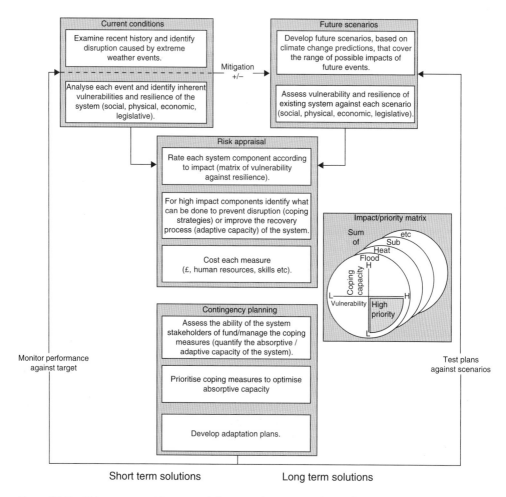

Figure 21.2 Risk assessment framework for assessing community resilience to EWEs.

of an extreme weather event scenario against a frame of reference that they are familiar with (e.g. if a similar frequency storm occurs in ten years' time, then, as a consequence of climate change, five times the number of houses are likely to be flooded than were flooded last time).

Once the currency of the scenarios has been established then the impacts of each extreme weather event (either as a consequence of antecedent conditions or from future scenarios) on the components (e.g. infrastructure, buildings and social networks) of the system can be assessed and those components (or subcomponents) that are highly vulnerable and have low inherent coping capacity can be identified. For each of these components interventions can be developed, either to reduce the vulnerability of the components to the impact of the event (e.g. for a flood scenario, install temporary flood barriers across doorways, seal airbricks etc.) or improve the coping capacity of the components (e.g. for a flood scenario, replace timber floors with concrete floors, raise electrical sockets above expected flood level etc.). The overall resilience of a system (e.g. house) can be obtained by summing the individual vulnerability and coping capacity of each component to each extreme weather event impact (e.g. flooding, overheating and subsidence). This summation process will not be linear as the priority given to each impact will vary depending upon the socio-economic and sociotechnical make-up of the community. A multicriteria decision making framework is used to produce an overall assessment of resilience that reflects local conditions and community expectations. Again, involving all community stakeholder groups in establishing the relative priority rating given to each impact will ensure that the community fully understands the implications of the risk assessment approach and do not have unrealistic expectations of the outcomes. When assessing priorities consideration needs to be given to both the social and physical aspects of the community as well as to the adaptive capacity that exists within the system. It is pointless to prioritise a range of interventions where neither the physical nor financial resources are available for their implementation. Such a process will simply raise expectations amongst community stakeholders and provide a false confidence in the resilience of the system.

Once priorities have been established and the effectiveness of the various interventions to raise the overall resilience of the community evaluated, those interventions that are highly prioritised and where the adaptive capacity exists to enact them are included as part of either the short term adaptation plan (where the event likelihood is imminent or already happening) or longer term adaptation plans to address the impacts of future climate change. In both cases the performance of the interventions needs to be measured (or in the case of the future scenarios tested) and the performance of the interventions fed back into the risk appraisal model as to inform the risk assessment process and provide the basis for ongoing review.

21.6 Final Thoughts

Whilst the new risk framework outlined above addresses many of the limitations associated with assessing the vulnerability, resilience and adaptive capacity of communities, there are still a number of factors that are not fully considered in the model. In particular a wider range of social, economic and political factors affect community resilience need to be examined and integrated into the risk appraisal and contingency planning part of the framework. Detailed examination of these factors is currently underway as part of the UK Engineering and Physical Sciences Research Council project into Community Resilience to Extreme Weather. This said, the risk assessment framework does provide an analytical tool for those responsible for advising built environment stakeholders on the potential impacts of extreme weather events on the physical resilience of communities. Of course, as with all such models, it's effectiveness to inform communities of their vulnerability, resilience and adaptive capacity to an extreme weather event will not be established until such an event occurs and the model validated.

Acknowledgements

The content of this chapter forms part of a multidisciplinary project into Community Resilience and Extreme Weather Events (CREW) being funded by the UK Engineering and Physical Sciences Research Council. The authors would like to acknowledge the contributions made by the following: Dr. G. Wood (Cranfield University), Dr. H. Fowler (University of Newcastle), Prof. G. Price (Glasgow University), Prof. L. Shao (De Montfort University), Dr. R. Few (University of East Anglia), Dr. F. Ali (University of Greenwich), Prof. D. Proverbs and Ms. C. Rose (University of the West of England), Dr. A. Wreford (Scottish Agricultural College), Dr. R. Soetanto and Mr. A. Mullins (Coventry University), Dr. D. Thomas (The University of Manchester) and Dr. B. Ingirige and Mr G. Wedawatta (University of Salford) for the general discussions that formed the background to this chapter.

References and Further Reading

AXA (2007a) Business continuity guide. Available from http://ww.thebci.org.

AXA (2007b) Business 4 Tomorrow. Available from http://ww.thebci.org.

BCI (2007) Expecting the unexpected – business continuity in an uncertain world. Available from http://ww.thebci.org.

Berkhout, F., Hertin, J. & Arnell, N. (2004) *Business and climate change: measuring and enhancing adaptive capacity.* Technical Report 11. Tyndall Centre for Climate Research, Norwich, UK.

Cabinet Office (2010) Draft Strategic National Framework on Community Resilience. Available from http://www.cabinetoffice.gov.uk/media/349129/draft-snframework.pdf.

Climate South East (2008) Climate change and small businesses. Available from http://www.climatesoutheast.org.uk/images/uploads/Climate_Change_and_Small_BusinessesFinal.pdf.

Cumbria Intelligence Observatory (2010) Cumbria Floods November 2009: Overview and Meteorological Circumstances. Available from http://www.cumbriaobservatory.org.uk/AboutCumbria/Floods/November2009.asp.

Cutter, S.L., Barnes, L., Berry, M., Burton, C., Evans, E., Tate, E. & Webb, J. (2008) A place-based model for understanding community resilience to natural disasters. *Global Environmental Change*, **18**, 598-606.

EQE International (2010) The European Storms Lothar and Martin December 1999. Available from http://www.absconsulting.com/resources/Catastrophe_Reports/Lothar-Martin%20Report.pdf.

IPCC (2007) The 4th Assessment Report: Working Group 1 – The Physical Science Basis, Chapter 3.8. Available from http://www.ipcc.ch/publications_and_data/ar4/wg1/en/contents.html.

Kovats, R.S., Johnson, H. & Griffiths, C. (2006) Mortality in southern England during the 2003 heat wave by place of death. *Office of National Statistics, Health Statistics Quarterly*, **29**, Spring.

OST (2004) Foresight: future flooding: executive summary. Available from http://www.foresight.gov.uk/Flood%20and%20Coastal%20Defence/Executive_Summary.pdf.

Paton, D. (2007) Measuring and monitoring resilience in Auckland. GNS Science Report 2007/18. GNS, Auckland.

Robine, J.M., Chueng, S.L., Le Roy, S., Van Oyen, H. & Herrmann, F.R. (2007) Report on excess mortality in Europe during summer 2003. EU Community Action Programme for Public Health, grant Agreement 2005114. http://ec.europa.eu/health/ph – projects/2005/action1/docs/action1_2005_a2_15_en.pdf.

Rose, C., Proverbs, D.G., Manktelow, K. & Booth, C.A. (2009) Psychological factors affecting flood coping strategies. In: *2nd International Flood Recovery, Innovation and Response Conference, Milan, Italy, 2010*. FRIAR, Milan, 305–312.

RMS (2000). U.K. floods, Nov 2000: Preliminary report of U.K. flood damage from increased rainfall in November 2000. Available from http://www.rms.com/publications/UKFLOODS_NOV2000.pdf.

RMS (2007) The Great Storm of 1987: 20-year perspective. Available from http://www.rms.com/Publications/Great_Storm_of_1987.pdf.

Tierney, K. & Bruneau, M. (2007) Conceptualizing and measuring resilience: a key to disaster loss reduction. TR News, **250**, May–June.

UKCIP (2010) Adaptation wizard. Available from http://www.ukcip.org.uk/index.php?option=com_content&task=view&id=147&Itemid=273.

The Socio-environmental Vulnerability Assessment Approach to Mapping Vulnerability to Climate

Fiifi Amoako Johnson, Craig W. Hutton and Mike J. Clarke

22.1 Introduction

Seeking solutions to climate change involves an understanding of how vulnerable we are to the hazards of climate change. This chapter attempts to measure the vulnerability to climate change induced flood hazards of the built environment of the Brahmaputra River Basin in the Assam State of India. There has been growing interest in quantifying and mapping vulnerability to climate hazards as a visual tool for planning and policy decision making (Sherbinin et al., 2007; Eakin and Luers, 2006; Morrow, 1999). Vulnerability maps can allow risk management professionals to target scarce resources to the most vulnerable in society. Maps can also indicate where improvement in built infrastructure can have the largest impact on reducing vulnerability. Two models that have informed vulnerability analysis are the risk-hazard (RH) model and the pressure and release model (Turner et al., 2003). From these models have emerged models for quantifying vulnerability to climate hazards, for example the vulnerability-resilience indicator prototype (VRIP) model proposed by Moss et al. (2001), the inference model by Alcamo et al. (2008) and the inference model based on political science perspective by Taenzler et al. (2008). In this study we use the socio-environmental vulnerability assessment (SEVA) approach to quantifying and map the spatial variations in vulnerability to flood hazards. The SEVA approach combines local knowledge with statistical techniques to quantify vulnerability to a specific hazard.

Climate hazards (e.g. floods, drought and river bank erosion, among others) have plagued the built environment of the Brahmaputra River Basin in the Assam State of India for decades with numerous social and economic consequences. The Brahmaputra River is a transboundary river with its

---

Solutions to Climate Change Challenges in the Built Environment, First Edition.
Edited by Colin A. Booth, Felix N. Hammond, Jessica E. Lamond and David G. Proverbs.
© 2012 Blackwell Publishing Ltd. Published 2012 by Blackwell Publishing Ltd.

origin in southwestern Tibet going through the Himalayas in the great gorges and into Arunachal Pradesh. It flows southwest through the Assam State of India (Upper Brahmaputra River Basin) and south through Bangladesh. This study focuses on the Assam Basin of the Brahmaputra River which stretches 720 km long and 90 km wide (Goswami 1985) and is home to more than 26 million people (Census of India, 2001). The channel of the river occupies about one tenth of the valley, with over 40 % of its area under cultivation.

The natural topography and physical make-up of the Basin renders it vulnerable to floods, making it an area of numerous social and economic disasters. The basin experiences the highest number of floods in India during the monsoon rains (June to September inclusive) and suffers flood damages on an annual basis. Historical records show that flooding has been part of the make-up of life in the basin since primeval times. The socioeconomic consequences are devastating. For example, the 2007 floods displaced 3 million people and contaminated most sources of drinking water (Bagchi, 2007).

The SEVA approach uses a participatory process involving residents, stakeholders and experts to:

1. assess the impact of climate hazards on the socioeconomic wellbeing of residents and their ability to mitigate or cope with the impacts or take advantage of opportunities that may arise;
2. identify relevant indicators for quantifying the impacts of climate hazards and coping capacity of residents; and
3. use appropriate statistical techniques which incorporate local knowledge to analyse spatial variations in socioeconomic vulnerability to hazards.

The SEVA approach profiles different aspects of the impacts of hazards and coping capabilities. The resulting output is then used to derive a combined vulnerability index which reflects the vulnerability of a given community to a specific climate hazard. Profiling the different aspects reveals the different drivers of vulnerability to a specific community. The resulting outputs are vital for policy decision making, planning and monitoring. Although the example outlined in this chapter is based on India, the proposed methodology can be replicated elsewhere.

22.2 The SEVA Approach

22.2.1 Conceptualizing Socioeconomic Vulnerability to Climate Hazards

The concept of vulnerability as a descriptor of the status of a society or community with respect to an imposed hazard or threat is deep rooted in a very broad research effort and its associated publications. The concept of vulnerability has been very widely treated in the literature, and the recent reviews by Villagrán (2006) and Birkmann (2006) draw together some highlights of a range of opinions. In regard to the assessment and reduction of socioeconomic vulnerability to climate hazards, different research and policy communities such as the disaster risk reduction, climate change adaptation, environmental management and poverty reduction community have taken up the discussion (Thomalla *et al.*, 2006). A common agreement among those communities has not been achieved yet, and discussions are still ongoing. Within the climate change community, divergent notions of vulnerability also exist. For example, the 'endpoint' definition sees vulnerability as the residual of climate change impacts minus adaptation (the remaining segments of the possible impacts of climate change that are not targeted through adaptation). In contrast, the 'starting point' views vulnerability as a general characteristic of societies generated by different social and economic factors and processes (Bogardi and Brauch 2005). The UN/ISDR (2004), on the other hand, classifies vulnerability in different dimensions or components (social, economic, physical and environmental). The present paper is not an attempt to further this high-level debate, but to focus on current developments and discussions as well as in-country local residents, stakeholder and expert knowledge to identify highly vulnerable areas.

For the purposes of this study, the optimum approach to vulnerability is to start with an influential established model – the IPCC (2001a) working definition – and then explore its ramifications in order to develop a set of working definitions and operational indicators for the study. Although, this may lack some of the sophistication of the conceptual literature, it provides a pragmatic route towards a realistic target for assessing vulnerability in data scarce regions. The starting point is to consider the core concept embodied in the IPCC's implicit definition of vulnerability as the degree to which a system is susceptible to, or unable to cope with, the adverse effects of climate change (IPCC 2001b). It is a function of the character, magnitude and rate of climate change to which a system is exposed, its sensitivity (degree to which a system is affected, adversely or beneficially, by climate-related stimuli) and its adaptive capacity (the ability of a system to adjust to climate change, moderate potential damages, take advantage of opportunities or cope with the consequences). The relation can be expressed as:

$$\text{Vulnerability} = f(\text{Hazard, Sensitivity, Adaptive capacity}) \quad (22.1)$$

The term *coping capacity* is often substituted for adaptive capacity, but appears *sensu lato* to have the same definition. The UK Department for International Development (White *et al.*, 2005) proposed:

$$\text{Vulnerability} = (\text{Exposure} \times \text{Susceptibility}) / \text{Coping Capacity} \quad (22.2)$$

This is another form of equation 22.1, as susceptibility has been substituted for sensitivity, and coping capacity substituted for adaptive capacity. For the purposes of this study, *adaptive capacity* is the preferable term as it relates most effectively to the concept of adaptive management and is more easily generalised to reflect society's ability to grasp opportunities as well as respond to threat. Both equations 22.1 and 22.2 emphasise the role of adaptive and coping capacity, but Villagrán (2001) moves further and identifies actual deficiencies in adaptive capacity as being pivotal:

$$\text{RISK} = f(\text{Hazard} \times \text{Vulnerability} \times \text{Deficiencies in Preparedness}) \quad (22.3)$$

This is an interesting formulation with a useful focus on deficiencies which might provide a basis for scoring scales for the various component attributes, but it is a much narrower viewpoint than that of equations 22.1 and 22.2 since preparedness is only one component of adaptivity and coping. The net outcome of this debate is that there is an encouraging convergence on the identification of the components involved in risk and vulnerability, but a remaining divergence on the relationship between them. Nevertheless, equation 22.1 is clearly the fundamental relationship that drives the vulnerability model, and suggests that vulnerability reflects the sum of the risks (hazards) to which a society or community is exposed, mitigated by its adaptive or coping capacity (its ability to respond effectively to risk) and compensated by the available alternative economic opportunities.

It is very important to note that climate change does not just bring threats; it also brings opportunities – new economic possibilities, improved environment or possibilities for enriching cultural change. If the new opportunities are more advantageous than the existing assets under threat, then it may not be necessary to 'protect' the asset or process at risk. New opportunities may compensate for threats and reduce the need for that society to defend the existing livelihood. It is helpful that this relatively simple view of vulnerability is entirely compatible with the general relationship:

$$\text{Process Risk (H)} = \text{Probability} \times \text{Magnitude} \qquad (22.4)$$

This is a simple and well-established statement, with a broad-based application and a fair degree of consensus. It is also easy to relate to human impact and response. This supposes that the terms process risk and natural hazard are one and the same, but in reality risk (like vulnerability and hazard) is defined in many different ways, so this equivalence may be disputed. Nevertheless, in the search for very simple relationships that do not contradict established approaches, equation 22.1 also appears to be compatible with the risk triangle (Granger 1998; Crichton 1999) that has been quite widely used to depict the interaction between the various drivers of risk. The risk triangle is defined below as depicted by GeoScience Australia (an online resource; 2007):

$$\text{RISK} = f(\text{Hazard} \times \text{Elements at Risk} \times \text{Vulnerability}) \qquad (22.5)$$

The concept of risk combines our understanding of the likelihood of a hazardous event occurring, with an assessment of its impact. Hazardous events can either be naturally occurring, such as earthquakes, tropical cyclones or coastal erosion; or they can be man-made, such as water pollution or a terrorist attack. Moreover, events can be sudden, as in the case of an earthquake; or they can occur over a period of time, as in the case of most environmental hazards. The impact of a hazardous event depends on the elements at risk, such as population or buildings, and their associated vulnerability to damage or change as a result of the event. Estimating risk is an uncertain science because it involves forecasting future events whose time and location of occurrence may be largely unknown. We capture this uncertainty mathematically in terms of probability.

Despite the huge range of possible formulations, equation 22.1, as derived from the IPCC, provides a viable initial working definition – but in practice it is difficult to implement locally in data-poor regions because it includes the full range of both biophysical and socio-economic factors (hazard and adaptive capacity). However, it can be suggested that the *hazard* term in equation 22.1 in effect serves mainly to scale the variability of vulnerability, spatially and temporally. Thus, for any one particular place and time, it may be possible to simplify the relationship to:

$$\text{Vulnerability} = f(\text{Sensitivity, Adaptive Capacity}) \quad (22.6)$$

In this form, the variability of vulnerability is seen to be driven locally mainly by socioeconomic factors. On this assumption, equation 22.1 represents the physical and social interrelationship underlying climate hazard induced risk as a whole, while equation 22.6 is appropriate as a basis for integrating the socioeconomic and environmental aspects of vulnerability.

22.2.2 Domains of Sensitivity and Adaptive Capacity

Sensitivity and adaptive capacity are not unidimensional but multidimensional because multiple factors contribute to a community being more or less sensitive or adaptive to climate-related hazard. This is because opportunities and access to resources (access to land, drinking water and health care, among others) and exposure to climate hazards (floods, drought, earthquakes, hurricanes, volcanic eruptions etc.) are not evenly distributed and vary spatially (Wisner *et al.* 2004). Some communities may be vulnerable because they are dependent on climate-sensitive livelihoods; for others, it is because their sources of drinking water are most exposed, they may lose essential ecosystems or they may be cut off from access to vital services. The ability to cope with climate hazards or take advantages of opportunities that may arise also varies between areas. For example, some communities may have easy access to economic alternatives, while others may have high human resource capacity. A combination of these factors determines a community's level of vulnerability. In the SEVA approach, we profile the different aspects of sensitivity and adaptive capacity. These are referred to as *domains* and are henceforth referred to as such.

22.2.3 Identification of Domains and Indication for Sensitivity and Adaptive Capacity

To avoid potential bias in the classification of domains and selection of indicators to quantify them (Jones and Andrey, 2007), a participatory process involving residents, stakeholders and experts is employed. The criteria for selection of domains and indicators are based on what residents, stakeholders and experts consider appropriate, relevant and robust. The domains and indicators for analysing sensitivity and adaptive capacity are identified through literature review, field observations and focus group discussion and one-on-one discussions with residents, stakeholders and experts working on both the hydrological and socioeconomic aspects of climate hazards.

The process of identifying domains and indicators for sensitivity and adaptive capacity for a particular hazard in a particular region begins with a review of literature on the key climate-related hazards, its impact and the socioeconomic background of residents in the region. From the review, an inventory of domains of sensitivity and adaptive capacity and the potential indicators to quantify them is developed. This is followed by focus group discussions with residents and one-on-one discussion with stakeholders and experts. The discussions focus on the key climate hazards, the degree to which resources or systems are affected adversely or beneficially and the ability of residents to adjust, moderate potential damages, take advantage of opportunities or cope with the consequences. Findings from the focus group and one-on-one discussions are used to revise the inventory. Field observations are then used to document physical evidence of the impacts of climate hazards and the results also incorporated into the inventory.

The next stage of the SEVA approach involves a stakeholder workshop aimed as a process of consensus building on the domains and indicators identified. The resulting inventory developed from findings from the literature, focus group and one-on-one discussions and field observations is used as a basis for a Delphi consensus process. The Delphi technique is a systematic and interactive technique for obtaining individual opinions and building consensus on a particular issue (Thangaratinam and Redman, 2005). A two-stage Delphi technique is used in this case to reach a consensus on the domains and the indicators selected to quantify them. In the case of this study, 64 government officials, representatives of nongovernmental organisations, academics and representatives of the basin's residents attended the workshop. The inventory of domains and indicators was presented to the participants at the workshop. The participants revised the inventory taking into consideration the severity and importance of each domain and indicator to that particular region. After each round a facilitator provided an anonymous summary of the participants' revised inventory and the reasons for the revision.

It is worthwhile mentioning that in most cases not all indicators identified by participants can be included in the final analysis due to data limitations. The indicators selected for the analysis, although they may not be exhaustive for assessing climate hazard induced vulnerability, they should be comprehensive and relevant for the practical assessment of vulnerability to climate hazards in the given location. The indicators may not be replicated exactly for other regions but could be modified to suit.

22.2.4 Ranking of Domains of Sensitivity and Adaptive Capacity

After consensus has been reached on the list of domains, a second Delphi is then employed in ranking the domains according to their importance and severity. In the case of this study, each participant at the workshop was asked to score the domains to sum up to a total score of 40, with the most important domain receiving the highest score and the least important receiving the lowest score. The scores were then averaged over the number of participants. The results were presented to participants and further deliberations undertaken to ensure that at least 95% of the participants agreed with the ranking. This approach was chosen as a way of validating the representativeness of the scores. The participants, also through a Delphi technique, indicated which domains represent sensitivity and which ones represent adaptive capacity. The domains, their rankings and how they are incorporated in the analysis are discussed in later sections.

22.2.5 Construction of Indices

A multidimensional matrix of indicators is often selected to represent each domain. To avoid scale dependence, it is important that all indicators are converted to standardised Z-scores. A maximum likelihood factor analysis is then used to derive a single factor score for each domain. The motivation for using maximum likelihood factor analysis is that in most vulnerability analysis indices are based on addition. However, variables used to quantify vulnerability in most cases are highly correlated, and therefore adding them up introduces a bias of double counting due to multicollinearity (Jones and Andrey, 2007). Factor analysis circumvents the problem of multicollinearity. The factor scores generated are then ranked, scaled to the range between 0 and 1 by R where $R = 1/N$ is the least sensitive or least adaptive community and $R = N/N$ is the most sensitive or most adaptive community, N being the number of areas.

To ensure comparability between the domain scores, it is appropriate that the domain scores are derived to have identical distributions with similar minimum and maximum values, with emphasis on the tail of the distribution. This helps to clearly distinguish the most sensitive and least adaptive communities. The exponential transformation has been widely used in this respect. The 'cancellation property' of the exponential transformation ensures that high scores in one domain do not cancel out low scores in others (Scottish Executive 2003). This property is highly desirable when combining scores from different domains. Equation 22.7 is used in this regard.

$$d_k = -23.026 * \log\{1 - R*[1 - e^{-\lambda/23}]\} \quad (22.7)$$

where d_k is the transformed domain score which ranges between 0 and 100 (d_i for sensitivity domains and d_j for adaptive capacity domains), −23.026 is a mathematical constant which gives a 10 % cancellation property, log is the natural logarithm, R is the ranked scores, e is the exponential transformation and the parameter λ controls the degree of progression of the score. In this case $\lambda = 100$. This transformation approach is employed for the analysis because it satisfies all the statistical requirements stated earlier (Scottish Executive 2003).

The next stage of the analysis requires combining the scores derived for each domain to create an overall score of sensitivity and also adaptive capacity. This process requires weighting the indices to reflect their severity or importance. Determining weights to attach to different indices to generate an overall index could be an intricate task. There are a number of propositions in the literature – theoretical, empirical, policy driven, consensus or purely arbitrary. To ensure that the weights applied to the scores in this study reflect the severity or importance of the domains based on the views of the basin's residents, stakeholders and experts working in the area, the Delphi scores were used to generate the weights – $w_k = t_k/n$, where t_k is the total score for domain k and n is the number of participants. In this regard, the weights reflect residents, stakeholders and experts knowledge and are based on consensus of people living and working in the area. Sensitivity and adaptive capacity domains are ranked together to understand their relative importance. The results of the ranking are presented in the results section (22.3). Overall exponentially transformed sensitivity and adaptive capacity scores are then derived using equations 22.8 and 22.9, respectively:

$$S = -23.026 * \log\left\{1 - R\left(\frac{1}{n_l}\sum_{l=1}^{l} d_i w_k\right) * \left[1 - e^{-1/23}\right]\right\} \quad (22.8)$$

$$AD = -23.026 * \log\left\{1 - R\left(\frac{1}{n_h}\sum_{h=1}^{h} d_j w_k\right) * \left[1 - e^{-1/23}\right]\right\} \quad (22.9)$$

where $\frac{1}{n_l}\sum_{l=1}^{l} d_i w_k$ and $\frac{1}{n_h}\sum_{h=1}^{h} d_j w_k$ are the sensitivity and adaptive capacity scores averaged over the respective number of domains, n_l is the number of sensitivity domains and n_h is the number of adaptive capacity domains.

Having derived the indices of sensitivity and adaptive capacity, the next stage of the analysis is to derive an overall index of vulnerability. Following the IPCC's conceptualisations of vulnerability (IPCC 2001a), we assumed an inverse relationship between sensitivity and adaptive capacity in deriving an overall index of vulnerability. Equation 22.10 is then used to derive an exponentially transformed index of socioeconomic vulnerability (V).

$$V = -23.026 * \log\left\{1 - R\left(\frac{S}{AD}\right) * \left[1 - e^{-1/23}\right]\right\} \quad (22.10)$$

where S and AD are the sensitivity and adaptive capacity scores. The indices generated from equation 22.7 are mapped to reveal the spatial variation in the domains of sensitivity and adaptive capacity and those from equations 22.8, 22.9 and 22.10 reflect the spatial variation in sensitivity, adaptive capacity and vulnerability to climate hazards respectively. The scores range from 0.001 to 100, with an increase in score showing increasing sensitivity, adaptivity and vulnerability. Thus, sensitivity, adaptive capacity and vulnerability are measured on a scale of 0 to 100.

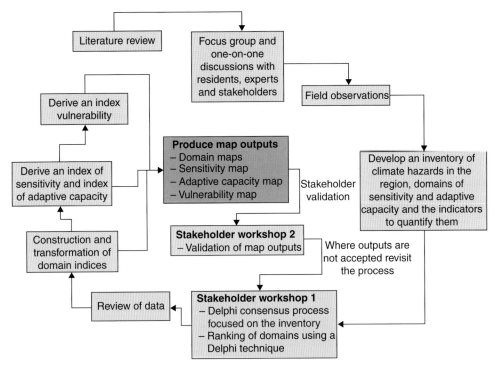

Figure 22.1 A diagrammatical overview of the SEVA approach.

22.2.6 Stakeholder Validation

After the map outputs have been produced, a stakeholder validation workshop is convened. This is to give the stakeholders the opportunity to validate the outputs and share opinions on the spatial disparities identified. In the case of this study, the validation workshop was convened in the Guwahati City in Assam and attended by 35 participants including government officials, journalists, representatives of nongovernmental organisations, academics and representatives of the basin's residents. Figure 22.1 is a diagrammatical overview of the SEVA process.

22.3 Results

22.3.1 Key Climate Hazards in the Basin

Flooding was identified as the most momentous climate hazard in the basin alongside droughts and river bank erosion. This finding concurs with findings from the literature (Jamir et al., 2008; Bagchi, 2007; Dhar and Nandargi, 2000; Chaudhari and Sinha, 1999; Sarma, 2005). The basin's residents reported that flooding is an annual occurrence with adverse socioeconomic consequences. The annual floods results in loss of livelihood mainly from agricultural products, contamination of drinking water, outbreak of diseases, loss of infrastructure (mainly housing, roads and bridges) and loss of ecosystems. Droughts in the basin are more erratic. They vary temporally and spatially. Prolonged dry spells in the basin as a results of deficits in the monsoon confronts families with unprecedented hardships. River bank erosion and its impacts were clear during the field observations. These include loss of agricultural land, houses and important infrastructure such as roads and health facilities. Excerpts from the focus groups and one-on-one discussions with the basin's residents, stakeholders and experts and field observations are presented in Table 22.1.

22.3.2 Domains and Their Ranking

Fourteen domains of sensitivity and adaptive capacity were identified and ranked using the Delphi technique. Figure 22.2 shows the domains, their classification and ranking in order of importance or severity. In all, ten of the domains were classified as sensitivity domains and four as adaptive capacity domains (Figure 22.2). The rankings in Figure 22.1 clearly shows that contamination to sources of drinking water, loss of livelihoods (mainly agriculture), lack of access to heath services, lack of economic alternatives and loss of housing are the major concerns of the basin's residents. Out break of disease, gender effects (impact on women), immigration, human capital and social networks although mentioned were not ranked high.

22.3.3 Review of Data

Having identified the domains and indicators to quantify them, there is the need to establish available spatial datasets that cover the area at the high resolution. A review of potential data sources for the study area was conducted. Three potential data sources with the required resolution were identified – the 2000–2001 Agricultural Census, 2001 India Population and Housing Census (Census of India, 2011) and 2001 Assam LANDSAT data. The data for the analysis comes from the 2001 Indian Population and Housing Census and 2001 LANDSAT data. The Agricultural Census data were not accessible to the researchers. The LANDSAT data include road density, agricultural land use, distance to main settlements and health centres. Road density and agricultural land use data were extracted using a 3 km buffer around each village or town point. Agricultural land use is expressed as the proportion of land use on the basis of commercial and noncommercial agricultural land. The distance measures (distance to main settlements and health facilities) are Euclidean distances. The data cover 14,775 villages or towns in the Assam Basin of the Upper Brahmaputra River.

Although 14 domains of sensitivity and adaptive capacity were identified, only six – loss of livelihood, access to health services, road infrastructure,

Table 22.1 Excerpts from discussion and observations with the basin's residents, stakeholders and experts.

Issues	Impacts	Who is most affected/benefit
Drinking water	In times of floods, drinking water sources submerge and get contaminated.	Communities with unimproved sources and those who have to travel long distances for water
Livelihood	Subsistence agriculture is the main source of income and livelihood, but highly susceptible to flooding. Loss of farm produce and equipment leads to loss of income. Often victims have to borrow (with high interest rates) to survive.	Communities with high dependency on agriculture
Sanitation	It was evident from field observations that most communities rely on poor drainage facilities which are not robust enough for the extent of floods experienced in the basin.	Communities with poor drainage facilities
Access to health and outbreak of diseases	The effects are both physiological and psychological. The indirect effects include – destruction to health infrastructure, loss of essential medicine and damage to referral systems and emergency services with detrimental immediate and long-term effects.	Communities with inadequate health care facilities and those exposed to communicable diseases
Destruction to housing	Loss of housing leads to massive displacements. They impacts both directly and indirectly on livelihoods, health, natural resource and the environment.	Areas where the quality of housing are poor the resilience of built up areas to floods are low
Road infrastructure	The basin's residents have high dependence on roads for sustaining incomes and livelihoods, e.g. transporting their farm produce to the market. Destruction to roads during floods exacerbates their socioeconomic stresses.	Communities with poor quality roads
Ecosystem loss	Floods affect the composition and distribution of ecosystems in the basin – access to food, medicine and water. High human (e.g. felling of trees for firewood and construction of houses) intrusion on the natural landscape of the basin was also blamed.	Communities with high dependence on ecosystems
Gender	Women have more responsibility for providing climate sensitive resources such as gathering firewood and water. Poor and disproportionate distribution of resources and opportunities also renders women very sensitive in the wake of floods.	Communities with marked gender disparities – participation in the labour market, education, health and decision-making

(continued)

Table 22.1 (Cont'd).

Issues	Impacts	Who is most affected/benefit
Immigration and urbanization	The influx of migrants put enormous pressure on already limited resources leading to conflicts. It also facilitates rapid loss of ecosystems due to construction of houses and dependence on forest resources.	Urban settlement and border communities
Availability of economic alternative	The availability of alternatives income and livelihood sources is vital for mitigating the effects of floods. In the aftermath of floods, some members (mainly men) of communities with close proximity to cities often migrate temporarily to engage in temporal economic activities such as trading pottering.	Communities with close proximity to cities most often have better alternative economic opportunities
Economic capacity	Economic capacity of individuals, households and communities are a key determinant of the ability to cope or adjust to the impacts of floods. The recurrent floods, livelihood insecurity and economic poverty have left most people in a cycle of poverty.	Communities less reliant on climate sensitive income and livelihood activities e.g. civil servants
Human resource capacity	The ability to cope with floods requires knowledge about options, the capacity to assess opportunities and the ability and skills to implement the most suitable and sustainable options. In most communities literacy, educational and employment levels are low.	Communities with high human capital have greater adaptive capacity.
Social networks	The ability to act collectively strongly determines the inherent capacity to adapt to floods in the basin. Stronger social networks enhance informal information sharing and are essential for enhancing capacity to cope or adjust to environmental stresses. Social networks serve as a vital information, education and communication tool for some communities in the basin.	Those endowed with strong social networks and civic associations are more likely to cope successfully with adverse situations.

gender effects, availability of economic alternatives and human capital – are analysed and mapped in this study due to data limitation. The indicators used to profile them are shown in Table 22.2. The indicators selected to represent each domain were what local resident, stakeholders and experts deemed appropriate. As mentioned earlier, not all indicators identified by participants are included in the final analysis due to data limitations.

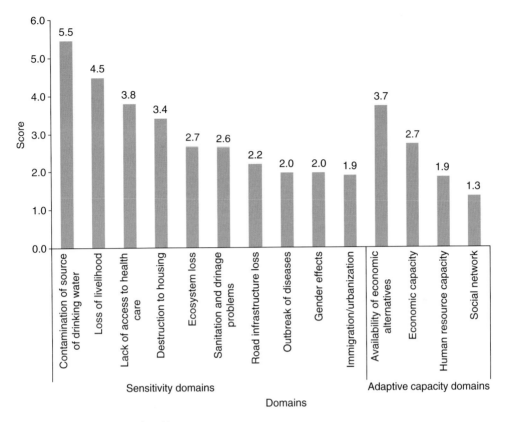

Figure 22.2 Domains and ranking.

22.3.4 Spatial Variations in the Domains of Sensitivity and Adaptive Capacity

Figure 22.3 shows the spatial variations in the domains analysed. The blue course shows the Brahmaputra River. The scores have been categorised into quintiles. Table 22.3 shows the categorisation of the scores. Figure 22.3 depicts clear variations in the domain scores. These variations are often masked when only one vulnerability score is created without considering its multidimensionality. Generally, Figure 22.3 shows that small communities (in terms of population size) are most sensitive and possess the least adaptive capacity to climate hazards. Considering livelihood sensitivity, Figure 22.3 shows a cluster of sensitive communities in the western part of the basin. With regard to health, road infrastructure and gender sensitivity, sensitive communities are more spread out but still with substantial clustering to the western part of the basin and in the Doom Dooma and Margherita Tehsils.

Considering spatial patterns in adaptive capacity, Figure 22.3 shows that there is a high concentration of communities with fewer opportunities to alternative economic activities within the southwestern part of the basin and also in the Subansiri (I and II), Dhemaji, Kadam and Dibrugarh West Tehsils

Table 22.2 Domains and indicators.

Domains	Indicators
Sensitivity domains	
Livelihood	1. Proportion of workers engaged in agriculture 2. Main to marginal agricultural workers (the ratio of agricultural worker who work for more than 6 months to those who work less than 6 months) 3. Subsistence crop land (in km^2) per 100 households 4. Commercial crop land (in km^2) per 100 household
Access to health care	1. Distance to nearest allopathic hospitals 2. Distance to nearest maternity and child welfare centres 3. Distance to nearest public health centres
Road infrastructure	1. Proportion of roads that are metalled 2. Proportion of roads that are national roads 3. Proportion of roads that are track paths 4. Proportion of roads that are unmetalled roads 5. Proportion of roads that are cart tracks 6. Proportion of roads that are foot paths
Gender	1. Female literacy rate 2. Proportion of female workers engaged in non-agricultural work 3. Sex ratio
Adaptive capacity domains	
Economic alternatives	1. Proportion of non-agricultural workers 2. Distance to the nearest city or town
Human Capital	1. Adult literacy rate 2. Proportion of the population (7+ years) who are working 3. Ratio of workers to nonworkers (dependency ratio) 4. Ratio of main workers to marginal workers

all to the north of the basin. There is also a high concentration of communities with low human resource capacity within the western part of the basin and the Doom Dooma and Margherita Tehsils.

22.3.5 Spatial Variations Sensitivity and Adaptive Capacity

Figure 22.4 shows the spatial variations in the sensitivity and adaptive capacity derived using equations 22.8 and 22.9, respectively. The categorisations are the same as shown in Table 22.3. Figure 22.3 clearly reveals that the most sensitive communities are more likely to be the least adaptive communities. Figure 22.4 reveals that the most sensitive and least adaptive communities are mainly in the western part of the basin and also along the river banks to the north. It can also be noted from Figure 22.4 that there is a high concentration of communities with low sensitivity and high adaptive capacity scores in the northeast and southeastern parts of the basin.

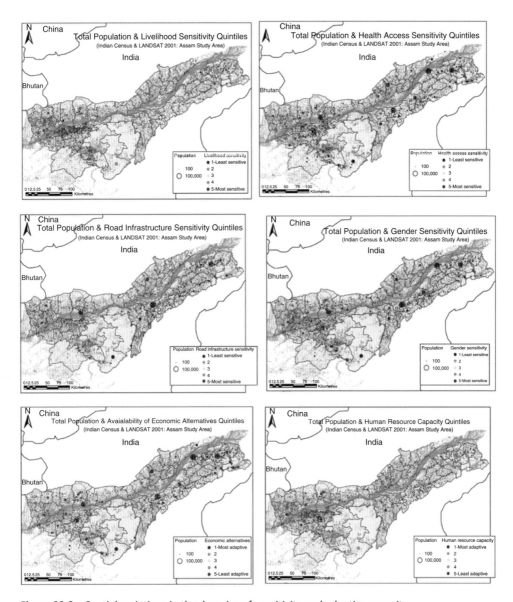

Figure 22.3 Spatial variations in the domains of sensitivity and adaptive capacity.

22.3.6 Spatial Variations in Vulnerability to Flood Hazards

Figure 22.5 shows the overall vulnerability map derived using Equation 22.10. The categorisations are the same as shown in Table 22.3. Figure 22.5 shows that the most vulnerable communities are in the western part of the basin and along the river bank to the north. It is also clear from Figure 22.5 that small communities are more vulnerable compared to main settlements.

Table 22.3 Categorisation of scores.

Quintiles	Score	Mean score	Standard deviation	Number of communities
1: Least sensitive / least adaptive	0.001–5.05	2.44	1.46	2953
2	5.06–11.55	8.15	1.87	2958
3	11.56–20.63	15.80	2.61	2955
4	20.64–35.85	27.41	4.35	2954
5: Most sensitive / most adaptive	35.86–100	54.64	15.54	2955
Overall		21.69	19.90	14775

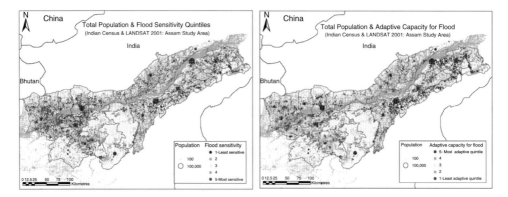

Figure 22.4 Variations in sensitivity and adaptive capacity to flood hazards.

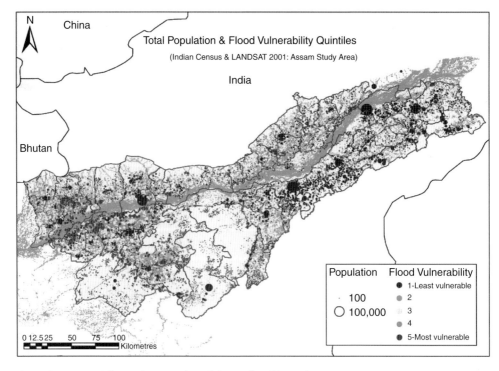

Figure 22.5 Spatial variations in vulnerability to flood hazards.

There is a high concentration of the least vulnerable communities in the northeast and southeast of the basin, where communities are not located along the banks of the river. Conversely where communities are located along the banks, it can be seen from Figure 22.5 that there is a concentration of vulnerable communities.

22.4 Conclusions

This paper illustrates the SEVA approach for quantifying and mapping vulnerability to climate hazards using statistical techniques which incorporate local knowledge. Domains and indicator selection are done through a participatory approach involving residents, stakeholders and experts. A Delphi process is used to arrive at a consensus on the domains and indicators identified. A Delphi technique is also used to derive weights to represent the severity or importance of domains, which are incorporated in the derivation of indices. Stakeholders are given the opportunity to validate the outputs and share opinions on the spatial disparities identified.

This process is adopted in this paper as the basis for a case study of flood hazards in the Upper Brahmaputra River Basin of the Assam State of India. Floods, droughts and river bank erosion were identified as the key environmental hazards in the basin. Floods were reported as an annual occurrence with adverse socioeconomic consequences – loss of livelihood mainly agricultural produce, contamination of sources of drinking water, outbreak of diseases, loss of infrastructure (mainly housing, roads and bridges) and loss of ecosystems, among others. In assessing the multidimensionality of vulnerability to flood hazards, 14 domains of sensitivity and adaptive capacity were identified and ranked using the Delphi technique. Contamination of sources of drinking, loss of livelihoods (mainly agriculture), lack of access to heath services, lack of economic alternatives and loss of housing were the major concerns of the basin's residents in times of floods. Although out break of disease, gender effects, immigration and urbanisation, human resource capacity and social networks were mentioned, they were not ranked high.

Spatial variations were identified in the domains of sensitivity and adaptive capacity. These variations are often masked when only one vulnerability score is created without considering its multidimensionality. The results show that, generally, small communities (with regard to population sizes) are the most sensitive communities and with the least adaptive capacity. The most sensitive and least adaptive communities are located mainly in the western part of the basin and also along the river bank to the north. Unsurprisingly, the results show that the most vulnerable communities are in the western part of the basin and along the river bank to the north.

Discussions with local experts associated with the Delphi process designed to weight the domains, provide a first order appreciation of the key elements of the spatial distribution of vulnerability shown in Figure 22.5. Firstly the predomination of high vulnerability associated directly with the river and wetlands (southwest and northwest) can be, at least in the main, attributed to the substantive influx of the Bangladeshi community who by virtue of their migratory status have a tendency to have much lower access to health care and education as well as being economically utilised to carry out the least well paid jobs in the region. Whilst it would be desirable to clearly identify this through the census, there is a pressure upon the immigrant population to identify themselves as of Indian origin and as such this data is therefore deemed unreliable. There is a clear region of low vulnerability to flood in the southeast of the basin which is thought to reflect the higher status afforded to the workers of commercial plantations in terms of access to education and health care as well as greater communication, substantive river defence networks and ties to the socio-economically low-vulnerability urban areas.

References

Alcamo, J., Acosta-Michlik, L., Carius, A., Eierdanz, F., Klein, R., Krömker, D. & Tänzler, D. (2008) A new approach to quantifying and comparing vulnerability to drought. *Regional Environmental Change*, 8(4), 137–149.

Bagchi, S. (2007) Disease outbreaks in wake of Southeast Asia floods. *CMAJ*, September 11, 177.

Birkmann, J. (ed.) (2006) *Measuring Vulnerability to Natural Hazards: Towards Disaster Resilient Societies*. United Nations University Press, Tokyo.

Bogardi, J. & H.G. Brauch (2005) Global environmental change: a challenge for human security – defining and conceptualising the environmental dimension of human security. In: *UNEO – Towards an International Environmental Organization – Approaches to a Sustainable Reform of Global Environmental Governance*, ed. Rechkemmer, A., Nomos, Baden-Baden, 85–109.

Census of India (2001) *Indian Population and Housing Census 2001*. Office of the Registrar General & Census Commissioner, India.

Chaudhari, S.N. & Sinha, S. (1999) Flood management in Assam – a viewpoint. *Journal of India Waterworks Association*, 31(2), 138–140.

Crichton, D. (1999). The risk triangle. In: *Natural Disaster Management*, ed. Ingleton, J., Tudor Rose, London, 102–103.

Dhar, O.N. & Nandargi, S. (2000) A study of floods in the Brahmaputra Basin in India. *International Journal of Climatology*, 20, 771–781.

Eakin, H. & Luers, A.L. (2006) Assessing the vulnerability of social-environmental systems. *Annual Review of Environment and Resources*, 31, 365–394.

GeoScience Australia (2007) What is risk? Available from http://www.ga.gov.au/urban/factsheets/risk_modelling.jsp.

Goswami, D.C. (1985) Brahmaputra River, Assam, India: physiography, basin denudation and channel aggradation. *Water Res.*, 21, 959.

Granger, K. (1998). *Geohazards Risk and the Community: Disaster Management: Crisis and Opportunity: Hazard Management and Disaster Preparedness in Australasia and the Pacific Region*. Centre for Disaster Studies–Cairns, James Cook University, Cairns, Australia.

IPCC (2001a) *Climate Change 2001: Synthesis Report*. World Meteorological Organization, UN Environment Programme, Geneva.

IPCC (2001b) *Climate Change 2001: Impacts, Adaptation and Vulnerability. Contribution to the Working Group I to the third assessment report of the Intergovernmental Panel on Climate Change (IPCC)*. Cambridge University Press, Cambridge.

Jamir, T., Gadgil, A.S. & De, U.S. (2008) Recent floods related natural hazards over West coast and Northeast India. *Indian Geophysical Union*, **12**(4), 179–182.

Jones, B. & Andrey, J. (2007) Vulnerability index construction: methodological choices and their influence on identifying vulnerable neighborhoods. *International Journal of Emergency Management*, **4**(2), 269–295.

Morrow, B.H. (1999) Identifying and mapping community vulnerability. *Disasters*, **23**(1), 1–18.

Sarma, J.N. (2005) Fluvial process and morphology of the Brahmaputra River in Assam, India. *Geomorphology*, **70**(3–4), 226–256.

Scottish Executive (2003) *Scottish Index of Multiple Deprivation: The General Report*. Scottish Executive National Statistics, Edinburgh.

Sherbinin, A.D., Schiller, A. & Pulsipher, A. (2007) The vulnerability of global cities to climate hazards. *Environment and Urbanization*, **19**(39), 39–64.

Taenzler, D., Carius, A. & Maas, A. (2008) Assessing the susceptibility of societies to droughts: a political science perspective. *Regional Environmental Change*, **8**(4) 161–172.

Thangaratinam, S. & Redman, C.W.E. (2005) The Delphi technique. *The Obstetrician & Gynaecologist*, **7**, 120–125.

Thomalla, F., Downing, T., Spanger-Siegfried, E., Han, G. & Rockström, J. (2006) Reducing hazard vulnerability: towards a common approach between disaster risk reduction and climate adaptation. *Disasters*, **30**(1), 39–48.

UN/ISDR (2004) *Living with Risk: A Global Review of Disaster Reduction Initiatives*. UN Publications, Geneva.

Villagrán, J.C.V. (2001) *La naturaleza de los Riesgos Asociados a Varias Amenazas en Poblados de Guatemala*. Technical report. SEGEPLAN, Guatemala.

Villagrán, J.C.V. (2006) *Vulnerability: A Conceptual and Methodological Review*. UN University Institute for Environment and Human Security, Bonn.

White, P., Pelling, M., Sen, K. Seddon, D. Russell, S. & Few, R. (2005) *Disaster Risk Reduction: A Development Concern*. DFID, London.

Wisner, B., Blaikie, P., Cannon, T. & Davis, I. (2004). *At Risk: Natural Hazards, People's Vulnerability and Disasters*. Routledge, New York.

Mitigation via Renewables

David Coley

23.1 Introduction

If we are to mitigate the effects of climate change, we need a rapid reduction in our production of greenhouse gases. Through the IPCC, attempts are being made to do this (e.g. Copenhagen 2009 and Kyoto 1997). Unfortunately, such reductions only include the developed world and the targets set have been small compared to what the climate science indicates is needed. Figure 23.1 shows the scale of the problem with most developed countries needing to cut their emissions by 80% or more. Figure 23.1 was created under the assumptions of *contraction and convergence*. This approach starts by setting a maximum safe atmospheric concentration for carbon dioxide, then estimates what level of global emissions gives rise to this, then apportions this to each country based on its population. Countries that currently produce more than this (largely the developed world) can buy the right to emit more from those that emit less than their quota (largely the developing world). Over time the total right to emit would be reduced until the safe level of emissions is reached. Contraction and convergence are seen by many as the only ethically sound ways of setting reduction targets that have a reasonable chance of gaining support from the world community.

In order to make the required reductions we will have to simultaneously reduce the amount of energy we use to achieve a set goal (e.g. keeping a building at a particular temperature), and decarbonise our energy production (i.e. increase the use of renewable energy). In the following we look at the plethora of renewable sources we can access; in Chapter 13 we saw how emissions can be reduced through energy efficiency, or the climate directly engineered. Chapter 13 also presented one of many possible combinations of changes to our energy supply that has the potential to make major carbon savings, together with an example abatement curve.

Solutions to Climate Change Challenges in the Built Environment, First Edition.
Edited by Colin A. Booth, Felix N. Hammond, Jessica E. Lamond and David G. Proverbs.
© 2012 Blackwell Publishing Ltd. Published 2012 by Blackwell Publishing Ltd.

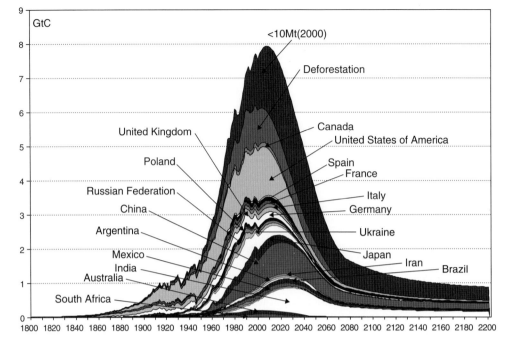

Figure 23.1 Contraction and convergence time series (from Coley, 2008; produced with software freely available from the Global Commons website).

23.2 Current World Sustainable Energy Provision

Thirteen percent of world primary energy is produced from renewable energy sources including hydropower, compared to 35% for oil, 24% for coal, 21% for natural gas and 7% for nuclear energy. This suggests that renewable energy should already be treated as seriously as other sources of energy. As Figure 23.2 shows, most of the flows of renewable energy we could tap into are far greater than our current world-wide use of energy (410×10^{18} joules per annum; or a continuous power use of 13×10^{12} watts, or 2 kW per person), so in theory we should be able to completely decarbonise our energy supply. The problem seems to be one of willingness to do this. Cost is often cited as a reason for not moving away from carbon based fuels to renewables. However, we do not apply this logic to other industrial pollutants such as radioactive emissions from nuclear power plants, where vast sums are spent on stopping pollution. We have yet to see carbon dioxide in this way, despite the impact we know it will have on the planet and its population. Thus it only makes sense to compare the cost to the cost of inaction.

Much (77.5%) of current renewable energy use is the noncommercial use of biomass, which represents 10.4% of world primary energy; the second largest is hydropower (16.3% of the renewable fraction, or 2.2% of primary energy). Geothermal represents 0.4% of primary energy. Solar, wind

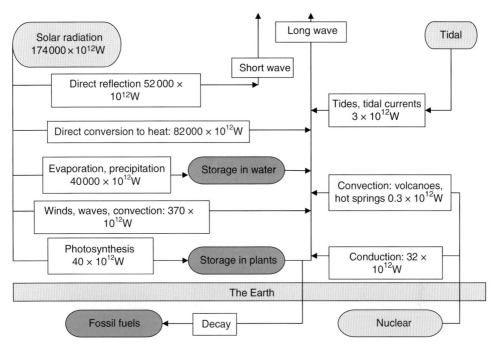

Figure 23.2 Power audit of the planet (data from Ristinen and Kraushaar, 1999).

and tidal represent less than 0.1% of the total world primary energy supply. Renewable energy use is growing at an average of 1.7% per annum, which is slightly greater than the increase in world energy demand (1.4%) (IEA, 2005). However, individual sustainable technologies are growing much faster than this. Wind power, for example, is increasing by 24% per annum.

Most solid biomass derived energy is produced and consumed within the developing world, where it is used for residential cooking and heating, with Africa producing 24% of the world's solid biomass supply (IEA, 2005). Because of this high level of biomass use, non-OECD regions account for 78% of renewable energy supply, while the OECD (the most developed nations) produce only 22% of the world's renewable energy, despite using 52% of world primary energy. Clearly, the developing world is ahead of the developed as regards sustainable energy provision. In non-OECD countries, 22% of energy use is on average from renewable sources, for OECD countries this falls to only 5.7%. Some would consider this inexcusable.

23.3 Solar Power

Every ten minutes, the surface of the Earth receives enough energy from the Sun to provide the primary energy needs of humankind for a whole year. It is, hence, unfortunate that we have based our societies on burning stored carbon rather than accessing this resource.

23.3.1 Passive Solar Heating

Space heating using solar gains has a long history. Pliny the Younger used windows built of thin sheets of mica to warm a room in northern Italy around 100 AD, thus saving on firewood. In addition, many Roman bathhouses used large south-facing openings to access this free source of heat. The eighteenth century saw the first widespread use of solar design with the introduction of conservatories for growing exotic fruits and as ways of creating a warm space before the general use of central heating. In the 1950s the architect Frank Bridgers designed the world's first commercial office building using solar space and water heating (the Bridgers-Paxton building in Albuquerque, United States).

The basic features of a solar heated space include large south-facing (north facing in the southern hemisphere) windows allowing sunlight to enter the room. This light is absorbed by the various surfaces in the room and then re-radiated as longer wavelength infrared radiation. Two, not so obvious, features are essential: the building must be constructed with a large amount of material that has a high thermal capacity, to store the heat, and the design must lose very little heat through the fabric. Low heat loss is essential because, compared to the amount of heat produced by a boiler, very little energy will be received from the sun during the all-important winter months. High thermal capacity is required because the sun will not shine every day or at night, and because we need the building to maintain an even and comfortable temperature. Without a high thermal capacity, the space will be too hot on sunny days and too cold on overcast ones.

23.3.2 Heat Pumps

One approach to solar energy collection is to extract heat from land, air or water which has been warmed by the sun. This warming need not be great and is carried out with the help of a heat pump (which operates in a similar manner to a fridge, but with the heat exchanger found on the back of fridge inside the building and the source of heat (normally the food in the fridge and heat ingress) being the outside world).

Heat pumps are unfortunately not a panacea for creating 'free' energy. Their performance will fall rapidly as the temperature of the source of heat falls. This can mean that in winter, when heating demand is greatest, their performance can be poor. A ground-source heat pump that uses the more constant conditions found underground is better in this respect but more costly to install.

23.3.3 Solar Water Heating

Solar water heating has a history as long as that of solar space heating. The Romans built water channels lined with black slate to preheat water for bath houses. 1891 saw Clarence Kemp of Baltimore, United States patenting the first solar water heater, and by 1897, 30% of homes in Pasadena, United States had Kemp's design fitted. In 1908 William J. Bailey invented a collector based on an insulated box containing copper coils – the basis of today's systems.

Several technologies exist for collecting solar energy and using it to heat water to modest temperatures. The simplest is just an unwanted domestic radiator painted black through which water flows and which is placed on a south facing roof. This will give a temperature rise of up to 10°C at mid-latitudes. By encasing the system in glass, heat loss can be greatly reduced and temperature rises of 50°C can be obtained. (Typically a series of copper pipes bonded to a black backing plate is used.) By using an evacuated glass tube containing an evaporator pipe, temperature rises of 150°C are possible. To reach higher temperatures mirrors or other concentrating systems are required and are more typically found in systems which raise steam for electricity generation.

23.3.4 High-Temperature Solar Power

Higher temperatures are achieved by gathering sunlight over a wide area and concentrating it on to a much smaller area for absorption. The Swiss scientist Horace de Saussure built the first solar collector in 1767, a design which was later used by the explorer Sir John Herschel in the 1930s to cook food on his travels in Africa. In the 1860s the mathematician Augustin Mouchot designed a solar-powered steam engine. During the 1870s and 1880s Mouchot and Abel Pifre produced solar stills, cookers and even a printing press.

In 1913 Frank Shuman demonstrated a 41 kW (55 horsepower) steam engine in Egypt that used five parabolic trough collectors, each 4 metres wide and 80 metres long, to focus the light on a cast iron pipe. Shuman also suggested that by building 52 000 km^2 (20 000 square miles) of collector in the Sahara desert more energy could be harvested than was being produced from coal at that time. 52 000 km^2 might seem a great area, but it would only have required 1000 sites each with a side of length around 7 km (4 miles) to achieve. Unfortunately, since 1913 our use of fossil fuels has spiralled and the area would now have to be very much greater.

Mouchot and Pifre's design has been improved by placing a small steam or Stirling engine, a thermoelectric generator or a cooled photovoltaic cell at the focus of the parabolic dish. The Barstow power plant in the United States contains 1818 heliostats (mirrors) of approximately 40 m^2 each, focusing on a 91 m tower and producing 10 MW$_e$ of electricity from high pressure steam at 560°C. Excess heat is stored (in molten salt) to allow 7 MW$_e$ of electricity production to continue at night.

OECD electricity production for thermal electric plants stands at 569 GWh per annum (IEA, 2005). The amount of passive solar energy provided by windows goes unestimated.

23.4 Photovoltaics

The sun is the ultimate source of renewable energy and electricity the highest grade of energy. The Holy Grail of alternative energy is probably the ability to turn sunlight into electricity with high efficiency, at low cost, using common materials that have structural characteristics that allow them to replace other components (e.g. roofing tiles). The search for this grail has been on for some time. In 1839, Edmond Becquerel noted that the voltage produced by a battery containing silver plates increased if the battery was exposed to sunlight. In 1833, Charles Edgar Fritts constructed the first true solar cell using slithers of selenium. His device had an efficiency of less than 1%. By 1954, Darryl Chapin and others from Bell Labs had produced a silicon solar cell with an efficiency of 6%. Improvements in design have now increased this to over 44% in the laboratory.

Unfortunately, approximately $30\,km^2$ of photovoltaics (PV) would be required to produce as much energy per annum as a $1\,GW_e$ fossil fuelled plant. This is a lot of land to cover in a human-made material and resistance from both the local population and environmentalists to such a facility might be expected. However, we are quite used to covering large areas of land with human-made materials when we build houses or other structures. If the roof of the building could be covered with PV cells, or even better, if the PV cells themselves could replace the roof structure, large amounts of electrical energy could be generated without overly impacting on the landscape (although the roofs would look somewhat different than traditional designs).

The average UK home uses 3880 kWh of electricity per annum (equivalent 440 W continuously) (UST, 2005). If covered in PV, a typical domestic roof could produce several times this amount of electricity over a year. It is worth pausing for thought in order to digest this result: much of the world's 'necessary' electricity use (i.e. that not used for heating) could probably be replaced by an alternative that would have little impact on the environment and without the building of any new large-scale structures – we just have to build our roofs from a different material. At present, the cost of PV cells is too high for this to happen unless large subsidies are in place. There are also additional questions that need to be addressed, such as the production of the required area of semiconductor and the safe use of the chemical compounds used in the process.

Worldwide PV installed capacity is around 16.6 GW (REN, 2009) and growing at nearly 50% per annum. This makes it the fastest growing sustainable energy technology.

23.5 Wind Power

Over 10 TW is flowing in the world's winds – enough to meet the world's primary energy demand. In comparison to technologies such as photovoltaics, the core technology is simple, and has been used by humankind for several thousand years. The earliest machines were used to raise water for irrigation, grinding corn and propelling boats. Many of these designs were found right across Asia, certainly from 250 BC and probably earlier. They were simple vertical axis machines with a central rotating shaft that incorporated a screen to ensure that only one side of the 'sails' felt the pressure of the wind. By the eleventh century AD, horizontal axis machines had been developed and introduced into Europe. The design was mainly used for grinding grain – hence the name *windmill*. By the eighteenth century up to 100 000 such mills were operating in Europe (Smil, 1994; DeZeeuw, 1978) and 10 000 were in England alone by the nineteenth century. It is interesting to note that an equivalent number of modern wind turbines each with a rated output of 2 MW$_e$ would give an installed capacity of 20 GW$_e$, 30% of current UK installed capacity. It could therefore be suggested that this figure might be taken as an historical minimum number that the landscape is capable of holding without undue degradation (although they would be taller than their historic counterparts). This would also return us to large-scale embedded generation, where there is less geographic separation between generation and use. This could be described as the *re-democratisation of supply* where we all experience both the benefits and the consequences of energy supply.

Ultimately there are limits to the percentage of a nation's electricity supply that can be generated by wind power because of the intermittent nature of the resource. However, if we move away from a carbon-based economy and toward a hydrogen-based one, then wind power could be used to make hydrogen by the electrolysis of water. This would avoid this limiting factor. The hydrogen could then either generate electricity when there was little wind, or be exported to be used as a transportation fuel.

Since the 1990s wind energy has developed in two directions: towards larger machines and to offshore siting. Larger machines are more efficient and allow much smaller sites to be used (one 1.5 MW$_e$ machine can replace five 300 kW$_e$ turbines, or thirty 50 kW$_e$ ones). They are also ideal for off-shore generation where visual impact and noise is less of a problem. Moving off-shore also allows substantial numbers of large machines to be grouped together in a windy environment.

Wind power has one of the smallest environmental footprints of any generating technology. No rivers need to be dammed, no radioactive waste is stockpiled and no oil spills can arise. However, there are concerns about how the look and sounds of the countryside can be altered by the construction of even a single wind turbine. Modern machines can be 70 m or more from ground to blade tip and need to be sited in locations with high wind speeds often near the top of slopes, which means they will be visible for a considerable distance. Because wind has a low energy density, wind farms need to be large and the requirement to space machines at least three to five rotor diameters apart to avoid problems with turbulence, means that they are even more extensive. Hence these is a clear visual impact. Other concerns arise over noise, scattering of radio, TV and radar signals and bird strikes. It is known that the disruption of some peat-based soils when forming the foundations of large turbines releases some of the carbon stored in the soil. On some sites the quantity of carbon released may equate to a reasonable fraction of the carbon saved by the turbine over its life.

Current worldwide wind installed wind power capacity is 120 GW$_e$, growing at around 25% per annum (REN, 2009). Growth is greatest in absolute terms in China and the United States.

23.6 Wave Power

Like wind power, wave power is an indirect form of solar power. Because water has a higher density than air, much higher energy densities are realised: 70 kW per metre of wave front or greater at some sites.

Many of the wave power devices prototyped so far would be placed a considerable distance from the shore and therefore it is probably the alternative energy technology with the least visual impact. Even designs which need to be on the seashore are likely to have scant impact on the marine environment or on nesting and feeding bird populations, as they have little effect on the general flow of water around them. The central issue for the development of wave power is not the size of the resource, nor possible environmental impacts, but the problem of harnessing it economically within the harsh marine environment. Tidal power faces a similar challenge, but has the advantage that the technology is likely to be deployed within sheltered estuaries and bays and hence in a less hostile setting. For wave power to work we need sites with regular, large waves.

A plethora of designs have been proposed with no one design dominating at either a theoretical or practical level, and it would be fair to say that the whole field is far less commercially developed than wind power. On-shore devices typically use small inlets in the coastline where waves can be captured and used to compress air and turn a turbine. Off-shore wave energy converters should be able to harvest more energy than their on-shore cousins as the size and reliability of the waves around them will be greater. Amongst the possibilities are the Sea Clam – a 60 m diameter toroid (think of a ring doughnut) consisting of 12 interconnected rubber cells. Waves passing the toroid compress cells forcing air through Wells's turbines into adjacent cells. An alternative version of the Clam is linear, with the air sacks following each other in a long line – 275 m for a 10 MW$_e$ device. Such devices would be laid in very large *energy fields* similar to a wind farm.

23.7 Large-Scale Hydropower

The use of falling or moving water to provide energy has a long history. The earliest use was probably as a simple way of raising water from a stream or river to adjacent fields for irrigation and was in use in the Middle East by 3000 BC. By 200 BC the technology had taken a leap with the invention of the vertical-axis mill for grinding corn, and then the horizontal axis geared system. This final design formed the basis of all arrangements for the next 2000 years. With only a few alterations, it proved so successful that it became the main source of mechanical power in much of the world and was used for pumping, iron working and paper manufacture.

The introduction of steam power during the industrial revolution reduced the need for water power for several reasons. The key factors being the greater power offered by steam and the severing of the link between the source location (the river) and the requirement location (the factory). The re-emergence of water power during the twentieth century was due to the simultaneous solution of these two problems. The first by the invention of more efficient systems with far greater hydraulic heads (traditional systems had heads no greater than the water wheel diameter), and the latter by the introduction of electricity distribution grids.

There are three technologies commonly in use for converting the potential and kinetic energy of water into electrical power: the reactive water turbine, the propeller and the impulse turbine, or wheel. Which is used depends mostly on the size of the scheme and the head (the vertical distance between the store of water and the outlet).

23.7.1 Environmental Impacts

Large-scale hydroelectricity has many benefits: it is cheap once the infrastructure is built, it uses no fossil fuels, it produces no direct carbon emissions and the reservoir may create a recreational facility. However, there are a series of negative impacts, some of which are considerable, and have led to major protests about the construction of new dams in various sensitive regions. Amongst the possible impacts are:

the need to resettle large numbers of people;
the loss of important archaeological remains;
loss of habitat;
loss of rare species;
major impacts on river wildlife and humans on the downstream side of the dam;
methane production from rotting vegetation in the flooded area;
high embodied carbon emissions if concrete used to build the dam;
loss of human life from dam failures; and
amplification of interstate tensions from diverting water resources.

Not all of these will apply to every project, and several of them apply in some form or other to many energy related or economic projects. However, it is probably true to say that the impacts are on a much larger scale. For example, the building of a new road might entail the compulsory purchase of land and the resettlement of a small number of people, but nothing like the 1.5 million people (CNN, 2005) who the construction of the Three Gorges hydroelectric scheme in China will require. This project submerged over 100 towns, and it is worth asking whether this would even be contemplated in the developed world.

If the construction of a hydroelectric scheme requires the flooding of a large area of forest, a substantial amount of biomass will be trapped beneath the surface. This will tend to decay anaerobically and produce methane, whereas the original forest would have decayed aerobically to produce carbon dioxide. Methane is 23 times more powerful as a greenhouse gas than carbon dioxide. Reliable figures are as yet unavailable, but there have been suggestions that some hydroelectric schemes might have an almost equivalent impact on climate change per kilowatt hour generated to that of conventional coal-fired generation. However, such claims remain controversial.

23.7.2 Pumped Storage

Unlike chemical energy, electrical energy is very difficult to store. This means that it is usually created on demand, which, as demand varies throughout the day and throughout the year, causes great problems for generating companies. Fortunately water turbines and generating sets work efficiently in reverse, allowing water to be pumped up to high-level reservoirs when demand is low. When demand rises, the upper reservoir can be drained back through the turbine into a lower reservoir. The overall efficiency of this process can be in excess of 70% and maximum generation can be reached in only a few seconds. The original electrical energy that has effectively been stored is usually from a source that finds it hard to match the rapid fluctuations in demand, nuclear or coal being obvious examples, or one where the efficiency drops rapidly when it is not running at full power. Another up-and-coming example is wind power, where there is little control over when the electricity is generated. If the upper reservoir also has a substantial catchment area, then the facility will also be a net generator of electricity.

Large-scale hydropower has all but reached its potential limit in most OECD countries. Capacity increased between 1990 and 2002 from 1,169.7 TWh to 1,230.5 TWh (i.e. an annual average increase of only 0.4%). The largest OECD hydropower producers are Canada, United States, Norway, Japan, Sweden and France. The world probably has the potential for producing about 100×10^{18} joules of electricity from hydropower – or one quarter of all the world's energy demand (current global consumption of hydroelectricity is 2631 TWh).

23.7.3 Small-Scale Hydropower

Due to their high initial cost and long construction times, large hydro schemes are amongst the most expensive forms of energy infrastructure. This means that hydropower is often deemed uneconomic. With small-scale hydropower, costs are reduced for schemes that do not require the creation of a dam and other large works – which solves the problem of methane emissions. If costs are low enough, individuals, communities or companies might be open to not making a strictly economic decision as the capital risked is small, and they may install a small hydro plant regardless. The reliability of small-scale hydro also makes it attractive for remote communities, particularly in the developing world, where fuel importation would be difficult (Nepal being an example).

23.8 Tidal Power

The regular rise and fall of the tides represent a shifting store of potential energy powered by the drag of the moon and the sun. This drag is gradually slowing the rotation of the Earth (by 0.016 seconds per thousand years; Brosche and Sündermann, 1978), but by any reasonable definition the resource is a renewable one. In order to make use of this alternative energy source we either need to somehow collect the water at high tide, then wait until the receding tide has created a sufficient head and drain the basin through a water turbine to generate electricity, or use the kinetic energy of tidal currents to spin turbines directly in much the same way as the kinetic energy of the wind is harvested by wind turbines. The loss of 0.016 s of rotational speed per thousand years equates to a loss of kinetic energy of 95×10^{18} joules per annum. This is equal to a quarter of current world primary energy use simply being converted into heat by friction on the sea bed. Much larger quantities could be extracted by slowing the tidal flow and slowing the rotation further – but still by an incredibly small amount.

Damming the Severn estuary (in England), as has been suggested several times, would create a 480 km² basin with an annual output of 61×10^{15} J. In such a system power can also be extracted during the filling of the basin if the incoming water is allowed to flow through the turbines. The head of water behind a tidal barrage will be much smaller than the head provided by a normal hydroelectric dam. This implies that the rotational speed of the turbines will be a lot slower (50–100 rpm). Another difference is that the whole basin needs to be emptied if possible in a single tidal cycle (approximately 12 hours), therefore a much larger number of turbines is required – 216 in the case of the proposed Severn barrage.

Importantly, tidal barrages could also operate as pump-storage facilities, with the turbines being used as pumps to increase the water level in the basin for later release. This suggests a natural symbiosis between wind and tidal power. Another possibility is to use multiple interconnected smaller basins. Water could then be pumped between basins and released sequentially in order to even out energy production.

Rather than trying to store large quantities of potential energy behind a barrage, it is possible to generate electricity from the kinetic energy of tidal currents near land masses. The general tidal flow is greatly altered by local coastal geography and this leads to higher than average water speeds past headlands and through constrictions. If these tidal streams have a speed greater than about 2.5 m/s (9 km/h) (mean springs peak flow) and are in water with a depth of about 20–30 metres, then it is possible to use a device very similar to a wind turbine to produce power.

With the exception of France which produces 536 GWh per annum and Canada which produces 311 GWh per annum (IEA, 2005), tidal power has yet to make an impact.

23.9 Biomass

Biomass can be considered as solar energy locked into the global carbon cycle. We can remove carbon temporarily from this cycle by growing biomass, then burn it to produce carbon dioxide and return it to the cycle. We have simply short-circuited a natural process and not increased carbon emissions.

Of all our energy feedstocks biomass has the longest history. It is thought that humankind had the ability to create fire at least 250 000 years ago (Goudsblom, 1992; James, 1989; Patel, 2005). The use of animals to pull loads and help with agriculture was common before 3200 BC when wheeled carts were introduced. Until the rise of coal during the industrial revolution, biomass, together with a small amount of water power, provided the majority of the world's energy. This situation is still true in some developing countries. Nepal and Ethiopia both meet the majority of their energy needs from biomass, and for others such as India and Brazil it is a more important source than oil. It is difficult to obtain reliable estimates of biomass use in much of the world: a great deal of it is produced locally often by the user, or is in the form of work done by draft animals, therefore there is no paper trail of trades from which to gather data. In the developed world, where the majority of biomass use is in the form of wastes or crops grown specifically as a feedstock and traded, the data are more reliable. Approximately one third of energy use derives from biomass in the developing world but this falls to only 3% in the industrialised nations. Because of the much greater per capita energy use in the developed world, this difference is reduced if one looks at the total energy derived or tonnage of biomass used. Developing world consumption is believed to be around 0.75 t per person per annum, and that of the developed world 0.3 t per person per annum – a much smaller difference. The United Kingdom and similar countries have lower levels of use than the developed world average; heavily forested nations such as the United States, Sweden, Austria and Switzerland have a greater than average use.

Biomass feedstocks come in a variety of forms and it is the form of the feedstock and the use to which the energy will be placed that dictates the degree of processing required and the form of energy conversion. Given reasonably dry biomass, one can simply burn it to provide domestic heating, or on a larger scale to raise steam for electricity generation. However, there are other possibilities which produce liquid or gaseous fuels for later combustion.

23.9.1 Combustion

Combustion is the simplest biomass conversion process, and is easily scaled to the size of the resource. It is still the major energy source in much of the world, particularly in rural communities. Specifically grown feedstocks such as willow trees can be used, as can industrial, agricultural or domestic waste.

The heat produced can be used within the industry itself, or exported for space heating, or if the scale of the operation is large enough, steam can be raised for electricity generation. The latter being a popular option for municipal waste incinerators.

23.9.2 Pyrolysis

This is the process used to produce charcoal from wood, but can be used with other feedstocks. By heating the biomass in the presence of insufficient oxygen, or a non-oxygen atmosphere such as nitrogen, the moisture and volatile compounds are driven off to leave a mix of carbon and inert materials collectively known as char. The volatile and liquid products are also energy sources and can either be captured or used to provide heat for the process itself.

23.9.3 Gasification

Like pyrolysis, gasification requires heating the biomass to high temperatures, but this time with a mix of steam and air or oxygen to produce a gaseous fuel. This was the process used to produce town-gas before the wide scale introduction of natural gas. Very pure fuels can be produced by such methods and if the system is designed to produce a mix of carbon monoxide and hydrogen called *synthesis gas* almost any hydrocarbon can be synthesised from the mix. This allows for the production of high-value liquid fuels.

23.9.4 Aerobic Fermentation

Given biomass with a high starch content, for example, potatoes, grains and cereals, or a high sugar content, such as beet and cane sugar plants, fermentation can be used to produce ethanol (C_2H_5OH), without many of the complications introduced by synthesis. This makes an excellent substitute for petrol (gasoline) and is highly popular in Brazil in particular. Oilseed rape can be converted to rape methylester, a substitute for diesel. The growing of crops for fermentation, rather than food raises questions of equity of land and changes to food prices. There are also questions about the destruction of natural landscapes in order to provide 'oil' for richer nations which have failed to instigate successful efficiency measures.

23.9.5 Anaerobic Digestion

A sludge of organic matter and suitable bacteria held in an airtight container (hence anaerobic) at a modest temperature of around 50°C will decompose to produce large quantities of biogas – a mix of 50–70% methane and carbon dioxide. This can be used on-site or burnt in internal combustion engines to produce electricity. A similar process will also occur in the semi-anaerobic conditions found in many landfill sites.

It has been suggested that by 2050 over 200×10^{18} J per annum could be provided by cultivated biomass. This equates to around half the world's primary energy consumption. In order to achieve this 400 million hectares of land would be needed (2.5% of the world's land area) and around half of all agricultural and forestry residues would need to be processed by energy recovery technologies. Much of this energy would come from North and South America, Africa and parts of Asia, suggesting a great change in the distribution of energy production, which could in turn have geopolitical consequences.

23.10 Geothermal

Like many energy resources, geothermal energy has a longer history than one might suppose. Roman spa towns were an early example as were Polynesian settlements in New Zealand 1000 years ago. Geothermal energy is mainly derived from radioactive decay of isotopes (principally Thorium-232, Uranium-238 and Potassium-40) deep within the Earth. The remaining one third is from heat left over from the original coalescence of matter that formed the Earth, where the kinetic and potential (gravitational) energy of the accreting particles was converted into heat. There are two ways that this immense store of heat can be accessed. We can either drill down closer to the radioactive sources that produce the heat (so called hot dry rock technologies), or use locations where the natural output of geothermal power is much higher than the average because of the presence of water (aquifers). This latter route is used in Iceland to satisfy much of its energy requirement. Iceland uses its geology to provide both hot water for domestic heating and to raise steam for electricity production and is hoping to produce enough hydrogen from the electrolysis of water to replace its use of oil as a transportation fuel. Such systems use up-wellings of magma to heat either water in the water table or pumped water to high temperatures and pressures. So cheap is this energy that it is not uncommon for heating and hot water costs to be unmeasured in domestic situations, but simply paid for by a fixed charge.

Such a happy coincidence of magma and people to use the heat is a rare thing. Usually either the energy produced must be turned into something more portable – electricity and then possibly hydrogen, or we must access deep hot rocks, usually granite, and pump water down to them.

Fifty-eight countries (*New Scientist*, 2003) use geothermal power; two example schemes being the Southampton and the Paris geothermal district heating schemes. In Southampton, United Kingdom, a bore hole was drilled to a depth of 1.8 km and brine at 70°C encountered within a permeable sandstone layer. Because this aquifer is naturally pressurised the brine rises unaided almost to the surface with pumping only required to complete the last 100 m. A heat exchanger is then used to extract 1 MW of heat which is increased to 2 MW by the use of heat pumps. The resultant heat is used within several city centre buildings for space and water heating.

Within the OECD geothermal electricity production grew from 28.7 TWh to 32.9 TWh between 1990 and 2002 (i.e. an annual average increase of only 1.1% – a very slow rate of growth). However, geothermal is still the third largest source of renewable energy. The largest producer is the United States, followed by Mexico, then Italy. Japan and New Zealand also have major facilities (IEA, 2005).

23.11 Nuclear: Fast Breeders and Fusion

At the current rate of use, easily accessible uranium deposits will only last for 70 years. This suggests that greatly expanding the use of traditional nuclear power cannot be considered a long-term mitigation strategy; however, it may have an important role in the short term. More important will be the potential switch to other forms of nuclear power.

23.11.1 Fast Breeder Reactors

Traditional (thermal) reactors can only make use of uranium-235, and this forms only 0.7% of natural uranium – the rest being uranium-238. However, a reactor can be built that relies on plutonium and uranium-238. The utilisation of uranium in such a reactor is about sixty times greater than that of a thermal reactor. This gain is such that concerns over the global reserve of uranium would be removed by the wide spread adoption of the technology.

The higher power density in the core of such reactors requires the rapid removal of substantial quantities of heat from the core and hence sodium has been used as the coolant in a series of fast breeder reactors. As anyone who has mixed sodium with water will know, the result of such an act can be explosive. As the coolant needs at some point to be used to raise steam to drive turbines this is a serious concern. Another concern is that large amounts of plutonium are required. Plutonium is toxic to humans and could be used by terrorists or others to pollute water supplies or possibly used within a nuclear device.

The first fast reactor was built at Dounreay, United Kingdom, in 1959. Since then such reactors have been built in France, Japan and the United States, but the technology has never been commercialised. India is also interested in developing the technology. It has been estimated that over $70\,000 \times 10^{18}$ joules could be created if all the accessible uranium was processed through such reactors. This should be compared to current world primary energy demand (409×10^{18} J).

23.11.2 Fusion

To make fusion work, we face the problem of creating temperatures similar to those in the centre of the sun and then confining the hot nuclei. As a temperature of many million degrees Centigrade would be required, the use of any physical container would be impossible since contact with a surface would lead to rapid cooling. In the sun this confinement problem is solved by gravity. On Earth we would be dealing with much smaller amounts of material and gravitational forces would be unable to offer confinement, but magnetic fields could be. The favoured shape for the containment vessel, or reactor, is a torus, or ring doughnut. Such a machine is called a 'tokamak'.

In theory, the successful and economic application of nuclear fusion to generate heat, raise steam and thereby generate electricity could solve all the problems associated with our reliance on fossil fuels. This is because the basic fuel, deuterium, could be provided in almost limitless quantities from sea water and nuclear fusion could provide four times more energy each per unit of fuel than nuclear fission. However, amongst physicists it has been a continuing joke for the last 40 years that fusion power will only take *another* 40 years to realise. And it is true that even today we have yet to build a reactor that can sustain fusion for a lengthy period or generate more energy than it takes in confinement and initial heating.

23.12 The Hydrogen Economy and Fuel Cells

A commonplace school science experiment is to extract hydrogen from water using electrolysis and then ignite the hydrogen with a match. The hydrogen burns in the presence of atmospheric oxygen to form water once more. This is true recycling. Water is converted to hydrogen and oxygen which burn to produce water! The process produces no pollutants during combustion except possibly nitrogen compounds formed at high temperatures from the abundance of nitrogen in air. The electrolysis could be carried out at central facilities and the hydrogen transported as a liquid at very low temperatures for final distribution to customers' vehicles or homes. For the process to make environmental sense, the original energy must be from a carbon-neutral source, for example wind power. An alternative source of hydrogen would be from fossil fuels, with the carbon fraction being burnt to generate electricity and form carbon dioxide. The carbon dioxide would then be sequestrated underground or in the oceans.

At three times that of petrol (gasoline), the energy density per unit mass of hydrogen also makes it look attractive. Apart from the problem of needing to increase the renewables base to provide the electricity to carry out electrolysis, the main problem that hydrogen faces is storage, as it is only liquid below −253°C. Another advantage of hydrogen is that it makes an ideal feedstock for fuel cells.

23.12.1 Fuel Cells

Fuel cells side step the need to convert chemical energy to thermal energy and then to rotational kinetic energy in a generating set in order to simply produce electricity. This is achieved by tapping directly into the electron exchange that occurs during a chemical reaction. By forcing the electrons through an external circuit part way through this exchange, useful work can be extracted from the flow of electricity. Central to the technology, and at the core of a fuel cell, is a membrane that will allow positive charges to flow through it, but not negative ones (electrons), which are then forced to take the long route via the external circuit – that is, via an electric motor or a lighting circuit for example.

William Grove invented the fuel cell in 1839 when he noted that not only could electricity be used to produce hydrogen and oxygen from water, but that these two elements could be combined to produce electricity. So, the technology has a history almost as long as many of our fossil fuel based technologies. Sporadic interest was shown in the idea throughout the late nineteenth and early twentieth centuries. In 1959 Francis Bacon demonstrated a 5 kW alkaline fuel cell which formed the basis of the fuel cells used by NASA for its Apollo missions.

Although low temperature fuel cells need to be powered by hydrogen, high temperature ones can make use of natural gas. Using natural gas means that carbon dioxide would still be released; however, it does mean that society would not need to invest in a hydrogen grid as the current gas grid could be used. Unfortunately the temperatures required are so high that such devices are not really suitable for rapid stop-start operation but need to be run continuously. This becomes practicable if they can export electricity from the home to form a national grid of base-load power stations. There is an obvious analogy here with the internet, where capacity is not based on a few large computers, but via a very large number of interconnected machines. British Gas and Ceres are currently (2010) starting to field-test 37 000 natural gas fuel cells in British homes.

23.13 Solutions

Current commercial worldwide energy consumption equates to a constant 13×10^{12} watts, or approximately 2 kW per person. One way to visualise this is that it is the same energy consumption as a 2-bar electric fire left on indefinitely for each person on the planet. In developed nations consumption might be closer to 4 bars, in the developing world less than one bar. It is clear that in order to move developing nations out of poverty, their level of energy consumption needs to become much closer to that of the wealthy nations. Thus we can expect worldwide energy consumption to rise, not fall. From a climate change perspective, this will only be a problem if such growth occurs via greater access to fossil carbon, such as oil, coal and natural gas.

As Figure 23.2 shows, many renewable sources have the ability to meet current demand. We will probably have to access a variety of these, and there is likely to also be an expansion of nuclear power in the coming decades. Solar alone has the ability in theory to meet demand 2 million times over and because photovoltaics, wind and biomass are now mature technologies they are likely to be at the centre of this new, solar-powered world. Because solar radiation is greater in some locations than in others it is likely that energy production will migrate towards these areas. North Africa is one such location, and the EU and others are researching the creation of a high-voltage direct current international electricity grid. Because the transportation of biomass has been shown to be relatively energy efficient, a similar trend is likely to occur with wood fuel, or biomass derived fuels, but in locations where water for plant growth is not in short supply. Pointing in the other direction, (i.e. towards small scale local generation, rather than remote mass generation) are solutions based on energy efficiency and urban generation, for example roof mounted photovoltaics. This could lead to a very different energy landscape where the developed world moves from approximately one power station per million people to possibly one micro-station per household.

Both these developments are likely to lead to a very different energy landscape and differing geopolitical and local tensions, the final result of which it is difficult to judge.

References

Brosche, P. & Sündermann, J. (eds.) (1978) *Tidal Friction and the Earth's Rotation.* Springer Verlag, Berlin.
Coley, D.A. (2008) *Energy and Climate Change.* Wiley, Hoboken, NJ.
CNN (2005) Available from http://www.cnn.com.
DeZeeuw, J.W. (1978) Peat and the Dutch Golden Age. *AAG Bijdragen*, **21**, 3–31.
Goudsblom, J. (1992) *Fire and Civilization.* Allen Lane, London.
International Energy Agency (IEA) (2005) *Renewables Information 2004.* International Energy Agency, Paris.
James, S.R. (1989) Hominid use of fire in the lower and middle Pleistocene: a review of the evidence. *Current Anthropology*, **30**(1), 1–26.
New Scientist (2003) 22 February, 40.
Patel, T. (1995) Burnt stones and rhino bones hint at earliest fire. *New Scientist*, 17 June, 5.
Renewable Energy Policy Network (REN) *World Renewables 2009.* Renewable Energy Policy Network, Paris.
Ristinen, R.A. & Kraushaar, J.J. (1999) *Energy and the Environment.* Wiley, New York.
Smil, V. (1994) *Energy in World History.* Westview Press, Boulder, CO.
University of Strathclyde (UST) (2005) Available from http://www.esru.strath.ac.uk/EandE/Web_sites/01-02/RE_info/hec.htm.

24

Complexities and Approaches to Managing the Adaptation of Climate Change by Coastal Communities

Annie T. Worsley, Vanessa J.C. Holden, Jennifer A. Millington and Colin A. Booth

24.1 Introduction – What's Special about the Coast?

The coastal zone is a unique environment. More specifically, it is a zone where a series of unique environments exist, based upon the localised condition of sediment supply, wind and wave conditions, local geology and geomorphology and anthropogenic activity. It is, very simply, the point where land and sea meet with the atmosphere and human communities. However, the boundary between terrestrial systems and marine systems is not an obvious one, the continuous movement of the tides results in an area of land that is neither terrestrial nor marine, but is shaped by the processes of both. In addition, the coastal zone marks the interface of terrestrial and marine systems with climate and human activity and is, therefore, the area where the greatest and most rapid changes can be seen, especially in 'soft' sediment coasts. This chapter will, therefore, concentrate on sedimentary coasts. For more information regarding definitions of the coastal zone, the reader is referred to McGlashen *et al.* (2005).

The importance of coasts should not be underestimated globally, of the current global population of over 6 billion, more than 50% live in the coastal zone (Haslett, 2000). Coastal zones cover some 8% of the earth's surface but contribute more than 25% to global biological productivity. Coasts are very important to the sustainability of living resources, providing up to 90% of the world's marine fish catches. The coastal zone is also a significant controller of climate, being source areas for key gases and elements to the atmosphere. In the United Kingdom, where the majority of major cities are situated in the coastal zone, the sea and the proximity to the coast are

---

Solutions to Climate Change Challenges in the Built Environment, First Edition.
Edited by Colin A. Booth, Felix N. Hammond, Jessica E. Lamond and David G. Proverbs.
© 2012 Blackwell Publishing Ltd. Published 2012 by Blackwell Publishing Ltd.

immensely important. Historically, communities have settled within the coastal zone, for hunting and fishing, for port development for import and export trade links, transport and, more latterly, tourism and recreation. Being a coastal and sea-faring nation is part of the British psyche, for nowhere in the United Kingdom is further than 75 miles from the sea. Here, coasts are major geographical features, more than 11,072 miles in length and containing an enormous variety of habitats and physical structures.

24.2 Coastal Landforms and Process

Soft sediment coastal landforms, such as sand dunes and salt marshes, are much more responsive to coastal changes, such as through erosion, than hard rocky coastlines. Soft sediment landforms can change dramatically within the space of a single high tide or storm event; whereas, many rocky features are worn away gradually until a process, such as undercutting causes a major rock fall or collapse.

Of all the coastal environments, salt marshes are the most susceptible to a rising sea level (Reed, 1990). Intertidal flats and salt marshes are of utmost importance both ecologically and socially. They are generally considered to be amongst the most biologically productive natural areas in the world (Nelson-Smith, 1977). As such, they provide an important habitat for many floral and faunal species. Since many areas of salt marsh are backed by land and buildings of economic value, they have a substantial value as a sea defence, with the vegetation being an important factor in wave attenuation (King and Lester, 1995).

Sand dune coasts are also very responsive to changes in sea level and climate. Formed by aeolian processes, they develop where the intertidal prism is large and shallow, where onshore winds are strong and where supplies of sand-sized offshore sediments are abundant. Like salt marshes, a range of habitats are plentiful (e.g. dunes ridges, slacks, heaths and grasslands) with dune systems often forming a 'barrier' to the sea, both physically (ridges may be many metres high, and dune fields several kilometres deep) and psychologically (they may be viewed by local populations as offering substantial protection to the built environment from waves and winds and, thus, to sea level changes). Both salt marsh and sand dune ecosystems respond to rising or falling sea levels by migrating seaward (as sea level falls) or landward (as sea level rises); thereby, being important forms of coastal defence, providing natural barriers to encroachment of land by the sea. This use of naturally occurring features within the landscape to manage the environment is known as 'soft engineering'.

Use of dunes and salt marshes as a soft coastal defence is usually very effective. However, in some cases, re-mobilization, or natural transgression processes are prevented by fixed landward boundaries (artificial or otherwise). This is termed 'coastal squeeze' and can result in direct loss of associated habitats (Doody, 2001). This occurs where an ecological zone (e.g. sand dunes or salt marsh) is backed to its landward side by a built structure or hard defence such as a sea wall. With rising sea levels, there is restricted opportunity for landward migration of the ecological zone – it cannot roll back – and so it is gradually squeezed into an ever decreasing area until the ecological zone can no longer exist. 'Coastal squeeze' often occurs at a greater rate than simple inundation (Orford and Pethick, 2006), not allowing sufficient time for natural recovery. This issue becomes especially complicated when the fixed boundary is a specific habitat valued for biodiversity (Natura, 2000).

24.3 Challenges Facing Coastal Communities

Since the end of the last glacial maximum (30,000 to 18,000 years ago), the sea has been gradually rising, by up to 100 metres to its present-day level (Valiela, 2006). On a geological timescale, the rise in sea level over a few hundred or even a thousand years is not particularly significant. However, major causes for concern are the potential acceleration in sea level rise due to anthropogenic factors; the increasing presence of high-value land (both in economic and social terms) close to the coastal zone; and the rising awareness of potential habitat infringement and loss due to human manipulation of the coastline in so many areas. In addition to the direct effects of sea level rise, there are implications for other consequences of climate change, such as increased storminess and rainfall, which can have significant impacts upon the coastal system in terms of tidal and wave action, and sediment delivery (including fluvial sources). The movement of sediment within the coastal zone is a key factor in the response of coastal landforms to climate change and sea level rise. If sufficient sediment is available to the system, the coastal landforms are more able to develop and adapt to the changing conditions. However, if sediment supply becomes limited, coastal erosion and, subsequently, coastal flooding will be exacerbated.

Following increased awareness of the trend during the mid twentieth century, as a result of coastal flooding in many areas globally, and extreme predictions during the 1980s, the science of predicting sea level rise has developed into a significant multidisciplinary body of data, models and forecasts. These are often led by governmental and international organisations, directed at determining the most accurate range of sea level rise scenarios in order to enable the most educated management strategies to be developed. More specifically, it is the degree to which the trend of rising sea levels is resulting from natural processes, and to which human activities have impacted upon and accelerated the trend, that is disputed.

Causes of 'contemporary' sea level rise are generally accepted and widely reported in various texts, including thermal expansion, crustal deformation, and changes in meteorological forcing factors (e.g. Bird, 1993; Gornitz, 1995; Diez, 2000; Pugh, 2004; Valiela, 2006). There are decadal scales of variation in sea level, notably the 18.6-year nodal modulation of the lunar tides (French, 2006), along with annual variations, such as maximum annual sea levels experienced around September, following heating and expansion of the water during the summer months (Pugh, 2004). Observed changes in the climate, its effects and its causes are covered in detail in the 2007 Intergovernmental Panel on Climate Change (IPCC) report (IPCC, 2007).

Sea level rise becomes an issue at the coast when land that is valued for either its monetary value (such as housing or industry) or its environmental value (such as habitat of an at-risk species) becomes under threat of tidal inundation. Increasingly, due to growing populations, and particularly higher proportions of an aged population, there are more developed areas of the coastline than ever before. This not only leads to the obvious impacts upon human society such as flooding, but also ecological impacts, notably 'coastal squeeze' (French, 1997, 1999; Haslett, 2000). There are clear conflicts in the management of this situation between the need to protect the anthropogenic infrastructure, whilst maintaining highly specialised ecological areas.

24.4 Ways of Managing Coastal Challenges

Previously, the risk of coastal flooding would have seen the building of defences and attempts being made to keep coastal water away from the land. However, there has been a fundamental change in thinking by policy makers, with a shift towards flooding being more 'controlled', with compromises made to re-create and encourage wetland environments to allow for the greater coastal inundations that are to be expected with rising sea levels and climate change. The presence of the soft engineering defences, such as salt marsh and sand dunes, are therefore of great consequence to the future planning and policy development, not only in terms of their value as a sea defence, but also as habitat for wildlife, and increasingly for the amenity value and 'green credentials' of what would otherwise require further hard engineered defences.

This move away from hard defences is reflected by developments in UK Government policy. As of 2009, the UK Government (Department for Environment, Food and Rural Affairs [DEFRA]) is developing an 'adaptation toolkit' aimed at assisting coastal communities to adapt to changes in the coastal environment, where hard engineered defences are not a feasible option. Between 2008 and 2011, up to £28 million is being made available to support the adaptation toolkit. The toolkit is designed to enable communities to better cope with coastal change, through changes in planning, building design, increased access to information and community engagement (http://www.defra.gov.uk).

There are issues and uncertainties surrounding the consideration of sea level rise over such relatively short time frames, such as a few hundred years. Natural variation around geological time can make short term variations appear largely insignificant. However, when considering policy and management, it is the timescales of decades that are most important and readily comprehensible when communicating management strategies to stakeholders. It is, therefore, appropriate to utilise current trends that are based around established data sets, such as the findings of the UK Climate Projections 09 (UKCP09), to provide the most practicable scenarios upon which to link other data sets and subsequently base conclusions.

The complex nature of managing coastal areas, when there are issues of divided ownership, a variety of land uses and stakeholder requirements, changing human influences, such as port operations, combined with 'natural' changes, such as sea level rise, underlines the need for an organised, comprehensive monitoring programme, with a sufficiently flexible management approach to incorporate changes in the natural systems as and when they become apparent.

Integrated coastal zone management (ICZM) emerged as a means of governance in the coastal zone in the late 1990s and has been well documented (Cicin-Sain and Knecht, 1998; Vallega, 1999; Ballinger 2005). In 2002, a formal recommendation by the European Commission encouraged all member states to adopt and implement ICZM. This demonstrated recognition of the importance of the coastal zone, that is, that area from the shallow marine environment to land surface affected by coastal processes, and the need to approach its management, in all its forms, from a holistic perspective. Today, adopting ICZM is widely understood and recognised. However, in reality it is challenging, since coastal zones are by their very nature complex, face diverse human and natural problems and can change rapidly in space and time. Many now appreciate that a geographical approach is preferable, since this is at once holistic and also incorporates physical and human systems. What is certain is that any approach to the management of built (anthropogenic) and physical (natural) environments needs to consider that change to coastal systems and processes is inherent (and often desirable) and, furthermore, management strategies must be collaborative (McFadden, 2008). ICZM is, therefore, a 'movable feast': both challenging and dynamic in itself.

24.5 Shoreline Management Plans

As part of ICZM, shoreline management plans (SMPs) have been nationally adopted to provide large-scale assessments and long-term policy frameworks associated with reducing the risk of the effects of coastal processes on both the natural and developed environment (http://www.defra.gov.uk). SMPs incorporate a littoral 'sediment cell' concept to coastal evolution management in England and Wales. An important objective of SMPs is to work in partnership with all interested organizations, such as local councils and the public, to enable sustainable future management decisions and to avoid fragmented attempts to protect one area at the expense of another. Such decisions need to take account of natural coastal processes and the interactions that occur within coastal systems at various temporal and spatial scales (Cooper and Pontec, 2006).

The shoreline of England and Wales was examined by Wallingford (1993) to provide initial guidance on suitable divisions of the coastline into littoral 'sediment cells', within which a strategic framework for the development of sustainable polices for coastal defences could be identified, based on natural process behaviour. This first generation of SMPs was successfully mapped and improved understanding of coastal processes operating on ~6000 km of coastline around England and Wales. SMP documents, applicable for the next 50 years, were put together on the basis of information that was known and available in 1998 and were designed to evolve as further information became available. Policy options were chosen from five possible alternatives: (1) do not provide any flood or coastal defence, (2) hold the existing defence line to maintain the shoreline in its present position and location, (3) advance the existing defence line to relocate the shoreline to seaward of its present position and location, (4) retreat the existing defence line to allow the shoreline to relocate landward of its present position or (5) use natural defence management to work with natural processes to minimize erosion and maximize accretion. The first generation of SMPs have been reviewed, accounting for increased awareness and scientific knowledge on future risk management challenges. Outcomes led towards producing SMP2s, which were completed in 2010 (http://www.defra.gov.uk).

24.6 Case Study: The North Sefton Coast

As a case study to the response of coastal communities to a changing environment over time, the Sefton coast is an excellent example. In 1990, its entire population of 300 000 lived within 15 minutes of the sea (Cox, 1990). On the southern shore of the Ribble estuary in north-west England, the north Sefton coast has a requirement for an integrated approach to its management, having a number of demands placed upon it based on the resources that it offers. An appreciation of both anthropogenic and natural factors is, therefore, imperative for the successful management of the area. Based upon a large collection of archived evidence (Figures 24.1 and 24.2), it is clear that the coastal zone has experienced sizeable human influence since the development of the town of Southport in the 1800s. Developments have included significant urbanisation with associated infrastructure development; land claim, predominantly for agriculture; port development with related estuarine channel training and dredging; tourism and recreation expansion and increases in industry and commerce, directly impacting upon the coastal zone. The tendency of the estuarine environment to seek to 'infill' with sediment has resulted in numerous land claim (or 'reclamation') schemes along the north Sefton coast. Many minor land claims were recorded throughout the nineteenth century (Gresswell, 1953), continuing through the twentieth century until 1980. Significant land claims were those to enclose a Marine Lake and construct the Coast Road at Southport (7 m OD), which enclosed salt marshes that are now brackish and an important Royal Society for the Protection of Birds (RSPB) reserve. This enclosure, however, now provides an engineered sea wall to the landward side of the active intertidal flats and salt marsh. This, therefore, has potential implications for allowing the process of 'coastal squeeze' to occur. Recent research along this stretch of coastline, however, has provided indications that currently the development of salt marsh is at a rate that is able to keep pace with local sea level rise (Holden, 2009).

To the south of the salt marsh, in an area fronting the central section of the town of Southport, no natural soft defences are present, requiring a

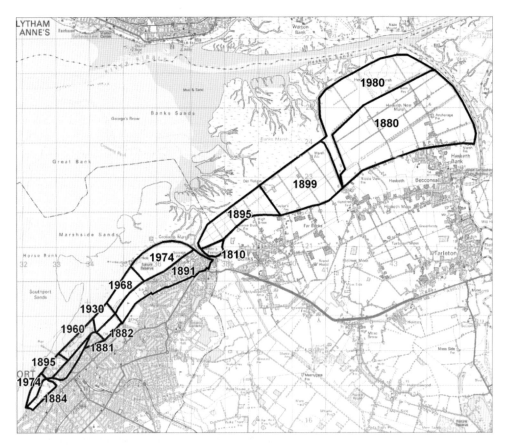

Figure 24.1 Extent of land claim on the north Sefton coast (image produced by and courtesy of Sefton Metropolitan Borough Council).

1.3 km section of sea wall to be completed (in 2002), raising the level of the existing sea defences to reduce the level of coastal flooding experienced, protecting the leisure and retail businesses, and infrastructure that had developed behind the existing sea wall.

In contrast to the anthropogenic pressures, the salt marshes of the Ribble Estuary are nationally and internationally recognised as being of major importance for their provision of habitats, both for flora and fauna, being designated as a Site of Special Scientific Interest (SSSI), a National Nature Reserve, a Special Protection Area (SPA), and a RAMSAR site. The area has a number of landowners, predominantly Sefton Metropolitan Borough Council and Natural England, but with many other stakeholders and interested parties (such as the RSPB and golf clubs) involved in the management of the area. The various bodies work together under the banner of the Sefton Coast Partnership, created in 2001 from the Sefton Coast Management Scheme, which was formally established in 1978 by a number of organisations with an interest in the entire Sefton coastline, either through management or land ownership, including Sefton Metropolitan Borough Council, National Trust,

328 Solutions to Climate Change Challenges in the Built Environment

Figure 24.2 Aerial photography showing the development of the sea wall and associated land claim fronting the town of Southport from (1) 1950s to (2) *ca*. 2000 (images courtesy of Sefton Metropolitan Borough Council, originator unknown).

Natural England (previously named English Nature), five private golf courses, and the Ministry of Defence (Atkinson and Houston, 1992). The primary objective, at the time, was to address issues of recreational pressure, habitat preservation and coastal defence. Before the Partnership, research on the coast had been fragmented and related only to specific issues. It was soon recognised, however, that a sound understanding of the processes occurring in the coastal zone was needed to allow informed sustainable decisions to be made. The Partnership now takes a more co-ordinated, long-term approach, with a task group in place specifically to address research issues.

24.6.1 SMPs on the Sefton Coast

An essential requirement to future management of the Sefton coast is to gain an appreciation of the scale of coastal retreat. The coast from Great Orme's Head to the Solway Firth represents cell 11 of the series of cells in existence around the United Kingdom (Figure 24.3). Five SMP divisions exist within cell 11, based on natural sediment transfer limits, rather than administrative boundaries (Lymbery, 1999). The Sefton coast falls within the two subcells, 11a and 11b, each with its own SMP (Figures 24.4 and 24.5). The Liverpool Bay SMP extends from Great Orme's Head to Formby Point and the Ribble Estuary SMP from Formby Point to the River Wyre.

Stage 1 of the Liverpool Bay SMP preparation identified the primary division of the shoreline into eight coastal process units (CPU) based on coastal process evaluation (Liverpool Bay Shoreline Management Plan, 1993), with CPU8 including part of the Sefton coast from Seaforth Dock to Formby Point. Currently, erosion of Formby Point is paralleled by increased accretion rates on northern and southern margins (e.g. Gresswell, 1953; Turner, 1984; Pye and Neal, 1994; Saye et al., 2005; Pye and Blott, 2008). The existing coastal defence policy for this area (Cell 11a, CPU8) is 'Hold the line', identified by the Liverpool Bay Shoreline Management Plan (1993), which remains the same course of action for future coastal defence policies, for both short-term and anticipated long-term management. This will require appropriate dune stabilization techniques, costing typically less than £50 000 per annum, to allow continued realignment processes, in order to absorb future coastal energy.

Evaluation of coastal process behaviour within sub-cell 11b identified eight CPUs based on coastal process evaluation (Ribble Estuary Shoreline Management Plan, 1999), with both CPU7 and CPU8 including part of the Sefton coast from Southport to Ainsdale and Ainsdale to Formby Point, respectively. The existing coastal defence policy for CPU7 is 'Natural defence management' and 'Hold the line', identified by the Ribble Estuary Shoreline Management Plan (1993), which remains the same course of action for short-term future coastal defence policies and is just 'Hold the line' for anticipated long-term management. The existing coastal defence policy for CPU8 is 'Natural defence management', which remains the same course of action for short-term future coastal defence policies and is combined with 'Managed retreat' for anticipated long-term management. Managed retreat of the

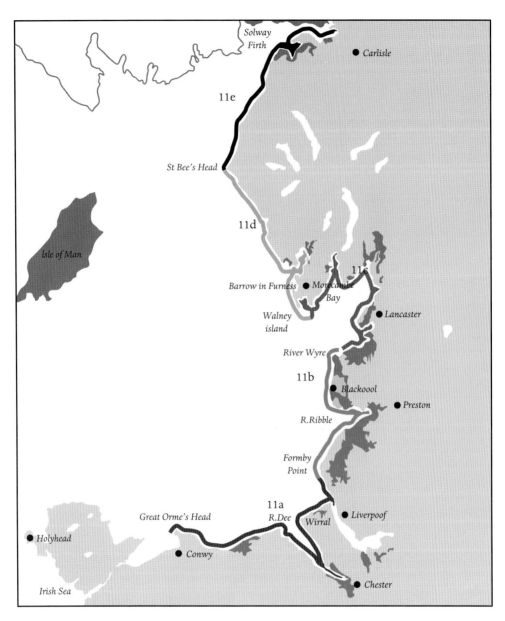

Figure 24.3 The eastern Irish Sea: Cell 11 (image courtesy of Sefton Metropolitan Borough Council).

shoreline includes the use of stabilization techniques, such as artificial planting, in an attempt to control the behaviour of the shoreline, allowing natural dune processes to continue relatively unhindered. For successful managed retreat, mobile dune widths must be maintained at present levels to encourage dune growth during environmentally stable periods, whilst allowing for dune erosion under unstable regimes.

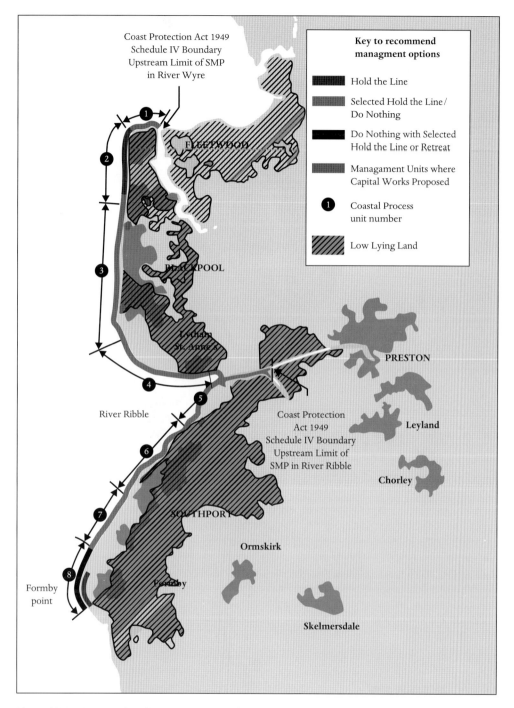

Figure 24.4 Current shoreline management plans for Sefton subcell 11b (images courtesy of Sefton Metropolitan Borough Council).

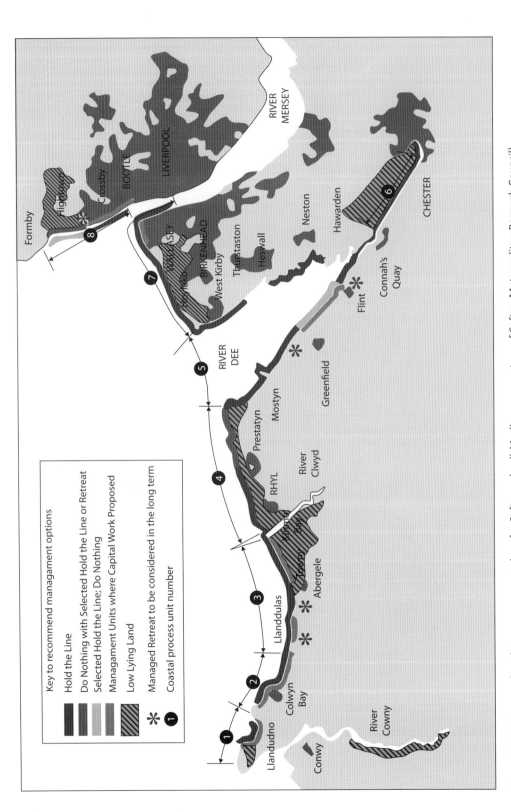

Figure 24.5 Current shoreline management plans for Sefton subcell 11a (images courtesy of Sefton Metropolitan Borough Council).

24.7 Solutions for Coastal Communities

Coastal environments vary considerably as do the approaches to their management. Significantly, soft sediment coasts, which are the most dynamic and responsive to changing climate and marine influences, present planners and coastal practitioners with some of the most difficult problems in preparing to protect the built environment. The speed of potential changes to sediment budgets and coastal alignments mean that planners need to be aware of the many natural phenomena that affect the coastal zone. In order to effectively prepare for projected coastal responses to various predicted climate change scenarios several suggestions are made for 'proactive *adaptive* management'.

Firstly, a strong coastal *partnership* is imperative to implement coastal defence strategies (both hard and soft engineering options). This enables strong links between land managers, government and non-government organisations and coastal professionals and practitioners. Secondly, the adoption of effective ICZM is imperative. Both ICZM and coastal partnerships benefit from involving local communities, not simply for the dissemination of information but also to garner support for initiatives and obtain valuable local knowledge. Key to coastal management is both detailed knowledge and understanding of the mechanisms of change but also how coastal zones will respond and how coastal communities will have to adapt. Adoption of strategies that can *adapt* to climate change challenges is therefore crucial. To do this, planners need to understand the needs of *end users* (communities, wildlife, industry and tourism) and adapt engineering solutions to the *specific* needs of local physical and human geographies.

References

Atkinson, D. & Houston, J.A. (1992) Towards a research strategy for the Sefton coast in north-west England. In: *Coastal Dunes*, ed. Carter, R.W.G., Curtis, T.G.F. & Sheehy-Skeffington, M.J. Rotterdam, Balkema.

Ballinger, R.C. (2005) A sea change at the coast: the contemporary context and future prospects of integrated coastal management. In: *Managing Britain's Marine and Coastal Environment: Towards a Sustainable Future*, ed. Smith, H.D. & Potts, S.J. Routledge, London, 186–217.

Bird, E.C.F. (1993) *Submerging Coasts*. John Wiley and Sons, Chichester.

Cicin-Sain, B. & Knecht, R.W. (1998) *Integrated Coastal and Ocean Management: Concepts and Practice*. Island Press, Washington, DC.

Cooper, N.J. & Pontec, N.I. (2006) Appraisal and evolution of the littoral 'sediment cell' concept in applied coastal management: experiences from England and Wales. *Ocean and Coastal Management*, 49, 498–510.

Cox, T.M. (1990) Coastal planning and management in a metropolitan area. In: *Planning and Management of the Coastal Heritage. Symposium Proceedings, Southport 1989*, ed. Houston, J. & Jones, C. Sefton Metropolitan Borough Council, Southport, UK.

Diez, J.J. (2000) A review of some concepts involved in the sea level rise problem. *Journal of Coastal Research*, 16, 1179–1184.

Doody, J.P. (2001) *Coastal Conservation and Management: An Ecological Perspective*. London, Kluwer Academic Publishers.

French, J.R. (2006) Tidal marsh sedimentation and resilience to environmental change: exploratory modelling of tidal, sea-level and sediment supply forcing in predominantly allochthonous systems. *Marine Geology*, **235**, 119–136.

French, P.W. (1997) *Coastal and Estuarine Management*. London, Routledge.

French, P.W. (1999) Managed retreat: a natural analogue from the Medway estuary, UK. *Ocean and Coastal Management*, **42**, 49–62.

Gornitz, V. (1995) Sea level rise: a review of recent past and near-future trends. *Earth Surface Processes and Landforms*, **20**, 7–20.

Gresswell, R.K. (1953) *Sandy Shores of South Lancashire*. Liverpool, Liverpool University Press.

Haslett, S.K. (2000) *Coastal Systems*. London, Routledge.

Holden, V.J.C. (2009) *Sedimentary accretion of the North Seftonc*. Unpublished Ph.D. thesis, University of Lancaster and Edge Hill University.

Intergovernmental Panel on Climate Change (IPCC) (2007) *Climate Change 2007: Synthesis Report. Contribution of Working Groups I, II and III to the Fourth Assessment Report of the Intergovernmental Panel on Climate Change*, ed. Pachauri, R.K. and Reisinger, A. IPCC, Geneva, 104.

King, S.E. & Lester, J. (1995) The value of salt marsh as a sea defence. *Marine Pollution Bulletin*, **30**, 180–189.

Liverpool Bay Shoreline Management Plan (1993) *Sub-Cell 11a: Great Orme's Head to Formby Point Plan Document*. Liverpool Bay Coastal Group, Liverpool.

Lymbery, G. (1999). Shoreline Management Plans, a partnership for coastal defence management. Available from http://www.seftoncoast.org.uk/articles/99winter_shoreline.html.

McFadden L. (2008) Exploring the challenges of integrated coastal zone management and reflecting on contributions to 'integration' from geographical thought. *The Geographical Journal*, **174**, 4, 299–314.

McGlashen, D.J., Duck, R.W. & Reid, C.T. (2005) Defining the foreshore: coastal geomorphology and British laws. *Estuarine, Coastal and Shelf Science*, **62**(1–2), 183–192.

Natura 2000 (2007) *Interpretation Manual of European Union Habitats*. European Commission DG Environment, Nature and Biodiversity, Brussels.

Nelson-Smith, A. (1977) Estuaries. In: *The Coastline*, ed. Barnes, R.S.K. John Wiley and Sons, Chichester.

Orford, J.D. & Pethick, J.S. 2006 Challenging assumptions of coastal habitat formation in the 21st century. *Earth Surface Processes and Landforms*, **31**, 1625–1642.

Pugh, D. (2004) *Changing Sea Levels: Effects of Tides, Weather and Climate*. Cambridge University Press, Cambridge.

Pye, K. & Blott, S.J. (2008) Decadal-scale variation in dune erosion and accretion rates: An investigation of the significance of changing storm tide frequency and magnitude on the Sefton coast, UK. *Geomorphology*, **102**, 652–666.

Pye, K. & Neal, A. (1994) Coastal dune erosion at Formby Point, north Merseyside, England: Causes and mechanisms. *Marine Geology*, **119**, 39–56.

Reed, D.J. (1990) The impact of sea level rise on coastal salt marshes. *Progress in Physical Geography*, **14**, 24–40.

Ribble Estuary Shoreline Management Plan (1999) *Sub-Cell 11b: Formby Point to River Wyre Plan Document*. Ribble Estuary Shoreline Management Plan Partnership, Sefton, UK.

Saye, S.E., van der Wal, D., Pye, K. & Blott, S.J. (2005) Beach-dune morphological relationships and erosion/accretion: an investigation at five sites in England and Wales using LIDAR data. *Geomorphology*, **72**, 128–155.

Turner, D.A. (1984) *A Guide to the Sefton Coast Data Base*. Engineer's Department, Metropolitan Borough of Sefton, UK.

Valiela, I. (2006) *Global Coastal Change*. Blackwell Publishing, Oxford.

Vallega, A. (1999) *Fundamentals of Integrated Coastal Management*. Kluwer Academic, Dordrecht.

HR Wallingford (1993) *Coastal management: mapping of littoral cells*. Report to MAFF, HR Wallingford Report SR328. Wallingford, UK, HR Wallingford.

Lessons for the Future

Jessica E. Lamond, David G. Proverbs, Colin A. Booth
and Felix N. Hammond

25.1 Introduction

Changes to world climates are predicted to result in an array of challenges to both natural and man-made habitats, chiefly because they have evolved or been designed to operate under different climate conditions. Humans, in particular, live complex lives governed by social, physical, economic and political systems, into which the introduction of change of any kind can be highly disruptive. Recognising the need for change, justifying the optimal direction of change and then enabling change are all necessary steps along the path to action. This is true of individuals, companies, nations and global organisations. Over the past decade, there has emerged a wide acceptance of the conclusions of climate science that the world will experience a shift in weather patterns and that it would be prudent to act now to prevent that shift from becoming catastrophic. This science is ably elucidated by Chapter 2, which demonstrates that warming has occurred in the last decades of the twentieth century and that the most likely scenarios from all types of climate models predict that warming will continue. Despite this acceptance, there has been very limited real action to either mitigate against climate change or adapt to it. As chapters in this book identify, the built environment has a large role in generating harmful emissions and, therefore, may be crucial in the reduction of greenhouse gases in the atmosphere. In particular, Chapter 4 discusses the scale of increased urbanism and its role in generating emissions, asserting that, by 2025, two thirds of the world's population will live in urban areas, with 90% of those residing in developing countries. In addition, the function of the built environment as a safe and healthy human habitat is threatened by the anticipated weather and ecosystem changes. Adaptation of the built environment to these predicted

Solutions to Climate Change Challenges in the Built Environment, First Edition.
Edited by Colin A. Booth, Felix N. Hammond, Jessica E. Lamond and David G. Proverbs.
© 2012 Blackwell Publishing Ltd. Published 2012 by Blackwell Publishing Ltd.

conditions can be central to policy to sustaining human communities into the future.

No one book could possibly cover in depth the vast range of research and practice in this area, but the chapters here have tackled a vast range of approaches, paradigms and adaptive responses. For these authors the challenges for the built environment go beyond simply selecting building materials and designing resilient structures, but encompass planning of urban landscapes and infrastructure and managing natural resources within urban spaces to maximise utility to humans, while minimising climate impacts. This chapter will consider three major themes, which are: (1) technological solutions; (2) working with the natural environment and (3) enabling change.

25.2 Technological Solutions

Emissions from the built environment during the operation of buildings principally stem from heating and cooling buildings to human ambient temperatures. Technological solutions for improved efficiency are discussed in Chapter 11 and Chapter 13. These solutions can be as simple as controlling heating systems via thermostats to a lower temperature or can be more advanced such as circulatory systems using the heat from outgoing stale air to warm fresh incoming air. Emissions can also be reduced by the use of renewable energy sources (Chapter 23) including solar energy heat pumps, wind power, hydropower and geothermal energy. It is seen that recent developments allow the greater use of renewables, many of which have the potential to reduce CO_2 levels (Chapter 11). Similarly, emissions from transport are another area where clean energy technology can limit the output of greenhouse gases (Chapter 16 and Chapter 13).

Waste (Chapter 17) is another result of not just the operation of buildings but also their construction, and can contribute to emissions not only through the energy requirements of processing materials but also through methane emissions from landfill. Technological approaches to reducing the impact of waste include improved recycling processes, mechanical biological treatment waste-to-energy incineration and anaerobic digestion. Chapter 11 discusses the ways in which embodied energy in building materials can be reduced and also covers emerging technologies in insulating materials. Building materials are also the topic of Chapter 12 which proposes that impact may be reduced by improvement of sustainability through decreased requirement for material, lower embodied energy, greater durability or reduced maintenance costs. Improved materials may also have advantages for adaptation to climate change that will increase their lifetime. The chapter considers a range of materials from concrete to composite materials and nanotechnology. The prevalence of concrete makes even small improvements in its manufacture highly significant, and it is clear that inclusion of recycled materials can not only preserve raw material but also, in some circumstances, improve material properties. Future trends seem to indicate that reducing recycled materials to nanoparticles could improve concrete and other materials, while reducing the quarrying and refinement of raw materials.

Further improvements in treatments for natural materials such as timber and bamboo will also be an important element in sustainable building materials. The efficient use of water may be an adaptation required during the increasingly warm summers expected in an era of climate change.

Improved technology in domestic water appliances can assist in saving water as can intelligent systems able to reuse water and harvest rainwater (Chapter 19). Flood proofing technology has progressed in the past decades through both resistant and resilient approaches, and many are described in Chapter 19. New materials may also be more flood resistant than more traditional building materials (Chapter 12).

25.3 Working with the Natural Environment

Mitigating against climate change is sometimes regarded as synonymous with preserving the natural environment, as for example the role of forests in carbon sequestration. However, this preservation is often seen as something that absolutely conflicts with urbanisation and the built environment. Urban spaces are often decried as barren but, in fact, Chapter 9 shows that the diversity of species and habitat in urban green space equals or surpasses equivalent tracts of rural land. Several other authors here demonstrate that attention to the use of natural resources within urban built environments is crucial and has multiple benefits. Greening of urban spaces not only reduces their contribution to emissions but also can protect against the impacts of climate change. Reduced emissions can be achieved through using the cooling power of vegetation and water to cool urban spaces, thus reducing the need for air conditioning (Chapter 14). Including green spaces within and close to urban dwellings provides leisure and recreational facilities and reduces stress for individuals (Chapters 10 and 14); it may also decrease the need for transportation either for leisure or for food distribution (Chapter 16).

The need to understand and quantify the totality of urban ecology rather than individual parcels is strongly presented here, as wildlife does not respect man-made boundaries and requires mobility across habitats. The removal of a seemingly insignificant section could, potentially, be devastating for the whole. The quality and nature of urban soils must be considered if their potential to mitigate or assist adaptation to climate change is weighed. The differences between urban and other soils is explored in Chapter 10, in which it is demonstrated that some concerns regarding the pollutant effects of urban life in terms of heavy metal contamination have been well documented. The characteristics of urban soils can make it unsuitable as an environmental resource and, therefore, protecting and enhancing urban soil characteristics may yield large dividends and increase the value greatly. For example, the role of river corridors in flood mitigation may also be compromised by soil quality if it is sterile, compacted and, therefore, less permeable. The value of urban green spaces can also be seen in the social and psychological benefits of leisure and exercise spaces and in carbon sequestration.

The further greening of urban environments is also advocated (Chapter 14) by the introduction and retrofitting of small pockets of vegetation for example as roofing material or on a wall. The chapter also argues that such strategies should be considered to mitigate the loss of biodiversity when formerly derelict but ecologically valuable areas are regenerated. It is also demonstrated that greening is an adaptation to extreme weather scenarios expected from climate change, such as flooding and the urban heat island effect. The cooling effect of vegetation through shading and transpiration effects is effectively explained and shown to be highly significant. Areas of water can also have a cooling effect and be part of an improved drainage and flood attenuation system. Additionally, the carbon sequestration potential of urban vegetation should not be ignored. This chapter presents case studies of urban greening installations suitable for retrofitting in a variety of urban and periurban settings.

Post-industrial landscapes feature in Chapter 18, which considers the little reported impact of climate change on man-made and natural slopes and waste deposits. Changes in precipitation patterns increased both wetness and dryness and could, potentially, destabilise previously well-engineered structures and cause dangerous landslides or leaching of toxins. The science of slope stability is admirably summarised in this chapter and the challenge to infrastructure installations demonstrated. The lack of recognition of the danger of climate impacts on such structures leads to a call for a risk assessment exercise.

Coastal zones are the residence of choice for half the world's population and are highly vulnerable to rising sea levels. While the coast itself is always changing, the value of the natural and built environments threatened by increased storminess and sea level rise are predicted to become a matter of grave concern. Although hard-engineered defences have served well in the past, current thinking involves a much wider toolkit for coastal adaptation. Chapter 24 discusses these holistic approaches with particular reference to the coastline of the northwest of England. Solutions cannot be prescriptive as the variability in coastal areas means that each solution will be unique but will involve a partnership between land managers, government and non-government organisations and coastal professionals and practitioners.

25.4 Enabling Change

The current debate on climate change is set within social, economic, political and natural science contexts much wider than the built environment perspective. To enable, or enforce, change may require a shift in perception and beliefs (Chapter 6). Climate modelling plays a large part in driving the acceptance of the need to change, and this is addressed in Chapter 3 which discusses the downscaling of models and the impact of global climate models on local weather predictions. Culture also contributes to the acceptance or otherwise of the necessity to adapt to climate change and facilitates or inhibits the implementation of climate-sensitive construction (Chapter 20). Efficient construction management can reduce energy expenditure, raw material demands and therefore waste (Chapter 11 and Chapter 17). The gap between the knowledge and implementation is addressed in Chapter 20, recognising that as an industry construction must be convinced of the business case for acting more responsibly in addition to having regulation imposed upon it.

The various initiatives, organisations and codes relating to both mitigation and adaptation are covered by Chapter 5, which reveals that the rate of change of climate in the past century necessitates changes in building technology at a more rapid rate than the previous evolution by trial and error allows. Science must intervene, but rapid change may also require cultural and social adaptation at an unprecedented rate. Not only technical but also political changes involving cooperation and effort from all stakeholders challenge us in the century to come (Chapter 21). Regulation has been introduced to control the disposal of waste (Chapter 17), transport emissions (Chapter 15 and Chapter 16), water directives (Chapter 19) and flood risk restrictions. Various initiatives on decent homes and the labour market are described in Chapter 6, which argues that international cooperation, such as carbon trading between rich and poor nations, may provide short-term respite from the need for all to change their lifestyles and allow time to develop other approaches.

Global organisations and national governments can encourage climate mitigation by incentives and they may also regulate, but they cannot act alone. Major changes in emissions and resilience will come from the sum of small actions under local circumstances. As discussed in Chapter 4, the knowledge required to convince local agencies of the need to adapt to and mitigate climate change must include the local consequences of not so doing. Chapter 3 discusses how climate models may be interpreted for this purpose, while Chapter 21 presents a risk framework developed to be used in evaluating risks for local decision makers. Chapter 22 shows how local factors and knowledge can be incorporated in maps of local vulnerability through the use of surveys and focus groups. By using such methods, the priorities of populations required to implement adaptation and mitigation policy will be aligned with the policies and, thus, those policies will stand a greater chance of success.

The contribution of the negative outputs of the built environment to climate change may seem obvious, and the avoidance of such outputs the highest priority. However, in the formulation of policy to combat these impacts, balance must be achieved between climate mitigation and economic prerogatives. Therefore, the impacts must be empirically captured and the value of urban and natural environments established. Traditional market forces-based economics has arguably failed to address the bigger issues of climate change (Chapter 6) and, therefore, new evaluation methods are needed. Two chapters in this book describe evaluation methods, which have often been seen as rivals in this respect: strategic environmental assessment (SEA) and cost–benefit analysis (CBA). However, they are both seen here as making useful, indeed complementary, contributions to increasing sustainable development. Chapter 7 demonstrates that SEA can absorb a wider range of stakeholder views and is designed to integrate fully into stakeholder engagement activity and policy formulation. Often it is seen as more open and transparent than the techniques of CBA. However, CBA allows more direct and robust comparison of alternatives. As discussed in

Chapter 8, methods for monetarisation of natural and environmental resources have developed over the previous decades. Despite their limitations, they make valuable contributions to project evaluation which can be widely applied. The challenge of climate change in these contexts is to incorporate increasing uncertainty and multiple-climate scenarios into these methods. The structured nature of approaches employed in CBA is seen to be amenable to such adaptation, where SEA can direct policy to the most advantageous scenarios. The future for project evaluation probably lies in a hybrid approach incorporating the strengths of each method.

While it is possible to monetarise or rank environmental resources using the methods described in Chapters 7 and 8, such analysis depends on deep understanding of these resources and their substitutability. So, for example, the value of green urban spaces will depend on the nature and potential of these spaces as leisure or environmental assets. The impact of climate change on such spaces is unpredictable, but urban gardens are a key tool in mapping and tracking changes in species better adapted to milder climates and, therefore, in measuring climate fluctuations. Understanding and preserving different types of urban green space are called for by these authors, in order that this diversity will be preserved and assist in the mitigation of climate change. Conflict may emerge between the value of some sites, such as post-industrial wastelands in ecological terms where the value is seen to be high and in aesthetic terms where urban planners may regard them as of low economic value and wish to gentrify neighbourhoods. Transportation is an example of urban activity that is difficult to tackle as an individual. Transport systems must be integrated if individuals are to be enabled to make different choices for necessary journeys (Chapter 16). Travel may also be reduced by planners ensuring that essential goods and services are available locally.

25.5 Final Remarks

Whichever approach, or approaches, to climate mitigation or adaptation are selected, a critical component is the knowledge and experience to select and apply it. As Chapter 19 identifies, a key enabling factor, therefore, is the training of the next generation of engineers, planners and politicians in the new technologies and approaches. Education of the wider public and their engagement in decisions is also crucially important, as illustrated by Chapter 16. Therefore, the hope of the editors of this volume is that the chapters included here will contribute to the greater awareness of solutions to climate change, and provide guidance to those seeking further practical knowledge or food for thought.

Index

abatement strategies 160–1
absorptive capacity 273
adaptation strategies 337–8, 341
   climate system 20
   coastal communities 321–35
   organisational culture 252–3
   policy initiatives 8, 50, 52–3
   strategic environmental assessment 83–4
   valuation of preferences 88
adaptive capacity 269, 271–9, 285–90, 293–8
advanced thermal treatment (ATT) plants 213–14, 218
aerobic fermentation 315
ageing populations 78
aggregate materials 142–4
agricultural soils management 160–1
air conditioning
   green infrastructure 170–1
   policy initiatives 50
   sustainable design 134–5
air pollution 37, 181–91
air transportation 195–6
airports 114
alien species invasion 108, 109
anaerobic digestion 315–16
anthropogenic climate change theory
   controversy and context 5–8
   emerging global trends 14–16
   primary cause of climate change 3–5
   strategic environmental assessment 75
appraisal mechanisms 198–9
Association of British Insurers 59
ATT *see* advanced thermal treatment
attribution methods 14

bamboo 145–6, 339
BC *see* black carbon
biodiversity 340
   green economics 58

   green infrastructure 167, 168–9
   socio-environmental vulnerability assessment 293, 295
   soil maps 113–14, 120–1
   urban environments 99–106
biofuels 57, 58, 159–60
biological waste treatment 212–13, 215, 218
biomass 114, 304–5, 314–16
black carbon (BC) 116–17
blastfurnace slag 143
Botanical Society of the British Isles (BSBI) 107
Brahmaputra River Basin 283–4, 291–300
BRE *see* Building Research Establishment
brick and masonry 144–5
brownfield sites 102–3, 109, 122
BSBI *see* Botanical Society of the British Isles
building construction *see* construction industry
Building Research Establishment (BRE) 136, 165–6
business continuity planning 276–7

cancellation property 289
capture technologies 155–6, 158
car fuel efficiency 157
carbon dioxide
   anthropogenic climate change theory 3–7
   climate system 11–13
   eco- and resilient materials 141, 148
   energy efficiency 155–7
   green economics 65
   green infrastructure 172–3
   renewable energy 303
   sustainable transportation 201–2
   urbanization 36–7
   waste management 207–10, 212–16
Carbon Emissions Reduction Target (CERT) 70
carbon footprints 65–6, 70–1, 82
carbon monoxide 37

Solutions to Climate Change Challenges in the Built Environment, First Edition.
Edited by Colin A. Booth, Felix N. Hammond, Jessica E. Lamond and David G. Proverbs.
© 2012 Blackwell Publishing Ltd. Published 2012 by Blackwell Publishing Ltd.

carbon neutral buildings
  energy efficiency 154–5, 158
  organisational culture 252
  valuation of preferences 88
carbon sequestration 155–7, 158, 172–3
carbon sinks
  strategic environmental impact assessment 81–2, 85
  urban soils 120
  valuation of preferences 87
carbon taxes 77
carpooling 201, 202
cathodic protection of steel 147
CBA *see* cost–benefit analysis
CBO *see* Congressional Budget Office
CDM *see* clean development mechanism
CE *see* choice experiments
census data 292
Central England Temperature record (CET) 17–18
CERT *see* Carbon Emissions Reduction Target
CFC *see* chlorofluorocarbons
CfSH *see* Code for Sustainable Homes
charrettes 135
China
  green economics 62, 66–7, 68–9
  urbanization 34, 36–7, 40–1
chlorofluorocarbons (CFC) 75
choice experiments (CE) 94
CHP *see* combined heat and power
circles of socialisation 262–3
circular rotation failures 228, 235
clay soil drying shrinking 47, 50
clean development mechanism (CDM) 217, 219
climate, definition 2–3
climate change, definition 1
climate change projections (UKCP09) 18
climate change theory 2–5
  see also anthropogenic climate change theory
climate engineering 155–7
climate system 11–22
  built environment 19–20
  emerging global trends 13–17
  emerging UK trends 17–19
  radiative forcing 11–14
  solutions 20–1
  strategic environmental assessment 83–4
closed-loop stepwise process 128–30
coal-based generation 158
coastal communities 340
  adaptation strategies 321–35
  case study 326–32
  challenges 323–4
  context 321–2
  integrated coastal zone management 324–5, 333
  landforms and process 322–3
  shoreline management plans 325–6, 329–32
  solutions 333

coastal squeeze 322–3, 326
Code for Sustainable Homes (CfSH) 131–3, 136–7
cogeneration *see* combined heat and power
cohesive forces 230, 235
combined heat and power (CHP) 79, 154, 214
combustion of biomass 314–15
committed warming 17
community resilience 273–5, 276–9
concrete 142–4, 147–8
congestion 37, 71–2
congestion charges 197, 201–2
Congressional Budget Office (CBO) 3, 6
conservation management 184–8
conservation sites 102
conservation tillage 160
constraints assessment 77–9
construction industry
  eco- and resilient materials 141–51
  flood risk 59
  green economics 57–62
  limited building lifespans 60–1
  organisational culture 251–67
  policy initiatives 47–9, 50–2, 54
  sustainable design 127–39
  unsustainable building practices 59–62
  urban and spatial planning 61–2
  valuation of preferences 87–8
consumer surplus 89–90
consumers driving green living 133–4
contaminated land 115–17, 119–20
contingent ranking (CR) 94
contingent valuation (CV) 94
contraction and convergence mechanism 65, 303–4
Copenhagen conference 46, 67
coping capacity 278–9, 285–6
cost of alternatives method 92
cost–benefit analysis (CBA) 88, 199, 341–2
CR *see* contingent ranking
cradle-to-cradle philosophy 128
created habitats 102
Curitiba (Brazil) 71–2
CV *see* contingent valuation
cyclic materials loops 129
cycling 201

DEC *see* displayed energy certificates
Decent Homes Standard 69–70
deforestation 75, 160
Delphi technique 288–90, 292, 299
detection methods 14
detention ponds 175, 244, 246
developing countries
  green economics 57–8, 65
  urbanization 34–5, 37–41
DEX *see* Dunfermline Eastern Expansion
diachronic environmental impacts 195–6

diesel particulates 182, 187
dinitrogen oxide *see* nitrous oxide
disaggregation of value 90
disaster resilience of place (DROP) model 273–5
dishwashers 240
displayed energy certificates (DEC) 136–7
district heating 154
Dongtan Green City project 68–9
dose–response method 93
drainage 122
DROP *see* disaster resilience of place
droughts
    green economics 58
    regional implications 29
    socio-environmental vulnerability assessment 292
    water resources 237
Dubai 62
Dunfermline Eastern Expansion (DEX) 175, 245
dynamical downscaling 26

early-successional habitats 102–3, 109
EC *see* elemental carbon
ECCP *see* European Climate Change Programme
eco- and resilient materials 141–51, 338
    brick and masonry 144–5
    concrete 142–4, 147–8
    future trends 149
    glass 145, 149
    nanotechnology 148
    polymer-based materials 147–8
    steel 146–7
    timber and bamboo 145–6
ecological footprint (EF) 82
ecological value
    brownfield and early-successional habitats 102–3, 109
    climate change 107–9
    context 99–101
    created habitats 102
    dimensions and boundaries 106–7
    garden habitats 105–6, 109
    habitat types 101–6
    post-industrial sites 103–5, 109
    seminatural habitats 101–2
    solutions to climate change 108–9
    urban environments 99–112
economic factors
    anthropogenic climate change theory 6, 7
    urbanization 34, 39–41
    valuation of preferences 87–98
    see also green economics
ecosystem services approach 63–4
eco-tech *see* green technologies
EF *see* ecological footprint
element risk 296
elemental carbon (EC) 182

energy efficiency 153–61
    abatement strategies 160–1
    carbon sequestration and climate engineering 155–7, 158
    carbon-neutral buildings 154–5, 158
    combined heat and power 154
    context 153–5
    sustainable development 157–61
Energy Performance of Buildings Directive (EPBD) 131, 136–7
Energy Performance Certificates (EPC) 131, 132–3, 137
energy usage
    climate system 19–20
    green economics 67, 70–1
    organisational culture 253
    policy initiatives 50–1, 54
    strategic environmental assessment 77, 79, 81
    sustainable design 128–30, 133–4
    urbanization 36, 40
engineered landfill sites 231–5
England and Wales Precipitation dataset (EWP) 18
English Flood and Water Management Bill 243
ensembles of opportunity 25–6
Environment Agency 59
environmental receptors 77–9
environmental sustainable transport (EST) 198
EPBD *see* Energy Performance of Buildings Directive
EPC *see* Energy Performance Certificates
EPOE *see* Extended Post Occupancy Evaluation
EST *see* environmental sustainable transport
European Climate Change Programme (ECCP) 48–9, 53
European Environment Agency 67
European Renewable Energy Directive 132
evaporative cooling 170–1
EWP *see* England and Wales Precipitation
Extended Post Occupancy Evaluation (EPOE) 135–6, 137
extensive green roofs 164–5
extreme weather events 269–81
    context and definition 269–70
    disaster resilience of place model 273–5
    green economics 59
    policy initiatives 52–3
    regional implications 29–30
    risk assessment framework models 276–9
    uncertainty and risk 270–1
    vulnerability, resilience and adaptive capacity 269, 271–9

fast breeder reactors 317
feed in tariffs (FIT) 137–8
fibre-reinforced plastic 147–8
fission 159
FIT *see* feed in tariffs

floating houses 149
flooding 339
  coastal communities 322–4
  green economics 59
  green infrastructure 167–8
  policy initiatives 50, 52–3
  socio-environmental vulnerability assessment 283–4, 292, 297–9
  strategic environmental assessment 78–9
  urban soils 121–2
  valuation of preferences 88, 90
  water resources 237, 242–7
fly ash 143
Foreseeable Natural Risk Prevention Plan (PPR) 52–3
forest fires 50
forest management 160
fossil fuels
  climate system 12, 20
  energy efficiency 157, 158
  renewable energy 318–19
  urbanization 36–7
French Climate Plan 53
frictional forces 229–30, 235
fuel cells 159, 318–19
fusion reactors 317–18
future costs and benefits 89, 90–1

gardens
  ecological value 105–6, 109, 118
  green infrastructure 175
gas-based generation 158
gasification 213–14, 315
GCM *see* general circulation models
GDP *see* gross domestic product
gender factors 293–6
general circulation models (GCM) 340
  climate system 17, 19–20
  downscaling 26
  ensembles of opportunity 25–6
  output interpretation 25
  parameterizations 24–5
  projections for future climate change 27–8
  regional implications 24–8
  urban soils 121–2
geo-engineering 64–5
geotechnical stability of landfill sites 223–36
geothermal power 304, 316–17
glass 145, 149
global climate models *see* general circulation models
global cooling 6–7
global warming 337
  controversy and context 5–8
  definition 1
  emerging global trends 13–17

  emerging UK trends 17–19
  greenhouse gases 3–5
  strategic environmental assessment 75
golf courses 327–8
government payments 93
granulated blastfurnace slag 143
green economics 57–74, 341–2
  construction industry 57–62
  costs of climate change 67–8
  Decent Homes Standard 69–70
  Dontan Green City project 68–9
  ecosystem services approach 63–4
  flood risk 59
  geo-engineering 64–5
  Homo economicus approach 63–4
  lifestyle changes 66, 67
  limited building lifespans 60–1
  limits to growth 62–3
  market mechanism 64
  project case studies 68–72
  raw materials 66–7
  Re-charge Scheme 71
  regulation 65
  strategic choices 64–6
  sustainable cities 71–2
  unsustainable building practices 59–62
  urban and spatial planning 61–2
  Warm Zone Project 70–1
Green Guide 136
green infrastructure 163–79, 339–40
  biodiversity 167, 168–9
  building efficiency 169–71
  carbon sequestration 172–3
  city centres 173–5
  context 163
  green roofs and walls 164–6, 173–4
  human health and wellbeing 171–2
  integration into built environment 163–7
  intercepting rainfall and reducing flood risk 167–8
  solutions for climate change 173–6
  suburbia 175
  urban heat islands 165, 169–71, 174
  urban periphery 176
  vegetated porous paving 166–7
  water resources 244
green roofs and walls 164–6, 173–4
green technologies 66–7, 71
greenfield sites 66, 122
greenhouse gases 338
  anthropogenic climate change theory 3–8
  climate system 11–13, 20–1
  eco- and resilient materials 141
  green economics 65
  policy initiatives 45, 46–7, 50
  regional implications 25–6
  renewable energy 303

Index

strategic environmental assessment 78
sustainable transportation 194–5, 201–2
urbanization 36–7, 40–1
waste management 207–16, 218–19
greywater harvesting 241–2
gross domestic product (GDP) 34, 57
ground granulated blastfurnace slag 143
ground stability effects 224–5
growth, limits to 62–3

habitat creation 120–1
habitat loss/destruction
    coastal communities 322–3
    ecological value 107
    green economics 58
    socio-environmental vulnerability assessment 293, 295
hay meadows 120–1
hazardous waste incineration directive 94/67/EC 217
heat pumps 306
heavy metal contamination 115–17, 119–20
hedonic pricing (HP) 95
high-temperature solar power 307
HIPS *see* home inspection plans
historic buildings 181–91
home inspection plans (HIPS) 133
Homo economicus approach 63–4
household water usage 239–40, 248
HP *see* hedonic pricing
human impacts *see* anthropogenic climate change theory
human resource capacity 294–6
hydrogen generation
    energy efficiency 155–6, 158, 159
    renewable energy 309, 316, 318–19
hydropower 304, 310–13
hypotaxia 61

ice cores 13
ICZM *see* integrated coastal zone management
ILO *see* International Labour Organisation
immigration 294–5, 299–300
incineration plants 212, 213–14, 217–18
industrial sites 114, 116
industrialization of construction 51, 54–5
infrared (IR) radiation 3–4
infrastructure
    socio-environmental vulnerability assessment 293, 295–6
    sustainable transportation 199–201
    water resources 242–6
    see also green infrastructure
inland water transport 196
Institute for Public Policy Research (IPPR) 134
insulation 154, 161

insurance policies
    extreme weather events 269
    green economics 59
    organisational culture 253
    policy initiatives 52
    water resources 248
integrated coastal zone management (ICZM) 324–5, 333
integrated planning 71–2
integrated solutions approach 209–16
intensive green roofs 164–5
Intergovernmental Panel on Climate Change (IPCC) 2–3, 5
    coastal communities 324
    policy initiatives 46
    socio-environmental vulnerability assessment 285–7, 290
    urbanization 35–6
International Labour Organisation (ILO) 69–70
IPCC *see* Intergovernmental Panel on Climate Change
IPPR *see* Institute for Public Policy Research
IR *see* infrared
isotope ratios 12

Kirklees (UK) Council 69–71
Kyoto Protocol 2, 64

land claims 326–7
land cover/use 2
    green economics 58, 61–2
    soil maps 113–15, 117
    strategic environmental assessment 81–2
land value tax 67
landfill 214–16, 219
    circular rotation failures 228, 235
    cohesive forces 230, 235
    context 223
    engineered landfill sites 231–5
    frictional forces 229–30, 235
    geotechnical stability 223–36
    ground stability effects 224–5
    organisational culture 265
    pore fluid suctions 230–1
    shallow slides 227–8, 233–6
    site design and components 232–3
    slope instability 225–8, 233–4
    soil shear strength 229–31, 233
    solutions and insights 235–6
    surface erosion 226–7
    vegetation 231, 233–5
landfill directive 1993/31/EC 216, 219
LANDSAT data 292
landslides 224, 227–8, 236
lapse rate feedback 16
large-scale hydropower 310–13
lawns 118

lifecycle assessment (LCA) 207–8, 209–10
lifestyle changes 66, 67
light-weight concrete 144
limestone soiling *see* particulate-induced soiling
limited building lifespans 60–1
limits to growth 62–3
London Bus Priority Network 202

marine transport 196
market mechanism 64
market price 89–90
masonry 144–5
maximum likelihood factor analysis 289
mechanical biological treatment (MBT) 212–13, 215, 218
metakaolin 143
metal contamination 115–17, 119–20
methane
 anthropogenic climate change theory 3, 5
 climate system 12
 strategic environmental assessment 78
 waste management 207–10, 212–16, 218–19
metropolisation 37–9
microgeneration 134
Millenium Development Goals 64
mitigation strategies 339, 341
 energy efficiency 153–61
 extreme weather events 273–5
 organisational culture 252–3
 policy initiatives 8, 50–1
 renewable energy 303–20
 strategic environmental assessment 83–4
 valuation of preferences 88, 92–3
motorway service areas 244–5
MSW *see* municipal solid waste
multi-GCM 25–6
municipal solid waste (MSW) 209, 212

nanotechnology 148, 338
National Adaptive Programmes for Action (NAPA) 40
National Environmental Policy Act (NEPA) 76
National Nature Reserves 327
natural fibres 143–4
natural impacts
 emerging global trends 14–16
 green economics 61–2
 greenhouse gases 4
 policy initiatives 50, 52–3
 strategic environmental assessment 83–4
natural resources 87–8, 95–6, 142–4
NEPA *see* National Environmental Policy Act
nitrous oxide
 anthropogenic climate change theory 5
 climate system 12
 waste management 207, 210, 212–15
North Sefton coast case study 326–32

nuclear power 159, 317–18
nutrient content of urban soils 116–19

ocean dead zones 61
opportunity costing 92
organisational culture 251–67
 achieving and sustaining culture of sustainability 257, 262–3
 case study 263–5
 climate change-driven construction 252–3
 context 251–2
 cycle of culture change 255, 263
 environmental and sustainability policies 257–61, 264
 framework for cultural change 262–3
 influences 262–3
 phenomenon and role of culture 253–5
 solutions 265
 training 257, 264
 UK construction industry 256–7
 web presence 257
ozone concentration 14

packaging and packaging waste directive 94/62/EC 217
paleo-climate reconstructions 13
parabolic collectors 307
parallel model paradigm 201
parameterizations 24–5
parking areas 122
particulate-induced soiling 181–91
 Bath case study 184–8
 classification systems 184–6
 climate change impacts 188
 context 181–2
 impacts and mechanisms 182–3
 urban particulate pollution 182
passive solar heating 306
Passivhaus buildings 154–5
PCMDI *see* Program for Climate Model Diagnosis and Intercomparison
pedagogical value
 characteristics of urban soils 115–17
 policy initiatives 119–22
 soil maps 113–19
 urban landscapes 113–26
 urbanization 113–15, 117
perturbed physics ensembles (PPE) 25–6
photovoltaics (PV) 159, 308
phytoremediation 120
planning strategies
 green economics 59, 61–2, 71–2
 strategic environmental assessment 80–2, 83–5
 sustainable design 135
 urban soils 121–2
 urbanization 38–41

policy initiatives 45–55
  adaptation policies 50, 52–3
  anthropogenic climate change theory 6, 8
  background 47–9
  built environment 45–7, 53–5
  construction industry 47–9, 50–2, 54
  extreme weather events 270
  mitigation policies 50–1
  solutions for the built environment 53–5
  strategic environmental assessment 83–4
  sustainable design 130–3
  sustainable transportation 197–8
  urban soils 119–22
  urbanization 41
  waste management 216–19
  water resources 238, 239–40, 248
polymer-based materials 147–8
polystyrene materials 148
ponds 244, 246
population growth 47, 113
pore fluid suctions 230–1
porous paving 166–7, 244, 246
post-industrial landscapes 103–5, 109, 224–5, 340
poverty 65, 319
power plant efficiency 158
PPE *see* perturbed physics ensembles
PPR *see* Foreseeable Natural Risk Prevention Plan
precipitation patterns
  emerging global trends 17
  emerging UK trends 18
  regional implications 28–9
  water resources 238
preferences *see* valuation of preferences
pressure and release model 283
process risk 286
product lifecycle 207–8, 209–10
professional indemnity insurance 253
Program for Climate Model Diagnosis and Intercomparison (PCMDI) 25
projections for future climate change 27–31
public goods 90
public participation 203
public transport 197, 201–2
pumped storage hydropower 312
PV *see* photovoltaics
pyrolysis 213–14, 315

radiative forcing 11–14
rail transport 196
rainfall *see* precipitation patterns
rainwater harvesting 166, 167–8, 241
rammed earth 145
RAMSAR sites 327
RCM *see* regional climate models
Re-charge Scheme 71

recycling 338
  eco- and resilient materials 146, 149
  waste management 211–12, 218
regional climate models (RCM) 26
regional implications 23–32
  characterisation of climatic risks 30–1
  climate modelling 23–8
  extreme weather events 29–30
  projections for future climate change 27–31
rehydration swelling 50
renewable energy 338
  biomass 304–5, 314–16
  context 303–4
  contraction and convergence mechanism 303–4
  current global sustainable energy provision 304–5
  energy efficiency 159–60
  environmental impacts 311–12
  geothermal power 304, 316–17
  green economics 71
  hydrogen generation and fuel cells 309, 316, 318–19
  hydropower 304, 310–13
  mitigation strategies 303–20
  nuclear power 317–18
  photovoltaics 308
  solar power 304–7, 319–20
  solutions 319–20
  strategic environmental assessment 78–9, 81
  sustainable design 129, 132, 134
  tidal power 304, 310, 313
  wave power 310
  wind power 304–5, 308–10
resilience to extreme weather events 269, 271–9
resilient building materials *see* eco- and resilient materials
re-use 211–12, 218, 240, 241–2
RH *see* risk-hazard
Rio de Janeiro Conference 2
risk assessment framework models 276–9
risk-hazard (RH) model 283
road pricing and taxing schemes 201–2, 203
road transport *see* transport and traffic
Royal Society for the Protection of Birds (RSPB) 326–7
runoff water 166, 167–8, 237, 244

salt marshes 322–3, 326–7
sand dunes 322–3, 329–30
SAP *see* Standard Assessment Procedures
satellite data 14
SBCI *see* Sustainable Buildings and Climate Initiative
SEA *see* strategic environmental assessment
sea ice 17

sea level rise
    coastal communities 322–4
    emerging global trends 16
    policy initiatives 47
seasonality 28–9
Sefton coast case study 326–32
self-compacting concrete 144
seminatural habitats 101–2
sensitivity 285–90, 293–8
SEVA *see* socio-environmental vulnerability assessment
sewage treatment plants 224–5
shadow project costs 93
shallow slides 227–8, 233–6
shear strength of soil 229–31, 233
shoreline management plans (SMP) 325–6, 329–32
significant possibility of significant harm (SPOSH) 116
simple climate models 24
single-occupant vehicles (SOV) 201, 202
Sites of Special Scientific Interest (SSSI) 327
slope instability of landfill sites 225–8, 233–4
sludge lagoons 224–5, 236
small-scale hydropower 312–13
Smith, Adam 89
SMP *see* shoreline management plans
snow cover 17
SOC *see* soil organic carbon
social inclusion 72
social learning 273–5
social networks 294–5
socialisation circles 262–3
socio-environmental vulnerability assessment (SEVA) 283–301
    approach 284–91
    Brahmaputra River Basin case study 283–4, 291–300
    conceptualizing socioeconomic vulnerability 284–7
    construction of indices 289–90
    context and background 283–4
    definitions of vulnerability 285
    domain identification and indicators 287–8, 292, 295–6, 299
    domain ranking 288–9, 292
    opportunities and threats 285–6
    results 291–300
    review of data 292–5
    sensitivity and adaptive capacity 285–90, 293–8
    spatial variations 295–9
    stakeholder participation/validation 291
socioeconomic factors
    strategic environmental assessment 76
    urbanization 39, 41
    vulnerability 284–7, 290, 293–6

soil maps
    characteristics of urban soils 115–17
    pedagogical value 113–26
    policy initiatives 119–22
    urbanization 113–15, 117
soil organic carbon (SOC) 116–18, 120
soil shear strength 229–31, 233
Soil Survey of England and Wales 114
soiling *see* particulate-induced soiling
solar power 304–7, 319–20
solar radiation reflection 157
solar water heating 306–7
SOV *see* single-occupant vehicles
SPA *see* Special Protection Areas
spatial planning 61–2
Special Protection Areas (SPA) 327
spontaneous successional habitats 102–3, 109
SPOSH *see* significant possibility of significant harm
SSSI *see* Sites of Special Scientific Interest
stakeholder participation/validation 84, 287–8, 291
Standard Assessment Procedures (SAP) 132–3
stated preference methods 94
statistical downscaling 26
steel 146–7
steel-reinforced concrete 142, 147
Stern Review 58, 64, 251
Stirling engines 307
storage technologies 156–7, 158
strategic environmental assessment (SEA) 75–86, 341–2
    addressing environmental issues 82
    analysing environmental problems 80
    baseline data 79–80
    climate change solutions 77–84
    constraints assessment 77–9
    context and historical development 75–7
    environmental receptors and interrelationships 77–9
    integrating environment with planning processes 80–2, 84–5
    limitations of strategic actions 77
    methods 82
    mitigation and adaptation policies 83–4
    stakeholder consultation 84
Strategy for Sustainable Construction 131
stratospheric cooling 13, 14–16
street trees 174–5
subcontractors 256
SUDS *see* sustainable urban drainage systems
surface erosion 226–7
Surface Water Management Team 243–6
Sustainable Buildings and Climate Initiative (SBCI) 48–9
sustainable cities 71–2

sustainable design 127–39, 338
  closed-loop stepwise process 128–30
  consumers driving green living 133–4
  context 127–30
  drivers 130–5
  energy and materials 127–30, 133–5
  innovation in technology and materials 127–8, 134–5
  policy initiatives 130–3
  rethinking construction 135–7
  solutions 137–8
sustainable development
  energy efficiency 157–61
  green infrastructure 169
  policy initiatives 46, 48–9
  strategic environmental assessment 76, 77
  urbanization 42
  valuation of preferences 90–1
sustainable mobility 201–2
sustainable procurement 66
sustainable transportation 193–205
  appraisal mechanisms 198–9
  barriers 199–200
  climate change impacts 194–5
  context and definitions 193–4
  diachronic environmental impacts 195–6
  indicators and levels of analysis 198–9
  infrastructure development 199–201
  perspectives 195–7
  policy initiatives 197–8
  public participation 203
  solutions 203–4
  sustainable mobility 201–2
  system development 197–203
  trade-off approach 195–7
sustainable urban drainage systems (SUDS) 79
swales 244–6
synfuels plants 159

taxation instruments 67, 77
thermal insulation 50–1
tidal power 304, 310, 313
timber 145–6
tokamaks 318
trade-off approach 195–7
training 257, 264
transport and traffic 342
  congestion 37, 71–2, 197, 201–2
  energy efficiency 157
  green economics 71–2
  organisational culture 253
  particulate-induced soiling 182–8
  renewable energy 316
  road traffic pollution 119
  strategic environmental assessment 83
  sustainable transportation 193–205
  urbanization 36–7, 40

travel cost method 94–5
tree rings 13
tropical cyclones 29–30

UHI *see* urban heat islands
UK Climate Impacts Programme (UKCIP) 53, 59
UKCP09 *see* climate change projections
ultraviolet (UV) radiation 3–4
UN-Habitat 33–4, 38–9
UNEP *see* United Nations Environment Programme
UNFCCC *see* United Nations Framework Convention on Climate Change
unfired-clay bricks 145
United Nations Environment Programme (UNEP) 46, 48–9
United Nations Framework Convention on Climate Change (UNFCCC) 2, 46, 48–9
United Nations Population Fund (UNFPA) 33–4
unsustainable building practices 59–62
urban environmental transitional model 38, 39–40
urban environments
  brownfield and early-successional habitats 102–3, 109
  climate change 107–9
  context 99–101
  created habitats 102
  dimensions and boundaries 106–7
  ecological value 99–112
  garden habitats 105–6, 109
  habitat types 101–6
  post-industrial sites 103–5, 109
  seminatural habitats 101–2
  solutions to climate change 108–9
urban flooding 237, 242–6
urban heat islands (UHI)
  green infrastructure 165, 169–71, 174
  policy initiatives 50
  urbanization 36, 37–8
urban particulate pollution 182
urban soils 339
  characteristics 115–17
  habitat creation 120–1
  nutrient content 116–19
  pedagogical value 113–26
  policy initiatives 119–22
  urbanization 113–15, 117
urbanization 33–43, 337, 339
  climate system 19–20
  coastal communities 326
  global trends 33–4
  impact on climate change 35–6
  mechanisms of climate change impacts 36–9
  pedagogical value 113–15, 117
  planning regulations 38–41
  policy initiatives 47

urbanization (cont'd)
    socio-environmental vulnerability
        assessment 294–5
    solutions for change 39–41
    water resources 237
utility 90
UV see ultraviolet

valuation of preferences 87–98
    categorization of human values 90–1
    context 87–8
    methods 91–5
    pricing methods 92–3
    solutions 95–6
    theory 89–91
    valuation approaches 93–5
vegetated porous paving 166–7
vegetation 231, 233–5
vernacular buildings 47–8, 52, 54, 135
volcanic eruptions 16, 62
vulnerability to extreme weather events 269, 271–9
    see also socio-environmental vulnerability assessment
vulnerability-resilience indicator prototype (VRIP) model 283

walking 201
Warm Zone Project 70–1
waste directive 75/692/EEC 217
waste incineration directive 2000/67/EC 217
waste management 38, 207–21, 338
    context 207–8
    hierarchy 210–16
    integrated solutions approach 209–16
    landfill 214–16, 219
    mechanical biological treatment 212–13, 215, 218
    policy initiatives 216–19
    prevention of waste 211, 218
    product lifecycle and lifecycle assessment 207–8, 209–10
    recycling and re-use 211–12, 218
    solutions 219

valuation of preferences 89
waste-to-energy incineration plants 212, 213–14, 218
Waste Strategy for England (2007) 219
waste-to-energy (WtE) 212, 213–14, 218
water butts 241
Water Framework Directive (WFD) 238
water meters 240
water resources 237–50, 339
    context 237–8
    flood-resilient measures 247
    flood-resistant measures 246–7
    flooding 237, 242–7
    greywater harvesting 241–2
    household usage and savings 239–40, 248
    present and future solutions 247–8
    property-level measures 246–7
    rainwater harvesting 241
    socio-environmental vulnerability
        assessment 293, 295
    supply shortages 237, 238–42
water vapour 3, 11–12, 16
wave power 310
WCBSD see World Council of Business and Sustainable Development
WCC see World Climate Conference
weather balloon data 14
wetlands 244–6, 324
WFD see Water Framework Directive
WHO see World Health Organizations
willing to accept (WTA) 94
willingness to pay (WTP) 89–90, 94
wind power 159, 304–5, 308–10
WMO see World Meteorological Organization
wood products 146
World Climate Conference (WCC) 48–9
World Council of Business and Sustainable Development (WCBSD) 58
World Health Organizations (WHO) 37
World Meteorological Organization (WMO) 2–3, 49
WTA see willing to accept
WtE see waste-to-energy
WTP see willingness to pay